Ecosustainable Polymer Nanomaterials for Food Packaging

Innovative Solutions, Characterization Needs, Safety and Environmental Issues

Ecosustainable Polymer Nanomaterials for Food Packaging

Innovative Solutions, Characterization Needs, Safety and Environmental Issues

Edited by
Clara Silvestre
Sossio Cimmino

CRC Press
Taylor & Francis Group
Boca Raton   London   New York

CRC Press is an imprint of the
Taylor & Francis Group, an **informa** business

CRC Press
Taylor & Francis Group
6000 Broken Sound Parkway NW, Suite 300
Boca Raton, FL 33487-2742

First issued in paperback 2016

© 2013 by Taylor & Francis Group, LLC
CRC Press is an imprint of Taylor & Francis Group, an Informa business

No claim to original U.S. Government works

Version Date: 20121115

ISBN 13: 978-1-138-03426-6 (pbk)
ISBN 13: 978-90-04-20737-0 (hbk)

Visit the Taylor & Francis Web site at
http://www.taylorandfrancis.com

and the CRC Press Web site at
http://www.crcpress.com

Contents

Preface

According to the latest official data, the global market of food packaging is worth $440 billion worldwide (about 2% of the gross national product of developed countries), with an annual increase of 10%. The food industry uses almost 65% of all packaging placed on the market, and the impact of the packaging on the retail cost of food is between 19% and 50%. Almost 40% of the total packaging is made of plastic. The growth rate of the market of plastic packaging in the last few years for both flexible and rigid plastic packages has been about 7%, the highest compared with other materials (glass 2%, metal 6%, paper 6%, and others, including wood, 3%).

These impressive numbers clearly explain why many of the world's largest food packaging companies are actively exploring the potential of polymer nanotechnology in order to obtain new food packaging materials with improved properties and also to be able to trace and monitor the condition of food during transport and storage.

Polymer nanotechnology, described as the next great frontier of material science in food packaging, with applications predicted to grow rapidly in the coming years, is currently developed to mainly improve barrier performance pertaining to gases, such as oxygen and carbon dioxide. Once perfected, sure from a safety point of view, and produced at a competitive ratio cost/performance level, the new Polymer Nano-material for Food Packaging (PNFP) will be very attractive for use in several food packaging applications.

The book aims at looking at the complete life cycle of the packaging based on polymer nanomaterials (raw material selection; production, structure, and properties characterization; analysis of interaction with food; and marketing, application, use, and disposal) through the contribution of several experts with the final objective of considering a balance between cost and performance, risk and benefit, and health and environmental issues, as well as the assessment of risk–benefit, and of identifying the barriers (in research and technology, safety, technology transfer, and communication) that prevent a complete successful development of the new technology and strategies to proceed further.

The book is divided into 13 chapters and covers the current developments in nanotechnology for food packaging application by providing a comprehensive review focusing on applications that are most likely to be accepted by consumers and attract regulatory attention in the immediate future.

Chapter 1 provides a general introduction to the topic, focusing on current issues and future trends. Chapter 2 introduces an "ethical design" as a concept that puts into practice key ideas like the precautionary principle and presents a model of accountability, responsibility, and ethical consideration commensurate with the current understanding of all sorts of risks and hazards, whether they are known, unknown, or unknowable. The book then delves into *science*: Chapters 3 and 4 are dedicated to the characterization of these novel materials, investigating the evolution of the rheology and of the structure and morphology with regard to processing conditions and kind of constituents. The interesting application of plasma technologies in the field of food packaging materials for production of barrier coatings on polymeric materials by nonequilibrium gas discharges is illustrated in Chapter 5. The main aspects of nanomaterials for food packaging from oil polymers (polyolefins) and from renewable resource polymers are discussed sequentially in Chapters 6 and 7 in order to better compare both kinds of materials. Chapters 8 and 9 deal with two specific innovative applications: the use of cellulose nanowhiskers for food biopackaging and edible nano-laminate coatings. The general issues linked to interactions of nanomaterials with food are covered in Chapter 10. Some examples of degradation under natural weathering exposure and recycling are provided in Chapters 11 and 12. Chapter 13 concludes the book with an overview on the rapidly evolving and expanding field of the usage of polymer nanocomposite materials for food packaging application. Each chapter contains a complete list of references related to the topic.

The book intends to show the reader that nanomaterials offer some exciting benefits to the food industry, including better materials for food packaging as well as safer foods on supermarket shelves that have lower incidences of contamination with chemical adulterants and potentially life-threatening microorganisms. The applications reviewed here were specifically chosen because they are the most likely nanofood products to be accepted by consumers in the short term. The science community must continue to be alert against the potential dangers that the use of this new technology will pose, and successful and safe implementation of these applications will require constant dialogue between scientists, technologists, and consumers. If these endeavors are successful, the benefits of nanotechnology may play an important role in making the world's food supply healthier, safer, and more plentiful.

This book is therefore addressed not only to researchers and engineers that actively work in the field of nanocomposites for food packaging but also to newcomers and students who have just started investigations in this multidisciplinary field of science.

We are indebted to all the authors for their worthy contributions; without their expertise, dedication, and hard work, the publication of this book would not have been possible. Many thanks are due to Dr. Marilena Pezzuto from ICTP-CNR, Italy, for her precious contribution in homogenizing all the manuscripts sent to the press and to the Taylor & Francis Group editorial staff and for their professionalism and wholehearted support.

All the authors have participated in the European activity COST Action FA0904 (2010–2014) "Eco-sustainable food packaging based on polymer nanomaterials," and they thankfully acknowledge the support of the action that is making possible the constitution of an impressive international scientific and technology network on issues related to the preservation, conservation, and distribution of high-quality and safe food through polymer nanomaterials.

Editors

Clara Silvestre, doctor in industrial chemistry, is a Senior Researcher Consiglio Nazionale delle Ricerche (CNR) at the Institute of Chemistry and Polymer Technology, Pozzuoli (Naples), Italy. She was visiting researcher at the University of Bristol (England) and associate researcher at the University of Massachusetts, Amherst, Massachusetts. She is the Chair of the COST Action FA0904 "Ecosustainable food packaging based on polymer nanomaterials," coordinator and responsible for national and international projects, and an EU expert evaluator/reviewer for 5/6/7FP projects. She is also a consultant for the International Atomic Energy Agency (IAEA) to develop a coordinated research project on the "Application of Radiation Processing Technology in the Development of Advanced Packaging Materials for Food Products."

Dr. Silvestre is the chair of the organizing committees of international conferences and is the author of over 140 papers and monographs as well as the holder of 3 patents. She is a member of the CNR Equal Opportunity Committee and she was also the EU Ambassador for Women and Science.

Sossio Cimmino, doctor in industrial chemistry, is Director of Research of Consiglio Nazionale delle Ricerche (CNR) at the Institute of Chemistry and Polymer Technology, Pozzuoli (Naples), Italy. He was a senior visiting researcher at the University of Massachusetts, Amherst, Massachusetts, and a visiting researcher at DSM-Geleen (the Netherlands). He was also a member of the Scientific Committee of the ICTP. Dr. Cimmino is scientific representative and grant holder of the COST Action FA0904 "Ecosustainable food packaging based on polymer nanomaterials." He serves as lecturer in international schools, meetings, conferences, and seminars; as coordinator and responsible for national and international projects; and as an evaluator/reviewer of FP6 and FP7 European projects. He is also a referee of international journals in polymer science, such as *Colloid and Polymer Science*; *Journal of Applied Polymer Science*; *Macromolecular Materials and Engineering*; *Polymer International*; *Thermochimica Acta*; *Journal of Polymer Science, Polymer Physics Edition*; *Journal of Macromolecular Science*; and *Macromolecular Symposia*.

Dr. Cimmino is the author of over 130 papers published in international journals and books, 140 congress communications, and holds 3 patents.

Drs. Sossio Cimmino and Clara Silvestre coordinate at ICTP-CNR, a research group whose activity focuses on the design of innovative polymer-based systems (homopolymers, copolymers, polymer blends, nanocomposites) through new mixing technologies, new formulations, and control of morphologies/structures for application in the area of food packaging, membranes, fire-resistant textiles, and biomaterials (www.ictp.cnr.it, www.napolynet.eu, www.costfa0904.eu).

Contributors

Zehra Ayhan
Department of Food Engineering
Mustafa Kemal University
Hatay, Turkey

Aida Benhamida
Faculty of Technology
Laboratory of Organic Materials
Engineering Department
Processes
University Abderrahmane Mira
Bejaia, Algeria

Hynek Biederman
Faculty of Mathematics and Physics
Department of Macromolecular
Physics
Charles University in Prague
Prague, Czech Republic

A.I. Bourbon
IBB–Institute for Biotechnology and
Bioengineering
Centre of Biological Engineering
Universidade do Minho
Braga, Portugal

Georgeta Cazacu
Department of Physical Chemistry
of Polymers
"P. Poni" Institute of
Macromolecular Chemistry
Iasi, Romania

M.A. Cerqueira
IBB–Institute for Biotechnology and
Bioengineering
Centre of Biological Engineering
Universidade do Minho
Braga, Portugal

Andrei Choukourov
Faculty of Mathematics and Physics
Department of Macromolecular
Physics
Charles University in Prague
Prague, Czech Republic

Sossio Cimmino
National Research Council
Institute of Chemistry and
Technology of Polymers
Pozzuoli, Italy

Raluca Nicoleta Darie
Department of Physical Chemistry
of Polymers
"P. Poni" Institute of
Macromolecular Chemistry
Iasi, Romania

Raluca Petronela Dumitriu
Department of Physical Chemistry
of Polymers
"P. Poni" Institute of
Macromolecular Chemistry
Iasi, Romania

Donatella Duraccio
National Research Council
Institute of Chemistry and
   Technology of Polymers
Pozzuoli, Italy

Yasemin J. Erden
Centre for Bioethics and Emerging
   Technologies
St Mary's University College
Strawberry Hill, United Kingdom

Lenka Hanyková
Faculty of Mathematics and
   Physics
Department of Macromolecular
   Physics
Charles University in Prague
Prague, Czech Republic

Evgeni Ivanov
Institute of Mechanics
Bulgarian Academy of Sciences
Sofia, Bulgaria

Mustapha Kaci
Faculty of Technology
Laboratory of Organic Materials
Engineering Department
   Processes
University Abderrahmane Mira
Bejaia, Algeria

Erich Kny
Austrian Institute of Technology
Vienna, Austria

Rumiana Kotsilkova
Institute of Mechanics
Bulgarian Academy of Sciences
Sofia, Bulgaria

Marek A. Kozlowski
Faculty of Environment
   Protection
Materials Recycling Center
   of Excellence
Wroclaw University
   of Technology
Wroclaw, Poland

Ekaterina Krusteva
Institute of Mechanics
Bulgarian Academy of Sciences
Sofia, Bulgaria

Ondřej Kylián
Faculty of Mathematics
   and Physics
Department of Macromolecular
   Physics
Charles University in Prague
Prague, Czech Republic

José María Lagarón
Novel Materials and
   Nanotechnology Group
Institute of Agrochemistry
   and Food Technology
Spanish Council for Scientific
   Research
Valencia, Spain

Amparo López-Rubio
Novel Materials and
   Nanotechnology Group
Institute of Agrochemistry
   and Food Technology
Spanish Council for Scientific
   Research
Valencia, Spain

Joanna Macyszyn
Faculty of Environment Protection
Materials Recycling Center
 of Excellence
Wroclaw University of Technology
Wroclaw, Poland

Antonella Marra
Department of Materials
 Engineering and Production
University of Federico II
 of Naples
Naples, Italy

Marta Martínez-Sanz
Novel Materials and
 Nanotechnology Group
Institute of Agrochemistry
 and Food Technology
Spanish Council for Scientific
 Research
Valencia, Spain

Geoffrey R. Mitchell
Centre for Rapid and
 Sustainable Product
 Development
Polytechnic Institute of Leiria
Leiria, Portugal

and

Centre for Advanced Microscopy
University of Reading
Berkshire, United Kingdom

Marilena Pezzuto
National Research Council
Institute of Chemistry and
 Technology of Polymers
Pozzuoli, Italy

A.C. Pinheiro
IBB–Institute for Biotechnology and
 Bioengineering
Centre of Biological Engineering
Universidade do Minho
Braga, Portugal

M.A.C. Quintas
IBB–Institute for Biotechnology and
 Bioengineering
Centre of Biological Engineering
Universidade do Minho
Braga, Portugal

and

Escola Superior de Biotecnologia
Centro de Biotecnologia e Química
 Fina
Porto, Portugal

Irina Elena Răschip
Department of Physical Chemistry
 of Polymers
"P. Poni" Institute of
 Macromolecular Chemistry
Iasi, Romania

Chérifa Remili
Faculty of Technology
Laboratory of Organic Materials
Engineering Department
 Processes
University Abderrahmane Mira
Bejaia, Algeria

H.D. Silva
IBB–Institute for Biotechnology and
 Bioengineering
Centre of Biological Engineering
Universidade do Minho
Braga, Portugal

Clara Silvestre
National Research Council
Institute of Chemistry and
    Technology of Polymers
Pozzuoli, Italy

Naima Touati
Faculty of Technology
Laboratory of Organic Materials
Engineering Department Processes
University Abderrahmane Mira
Bejaia, Algeria

Cornelia Vasile
Department of Physical Chemistry
    of Polymers
"P. Poni" Institute of
    Macromolecular Chemistry
Iasi, Romania

Antonio A. Vicente
IBB–Institute for Biotechnology and
    Bioengineering
Centre of Biological Engineering
Universidade do Minho
Braga, Portugal

Lynda Zaidi
Faculty of Technology
Laboratory of Organic Materials
Engineering Department
    Processes
University Abderrahmane Mira
Bejaia, Algeria

1

Polymer Nanomaterials for Food Packaging: Current Issues and Future Trends

Clara Silvestre, Marilena Pezzuto, Sossio Cimmino, and Donatella Duraccio

CONTENTS

1.1 Introduction

Since its starting in the nineteenth century with the invention of canning, modern food packaging has made great advances as a result of global trends, technological improvements, and consumer preferences. The importance of canning as a food preservation and containment method was soon recognized and still is. In 1937, Prescott and Proctor of the Massachusetts Institute of Technology[1] described the importance of canning as follows: No technologic advance has exerted greater influence on the food habits of the civilized world than the development of heat treatment and the use of hermetically sealed (air-tight closure) containers for the preservation of food. In 2011, John D. Floros, head of the Food Science Department at Penn University, stated that the invention of the can and canning has truly helped society resolve major issues of hunger and diseases connected directly to lack of food or nutrients.[2]

Modern food science and technology has extended, expanded, and refined the traditional packaging methods and added new ones. Moreover with the move toward globalization, safety and longer shelf life are required, along with the monitoring of safety and quality based upon international standards. Nanotechnology, the manipulation of matter at the nanoscale, i.e., dimensions of approximately 1–100 nm, can address all these requirements and extend and implement the principal packaging functions—containment, protection and preservation, marketing, and communication.[3,4]

Nanotechnology is already having a dramatic effect on the entire food industry. In 2010, the nano-food industry reached a value of $20.4 billion, from $2.6 billion in 2006 and there are already over 200 companies pursuing research in food nanotechnology.[5] The nanomarket in 2020 is predicted to reach $3 trillion with 6 million workers involved.

Packaging based on nanotechnology, increasing the performance of the packaging, can also contribute to reducing the huge amount of food lost between the points of production and consumption, which is estimated to be about 30%–40% of the total and cannot be tolerated taking into consideration the 2009 report of the World Summit on Food Security, where it was recognized that by 2050 food production must increase by about 70%–34% higher than it is today to feed the anticipated 9 billion people (FAO 2009).[6]

In addition, as energy and resource conservation is becoming increasingly critical, there is a need for the future to provide sufficient food for everyone in a sustainable and environmentally responsible manner, without compromising the precious natural resources. Food manufacturing systems must therefore become more efficient, use less energy, generate less waste, and produce food with extended shelf life. Nanotechnology can address also this additional challenge.[7–18] Despite the large amount of research being undertaken in industry and academia, advanced technology for food packaging is still in a developmental stage. Successful technical development of materials for food packaging has to overcome several multidisciplinary barriers (science and technology, safety regulation, standardization, trained workforce, and technology transfer) in order that commercial products can benefit from the global market potential and requires therefore a high degree of multidisciplinarity.

The future of packaging based on nanomaterials is also contingent upon the way this emerging technology is handled by regulatory agencies. The enormous potential benefits offered by nanotechnology must be weighed against the potential risks of use and abuse of nanomaterials and in large part these risks are still being evaluated. When it comes to food and food packaging materials incorporating nanoscale materials, there are numerous data gaps that need to be filled in order to demonstrate product safety to a wary public. These data gaps include a lack of information regarding nanomaterial migration through packaging materials; the interaction of nanomaterial biomolecules and cellular components; the value of mass-based definitions of dosage in the context of nanomaterials; the interrelationships between nanoparticle characteristics (size, shape, surface charge, etc.) and toxicity or pharmacokinetic properties; appropriate and consistent methods to identify, characterize, and quantify nanomaterials in complex food matrices; chronic toxicity of nanomaterials or toxicity following oral routes of exposure; and biodegradability of nanomaterials or the toxicity of nanomaterials to ecologically important organisms.

The main drivers for most of the innovations are consumer and food service needs and demands mainly related to the global and fast transport of food. It is also required sustainability. According to the Sustainable Packaging Coalition®,[7] an international consortium of more than 200 industry members, sustainable packaging is characterized by the following main criteria:

- It is beneficial, safe, and healthy for individuals and communities throughout its life cycle.
- It meets market criteria for both performance and cost.
- It is sourced, manufactured, transported, and recycled using renewable energy.
- It optimizes the use of renewable or recycled source materials.
- It is manufactured using clean production technologies and best practices.
- It is made from materials healthy throughout the life cycle.
- It is physically designed to optimize materials and energy.
- It is effectively recovered and utilized in biological and/or industrial closed loop cycles.

1.2 Polymer Nanomaterials for Food Packaging

The introduction of polymers as food packaging materials began in the 1930s, assuming a continuously increasing role.[8–19] The first package made of polymer material, a pouch of vinyl-coated cellophane, proved useful in protecting food for a short period of time while providing flexibility, lightweight, and transparency needed to view the content. Due to the good mechanical, physical, and chemical properties of the polymers, as well as the wide range of properties that can be tailored by modifying the formulation and the processing conditions, the use of plastics as packaging materials increased and substituted many applications of metal, glass, and paperboard. In 1950s, polyvinyl chloride (PVC) and polyolefins emerged: PVC rapidly replaced glass for large containers for a variety of liquids enclosing liquors, but because of the finding of the migration of harmful residual monomer into the beverage, the use of PVC as material for liquid containers was stopped. Polyolefins became soon very popular due to their characteristics: resistance to several foodstuffs, resistance to impact, barrier to moisture and gas, easy moldability, and safety for food contact.[20–22] From then several types of polymers have been used as packaging materials. Thermoplastics represent the greatest share of polymers used in food packaging because they can be easily molded into any shape needed to fulfill the packaging function, moreover they have good characteristics in recycling and waste to energy conversion. Thermosets are also used as packaging materials, mainly used in closure and trays rather than primary food packages.[12,13] In the polymer global market that has increased from some 5 million ton in the 1950s to nearly 100 million ton today, almost 40% is covered by packaging, with the packaging industry itself worth about 2% of the gross national product in developed countries.[23] Polymer packaging provides many properties including strength and stiffness, barrier to oxygen and moisture, resistance to food component attack, and flexibility. Though polymers have revolutionized the food industry and possess numerous advantages over conventional materials, their major drawback is a generally low permeability to gases and other small molecules. In Figure 1.1, the permeability for water vapor and oxygen of several polymers is reported.[24]

Polymer nanotechnology can solve the aforementioned problems. Polymer nanotechnology involves the design, the processing, and the application of polymer matrices filled with particles and/or devices that have at least one dimension smaller than 100 nm.

Actually, polymers that are most frequently used for food packaging include, but are not limited to, polyolefins such as polypropylene (PP) and various grades of polyethylene (high-density polyethylene [HDPE], low-density polyethylene [LDPE], etc.), polyethylene terephthalate (PET), polystyrene (PS), and PVC.[25–29] Filler materials can include clay and silicate nanoplatelets, silica (SiO_2) nanoparticles,[30–32] carbon nanotubes,[33–39] graphene[39,40–42], starch nanocrystals[43,44], cellulose-based nanofibers or nanowhiskers,[45–53] and other inorganic nanoparticles.[54–57]

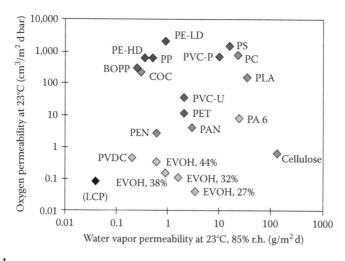

FIGURE 1.1
Permeability for water vapor and oxygen at 23°C and 100 μm film thickness of commonly used thermoplastics and special packaging type polymers.

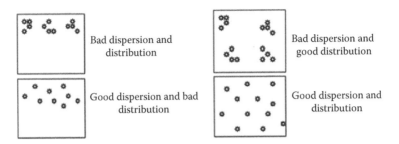

FIGURE 1.2
Schema of dispersion and distribution of nanoloads in a polymer matrix, where dispersion and distribution are the level of nanoloads' conglomeration and homogeneity in the matrix, respectively.

Polymer nanomaterials for food packaging (PNFP) present not only enhanced barrier properties but also several other improved properties such as, for example, mechanical properties[33,58–70] and flame resistance,[71–77] with respect to the unfilled polymers.

The properties of the polymer nanomaterials and their improvement with respect to traditional composites depend on the structure and morphology developed during the processing step, in particular the dispersion and distribution of nanoparticles in the matrix, the interfacial adhesion, and the influence of the nanoparticles on the polymer matrix. Moreover, in order to select the best packaging materials, the food processing methods (e.g., thermal/nonthermal processing methods, irradiation) should be taken into account, as stated by a recent review on the benefit of processing and packaging.[78]

In turn, structure and morphology depend on processing conditions, composition, and constituent properties. Without a good dispersion and distribution of nanoloads, the properties of nanocomposites are limited (Figure 1.2).

Good dispersion and distribution are the key factors for improved materials, and several methodologies have been applied depending on the shape of the nanoparticles and their compatibility with different polymer matrices in order to reach homogeneous morphology. The modification of the nanoparticle surfaces and/or the polymer, the optimization of the processing conditions (Figure 1.3), and the use of interfacial agents are the most used methodologies to achieve the best morphology.

For platelet-like nanoparticles (layered silicates are only compatible with polar polymer clay), the objective is to achieve an exfoliated structure with layers of the silicate completely separated and homogeneously distributed throughout the matrix. To make layered silicates (LS) compatible with nonpolar polymer matrices, the normally polar surface of the clay can be converted into an organophilic one, for example, by using alkyl ammonium chains.

For three-dimensional nanoparticles, the increase in compatibility can be obtained by modifying the nanoparticle surface, for example, with a grafting agent (covalent functionalization of nanoparticles) or coating agent (agent physically absorbed on nanoparticles).

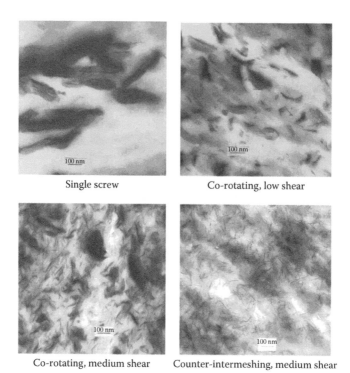

Single screw   Co-rotating, low shear

Co-rotating, medium shear   Counter-intermeshing, medium shear

FIGURE 1.3
Example of influence of processing conditions on morphology. Transmission electron photomicrographs using different extruders and screw configurations. (Reprinted from *Polymer*, 42, Dennis, H.R., Hunter, D.L., Chang, D., Kim, S., White, J.L., Cho, J.W., Paul, D.R., 9513. Copyright 2001, with permission from Elsevier.)

1.3 Innovative Food Packaging Solutions

Food packaging's principal aim is to ensure safe delivery of food to the consumer in sound condition.

Packaging performs a series of different functions. The principal functions are containment, protection and preservation, marketing, and communication.[9–11,17,79–82]

The containment function seems obvious, but without this function product loss and pollution would be widespread. The protection and preservation function, often regarded as the primary function of the package, is needed to maintain quality and safety of packaged food, to retard deterioration, and to extend the shelf life.

The communication function serves as a link between producer and consumer providing mandatory information such as weight, source, ingredients, and expiration date. This function can also contribute additional information to add appeal to consumers, maximizing sales and hence profits.

In order to select the best packaging material, the food processing methods, for example, thermal/nonthermal processing methods and irradiation, should be taken into account, as stated by a recent review on the benefit of processing and packaging.[78]

The use of polymer nanotechnology can extend and implement all the principal functions of the package.[9–11,79–82]

Food packaging based on polymer nanomaterial (PNFP) results in light materials with enhanced mechanical and thermal properties and can ensure better protection of the food. Once perfected, sure from a safety point of view, and produced at a competitive ratio cost/performances, the new material will be very attractive for use in several food packaging applications. This is the reason why many of the world's largest food packaging companies are actively exploring the potential of nanotechnology in order to obtain new food packaging materials with improved mechanical, barrier, and antimicrobial properties and to be able to trace and monitor the condition of food during transport and storage.

Three main applications for PNFP can be defined: "Improved" PNFP, "Active" PNFP, and "Intelligent/Smart" PNFP.

1.3.1 Improved PNFP

The possibility to increase the performances of the polymer for food packaging by adding nanoparticles has led to the development of a variety of polymer nanomaterials. The presence of nanoparticles in the polymer matrix improves the properties of the packaging materials by increasing flexibility, gas barrier properties, temperature/moisture stability, and biodegradability.[17,83–87]

PNFP are actually developed mainly to improve barrier performance pertaining to gases such as oxygen and carbon dioxide and water vapor.

Polymers incorporating clay nanoparticles are among the first polymer nanomaterials to emerge on the market as improved nanomaterials for food packaging.

The presence of nanofillers into a polymer matrix affects the barrier properties creating a tortuous path and modifying the polymer matrix at the polymer particles interface mainly in the case the interactions polymer-particles are favorable (increasing the rigidity of the material around the particles). The presence of nanoparticles modifies the pathway for gas diffusion impacting the barrier properties. Moreover, the nanoparticles can change the matrix characteristics (for example, crystallinity, structure, and morphology) from which the barrier properties are strongly dependent).[88–90]

Several examples are reported in literature; a nonexhaustive table reports examples of polymer–clay nanocomposite systems and the improvement on oxygen and water vapor permeability (see Table 1.1).

From all the literature papers it was stated that the barrier properties depend on processing method used and type of organic compatibilizer used. It was reported that increasing the number of octadecyl substituents in ammonia modifiers (that increase the d-spacing between individual clay platelets), the oxygen permeability of a linear HDPE decreases of almost 40%[108] Figure 1.4.

TABLE 1.1

Some Representative Polymer–Clay (Montmorillonite, MMT) Nanocomposite Systems and Their Improvement on Oxygen and Water Vapor Permeabilities, Expressed as Improvement Ratios: The Ratio of the Gas Permeability or Transmission Rate of the Virgin Polymer to the Gas Permeability or Transmission Rate of the Polymer Composite, Measured at the Same Conditions

Polymer	MMT (wt.%)	$P(O_2)$	$P(H_2O)$	References
Poly(imide)	8	13.0	7.4	[91]
	2	19.8	—	[92]
Poly(styrene)	16.7	2.8	—	[93]
Poly(amide)	5.5	>1100	—	[94]
Poly(ethylene terephthalate)	5	15.6	1.2	[95]
	5	2.23	1.15	[96]
Poly(urethane)	6	0.7–1.3	1.6–1.7	[97]
Poly(methylmethacrylate)	5	1.83	1.7	[98]
Poly(lactic acid)	5	1.16	1.21	[96]
	5	1.2–1.9	1.7–2.0	[99]
Poly(hydroxybutyrate-co-valerate)	5	1.36	2.16	[96]
Poly(caprolactone)	12	—	4.87	[100]
Poly(vinyl acetate)	6	—	~3	[101]
	20	>21	—	[102]
Poly(propylene)	5	~1.4	~1.7	[103]
	1	1.53	—	[104]
	1	1.24	—	
High-density poly (ethylene)	4	1.2–1.7	—	[90]
	5	2.8–2.9	1.8–2.4	[105]
Low-density poly (ethylene)	4.76	2.2	—	[106]
Thermoplastic starch	10	—	~1.7	[107]

Recently, by using a layer-by-layer assembly of a positively charged polymer and a negatively charged clay, a new nanomaterial was formed with a brick wall structure possessing ultrahigh barrier properties Figure 1.5.

In this method, the morphology is optimized and the clay platelets are organized in monolayers oriented in a direction perpendicular to the gas direction. These new nanocomposites have also shown extensive improvement in flame retardancy and strengths.[109,110]

Several different polymers and clay fillers can be used for obtaining improved clay–polymer nanomaterials. The nanoclay mineral generally used in these nanomaterials is montmorillonite, which is a relatively cheap and widely available natural clay derived from volcanic ash/rocks. Nanoclay has a natural nano-layer structure and when well dispersed in the matrix, it provides substantial improvements in gas barrier properties. Such improvements have led to the development of nanoclay–polymer nanomaterials for potential use in a variety of food packaging applications, such as processed meats, cheese, confectionery, cereals, boil-in-the-bag foods, as well as

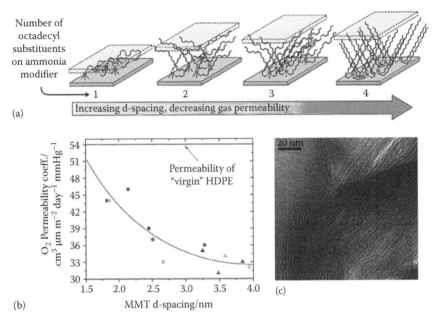

(a)

(b)

(c)

FIGURE 1.4
(a–c) Oxygen permeability of a linear HDPE as a function of d-spacing between individual clay platelets. (Reprinted from *Journal of Materials Chemistry*, 15, Maged, A.O., Rupp, Jörg, E.P., and Suter, Ulrich W., 1298, Copyright 2005, with permission from Royal Society of Chemistry.)

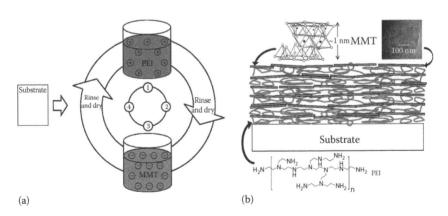

(a)

(b)

FIGURE 1.5
Brick wall structure possessing ultrahigh barrier properties obtained by using a layer-by-layer assembly of a positively charged polymer and a negatively charged clay. (a) Schematic of layer-by-layer assembly with cationic polyethylenimine (PEI) and anionic montmorillonite (MMT). (b) Crossectional illustration of the resultant nanobrick wall microstructure with the structures of MMT and a TEM image of the surface of a single clay platelet. (Reprinted from *Langmuir*, 27, Priolo, M.A., Holder, K.M., Gamboa, D. and Grunlan, J.C., 12106. Copyright 2011, with permission from American Chemical Society.)

in extrusion-coating applications for fruit juices and dairy products, or co-extrusion processes for the manufacture of bottles for beer and carbonated drinks. Examples of available nanoclay composites include (a) multi-layer PET bottles and sheets, polymerized nanocomposite nylon film for food and beverage packaging to minimize the loss of CO_2 from the drink and the ingress of O_2 into the bottle, thus keeping beverages fresher and extending their shelf life and (b) new hybrid plastics, which comprise polyamide (PA) and layered silicate barriers. The resulting films present increased barrier properties and enhanced gloss and stiffness. It is intended for use in applications where conventional PA is too permeable and ethylene vinyl alcohol (EVOH) coatings are too expensive, for example, paperboard juice containers. The current industrial applications of nanoclay in multilayer film packaging include beer bottles, carbonated drinks, and thermoformed containers.

Moreover, the incorporation of nanoclays into packaging offers several additional advantages:

- Reduction in raw materials. Improved stiffness enables the use of fewer raw materials, and down-gauging by 20% can be achieved. Lighter packaging may lead to savings in the cost of transportation, storage, and recycling.

- Less dependence on speciality products. Polymer–clay nanocomposites can be alternatives to expensive speciality materials.

- Elimination of secondary processes. High-cost operations such as laminations for barrier packaging or mechanical surface finishing can be eliminated.

- Less complex structures. Nanocomposites may have, for example, less complex structures than multilaminates and this can lead to easier recycling.

- Reduction in machine cycle time. By changing the physical and thermal properties of polymers, it is possible to reduce pack production times.

1.3.2 Active PNFP

Active polymer nanomaterials for food packaging are intended to maintain or improve the condition of packaged food. They are designed to deliberately incorporate components that would release or absorb substances into or from the packaged food or the environment surrounding the food. Polymer nanocomposites incorporating metal or metal oxide nanoparticles have been developed for antimicrobial "active" packaging, abrasion resistance, UV absorption, and/or strength. The nanomaterials used as UV absorbers (e.g., titanium dioxide) can prevent UV degradation in plastics such as PS, PE, and PVC. The metal and metal oxide nanomaterials commonly used are silver, gold, zinc oxide, silica, titanium dioxide, alumina, and iron oxides. Other semi-conductor nanoparticles (e.g., cadmium telluride/gallium arsenide) have also been used in the development of nanocomposites. A number of "active" PNFP have been developed based on the antimicrobial action of nanosilver, which is claimed to preserve the food materials for longer by inhibiting the growth of microorganisms. The antimicrobial properties of nano-zinc oxide and magnesium oxide have recently been discovered. Compared to nanosilver, the nanoparticles of zinc oxide and magnesium oxide are expected to provide a more affordable and safe food packaging solution in the future. Nanomaterials containing nano-zinc-oxide-based light catalyst, claimed to sterilize indoor lighting have been recently introduced.

Active packaging performs several tasks: In dependence of the task required, the mechanisms of the action as well as the substances used are very different; see examples in Table 1.2.[111,112]

TABLE 1.2

Mechanism of Action for Different Type of Active Packaging System

Type of Active Packaging System	Substances Used and Mode of Action
Oxygen scavengers	Enzymatic systems (glucose oxidase–glucose, alcohol oxidase–ethanol vapor)
	Chemical systems (powdered iron oxide, catechol, ferrous carbonate, iron–sulfur, sulfite salt–copper sulfate, photosensitive dye oxidation, ascorbic acid oxidation, catalytic conversion of oxygen by platinum catalyst)
Carbon dioxide absorbing/ emitting	Iron powder–calcium hydroxide, ferrous carbonate–metal halide
Moisture absorbing	Silica gel, propylene glycol, polyvinyl alcohol, diatomaceous earth
Ethylene absorbing	Activated charcoal, silica gel–potassium permanganate, Kieselguhr, bentonite, Fuller's earth, silicon dioxide powder, powdered oya stone, zeolite, ozone
Ethanol emitting	Encapsulated ethanol
Antimicrobial releasing	Sorbates, benzoates, propionates, ethanol, ozone, peroxide, sulfur dioxide, antibiotics, silver-zeolite, quaternary ammonium salts
Antioxidant releasing	Butylated hydroxyanisole (BHA), butylated hydroxytoluene (BHT), tertiary butyl hydroquinone (TBHQ), ascorbic acid, tocopherol
Flavor absorbing	Baking soda, active charcoal
Flavor releasing	Many food flavors
Color containing	Various food colors
Anti-fogging and anti-sticking	Biaxially oriented vinylon, compression rolled oriented HDPE
Light absorbing/regulating	UV blocking agents, hydroxybenzophenone
Monitoring	Time–temperature indicators
Temperature controlling	Non-woven microperforated plastic
Gas permeable/breathable	Surface treated, perforated, or microporous films
Microwave susceptors	Metallized thermoplastics
Insect repellant	Low toxicity fumigants (pyrethrins, permethrin)

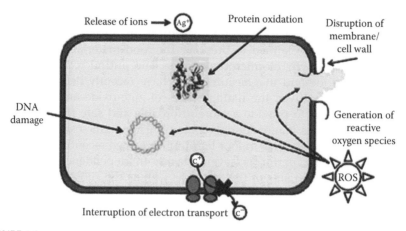

Release of ions → Ag⁺   Protein oxidation   Disruption of membrane/cell wall

DNA damage

Generation of reactive oxygen species

ROS

Interruption of electron transport

FIGURE 1.6

Various mechanisms of antimicrobial activities exerted by nanomaterials. (Reprinted from *Water Res.*, 42, Li, Q., Mahendra, S., Lyon, D.Y., Brunet, L., Liga, M.V., Li, D., and Alvarez, P.J.J., 4591, Copyright 2008, with permission from Elsevier.)

In the case of antimicrobial nanoparticles, several mechanisms of action have been proposed. The nanoparticles can cause cell damage directly interrupting the transport of the electrons, disrupting/penetrating the cell membrane, or oxidizing cell components or indirectly producing secondary products like reactive oxygen species or dissolved heavy metal ions. The schema of the various mechanisms of antimicrobial activities exerted by nanomaterials is reported in Figure 1.6.[113]

1.3.3 Intelligent/Smart PNFP

Intelligent food contact materials are intended to monitor the condition of packaged food and/or the environment surrounding the food. Polymer films incorporating nanodevices (sensors or antigen-detecting biosensors) are being developed for use in "smart" packaging. First developments were based on independent devices incorporated with the product in a conventional package. The embedded sensors in a packaging film are able to detect food-spoilage organisms and trigger a color change to alert the consumer that the shelf life is ending or has ended. Examples include devices that will provide a basis for intelligent preservative packaging technology that will release a preservative if food begins to spoil. Nanoscale-sensing devices are also under development that, when attached to packaging, will enable the food or food ingredients to be traced back to the source of origin. Further developments in this field include the so-called "Electronic Tongue" technology that is made up of sensor arrays to signal condition of the food. DNA-based biochips are also under development, which will be able to detect the presence of harmful bacteria in meat or fish or fungi affecting fruit. Other advances in this field include

TABLE 1.3

Intelligent Packaging Systems for Food Applications

Indicator	Principle/Reagents	Information About	Application
Time–temperature (external)	Mechanical Chemical Enzymatic	Storage conditions	Food stored under chilled and frozen conditions
Oxygen (internal)	Redox dyes pH dyes Enzymes	Storage conditions Package leak	Food stored in modified atmosphere (MA) packages
Carbon dioxide (internal)	Chemical	Storage conditions Package leak	MA or controlled atmosphere (CA) food packaging
Microbial growth (internal)	pH dyes All dyes reacting with certain metabolites (volatiles or nonvolatiles)	Microbial quality of food	Perishable food

nanostructured silicon with nanopores for potential applications in food packaging, with the aim to enable detection of pathogens in food and variations of temperature during food storage. Examples of application and intelligent packaging technologies used are summarized in Table 1.3.

1.4 Risks and Benefit of the Use of PNFP for Environment and Health

Although there are great performances of PNFP, the industrial applications are coming relatively slowly, with few large corporations like Honeywell, Mitsubishi Gas and Chemical, Bayer, Triton Systems, and Nanocor currently acting as pioneers. In general, it appears that there is a reluctance to embrace this new technology due to cost and variability in the quality of some of the products, drawbacks in the production of PNFP. Optimization of the processing conditions becomes a crucial point in the production, and it can be an expensive and low cost-competitive initiative. To use different materials with different properties (mainly flow characteristics and crystallization/solidification rate), it is necessary to have equipment conversion that accepts new material through recalibration. This is a big investment for converters to afford and cannot be wasted.

The use of PNFP is also posing concerns on the health due to indirect sources of food contamination by nanoparticles. While nanoscale materials possess more novel and unique physicochemical properties, they also have an unpredictable impact on human health. The challenge is to translate research and training into interim guidance for exposure assessment and exposure minimization and to fill the knowledge gap concerning the impact of nanomaterials on the human body, their interactions with biological systems, providing up-to-date information on the potential toxicological effects, and safety evaluation of the nanostructured materials on human health for a wide audience and their risk assessments. The main risk of consumer exposure to nanoparticles from food packaging is likely to be through potential migration of nanoparticles into food and drinks. However, only a few reports deal with migration data, despite the fact that a number of PNFP are already available and in commercial use in some countries.

Moreover, the penetration of the new products in the market is also driven by the consumer perception. Public perception is a major factor determining the commercial success of a new material/technology. Some papers and reports on public perception about nanotechnology stated that European consumers[114–117] are sceptical of the use of nanoparticles in food, probably because of the little awareness of nanotechnology, so the new products must be associated with a documented security that benefits will definitively outweigh the risks, bringing also safety and economical and environmental advantages. Moreover, recently, a study showed that when nanotechnology is briefly explained, nanotechnology applied to packaging is perceived as being more beneficial than nanotechnology-engineered food.

Appropriate information on risk/benefit assessment, as well as standardization in specific areas, can vitally contribute to effective dissemination of new knowledge, turning scientific curiosities into industrial powerhouses.

The foreseeable extensive use of nanotechnology by food packaging industry, as well as by any other sector using nanotechnology, is stirring up environment and health safety concerns. Nanomaterials can enter the environment in the course of their life cycle. How long they survive there, and in which form, i.e., how long they persist, is still a matter of investigation. Moreover, as new particles and applications are developed, and as more information becomes available on fate and behavior, routes of uptake and entry into the atmosphere must be updated. Several steps must be followed to determine the fate of the starting products of nanoscale substances and their transformation products resulting from production, processing, and use in the target compartments: identification of the nanoparticles that are persistent and accumulate in the environment through suitable measurement methods for the identification in water, soil, and sediment; analysis of the behavior of the nanomaterials after use, during disposal, land filling, incineration, or reutilization; and testing of ecotoxicity during the entire life cycle. A crucial factor for the determination of a risk of exposure to nanomaterials is the stability of these nanoparticles. The knowledge of the behavior and effects of nanoparticles in the environment and living organisms is increasing almost exponentially, caused by a massive interest in the scientific community and increased funding. However, the field is far from mature although current predictions suggest that the contribution of nanoparticles used in food packaging to the total of the environmental concentration seems to be trascurabile. Moreover nanoparticles can have also a positive effect because they can modify the durability of the polymer influencing degradation process in dependence of different environmental conditions.

Many people fear risk of indirect exposure due to potential migration of nanoparticles from packaging. Growing scientific research evidences that free nanoparticles can cross cellular barriers and that exposure to some of these nanoparticles may lead to oxidative damage and inflammatory reactions.

Three different ways of entrance penetration of nanoparticles in the organism are possible: inhalation, entrance through skin penetration, and ingestion.

For food packaging nanomaterials, the inhalation and the entrance through skin penetration is relative mainly to the workers in the factories producing the nanomaterials.

For the final consumers of food packaged with nanomaterials, the first concern is to verify the extent of migration of nanoparticles from the package into the food and then, if this migration happens, the effect of the ingestion of these nanoparticles inside the body from the mouth to the final gastrointestinal tract. There is a crucial need to understand how these particles will act when they get into the body, how the nanoparticles are absorbed by different organs, how the body metabolizes them, and how and in which way the body eliminates them. Few studies are present in the literature on the migration of nanoparticles from the package to the food. Although these cases seem to give some reassurance about safety, the number of tests on migration is too limited and further investigation needs to be performed before using these materials. The risk assessment of nanomaterials after ingestion has been studied only for few of the nanoparticles used in food packaging. Some results on TiO_2, Ag nanoparticles, and carbon nanoparticles/nanotubes show that nanoparticles can enter circulation from the gastrointestinal tract. These processes are likely to depend on the physical–chemical properties of the nanoparticles, such as size, and on the physiological state of the organs of entry. After the nanoparticles have reached the blood circulation, the liver and the spleen are the two major organs for distribution. For certain nanoparticles, all organs may be at risk including the brain, the reproductive system, and the fetus in utero. Conversely nanoparticles (clay) can also be used to slow down the rate of migration of potential harmful additives into food as it was reported in the case of substances from the matrix polymer (PA) into the food up to 6× as reported in reference.[118]

1.5 Regulatory Compliance for PNFP

As developments in nanotechnology continue to emerge, its applicability to the food packaging industry is sure to increase. The success of these advancements will be strictly dependent on exploration of regulatory issues.[119–121,122,123]

The majority of legislations dealing with nanotechnologies tend to be cautious toward potential risks posed by the new applications. Interesting is the case of Taiwan where they have introduced the Nano Mark System: This is a quality-like symbol of assurance to consumers, which certifies that a product uses a genuine nanotechnology.[124] In Australia, nanotechnologies are regulated by the NICNAS 2010 legislation,[125] which regulates chemicals for the protection of human health and the environment: This has recently introduced new administrative processes to address nanotechnology. NICNAS determines volumes, types, and data holdings of nanomaterials being used in Australia and has the responsibility of determining if legislation is sufficient to protect people from potential risks arising from nanotechnology. In this way, NICNAS monitors changes in industrial usage and can legislate accordingly to ensure legislation remains at the forefront of developments and to ensure emerging challenges in industrial chemical regulation, including the challenge of nanotechnology, are under control.

In the United States, multiple federal agencies regulate products associated with nanotechnologies and nanomaterials, but there is no regulatory framework that provides consistent and comprehensive screening and protections for consumers. The United States Food and Drug Administration's (FDA's) regulatory framework is challenged by the complexities of nanotechnologies and it is thought that risk assessment research is not progressing at a sufficient rate to deal with advancements in nanotechnologies.

The FDA issued in July 2007 its Nanotechnology Task Force Report. Anticipating the potential for rapid commercialization in the field, the FDA report recommended consideration of agency guidance that would clarify what information the industry needs to provide to the FDA about nanoproducts and also when the use of nanoscale materials may change the regulatory status of products. More information is available at http://www.nano.gov. Additional information may also be found at the National Cancer Institute website http://nano.cancer.gov; http://www.fda.gov/ScienceResearch/SpecialTopics/Nanotechnology/NanotechnologyTaskForceReport2007/default.htm. Draft guidance has been prepared by the FDA in June 2011,[124] intended for manufacturers, suppliers, importers, and other stakeholders. The guidance describes FDA's current thinking on whether FDA-regulated products contain nanomaterials or otherwise involve the application of nanotechnology. It states that there is a critical need to learn more about the potential role and importance of dimensions in the characteristics exhibited by engineered nanomaterials that may be used in producing products regulated by FDA. Premarket review, when required, offers an opportunity to better understand the properties and behavior of products that contain engineered nanomaterials or otherwise involve application of nanotechnology. And where products applying nanotechnology are not subject to premarket review, the agency urges manufacturers to consult with the agency early in the product development process. In this way, any questions related to the regulatory status, safety, effectiveness, or public health impact of these products can be appropriately and adequately addressed. It is also stated that FDA may issue additional guidances to address considerations for specific products or classes of products, consistent with the Principles for Regulation and Oversight of Emerging Technologies released in March 2011 as well as the Policy Principles for the U.S. Decision-Making Concerning Regulation and Oversight of Applications of Nanotechnology and Nanomaterials released in June 2011.[125,126]

The conclusion of this document reports that nanomaterials should not be deemed or identified as intrinsically benign or harmful in the absence of supporting scientific evidence, and regulatory action should be based on such scientific evidence. In general, however, and to the extent consistent with law, regulation should be based on risk, not merely hazard, and in all cases the identification of hazard, risk, or harm must be evidence based. In applying these principles, regulators should use flexible, adaptive, and evidence-based approaches that avoid, wherever possible, hindering innovation and trade while fulfilling the Federal Government's responsibility to protect public health and the environment.

In conclusion, it seems that the U.S. approach gives considerable credibility to the idea that "the dose makes the poison" so that toxicological justification is not needed or is greatly minimized by exposure assessments: In Europe, the approach is based on the theory that all materials should be explicitly cleared and publicized in regulations and that all clearances must be based on a toxicological evaluation of the listed substances, while the European approach starts from the principle that there must be toxicological data on all substances regardless of the level of anticipated exposure.

The main EU regulatory framework related to use of food contact materials, Regulation (EC) 1935/2004, states that any material intended for food contact must be suitable and inactive to avoid that the substances are transferred to products in such quantities to harm human health or to bring about an unacceptable change in food composition or properties. This rule applies to the migration of "nanocomponents" from packaging and to the use of "active packaging" and "intelligent packaging," that they are not inert by design, and, therefore, addresses the main requirements for their use. So any nano-sized ingredient intended to be released would have to be evaluated as a direct food additive. Restrictions on these substances take the form of limits of their migration into foodstuffs or limits on the composition of the materials. These rules are relevant also for nanomaterials but the safe maximum migration limits that have been determined for macrocomponents cannot be applied to their nanoequivalents, due to possible differences in their physical, chemical, or biological properties. EU regulation describes also test procedures. It has to be determined if the current test procedures are valid also with respect to the possible transfer of nanoparticles from materials into food. A working document is currently being discussed by the European Commission and Member States. Until such legislation is completed and adopted, nanomaterials will continue to be dealt with by a combination of general EU food law and more specific controls on particular materials.

The European Food Safety Authority (EFSA)[127] in February 2009 has concluded its assessment of the potential risks of nanotechnology for food and feed, providing a scientific opinion stating that a cautious, case-by-case approach is needed as many uncertainties remain over safe use. In particular, current uncertainties for risk assessment and the possible applications in the food arise from difficulty to characterize, detect, and measure nanoparticles and limited information available in relation to aspects of toxicology. The available data on oral exposure to specific nanoparticles and any consequent toxicity are extremely limited and the majority of the available information on toxicity of nanoparticles is from in vitro studies. In May 2011, the EFSA[128] published the guidance on the risk assessment of the application of nanoscience and nanotechnologies in the food and feed chain with the aims of giving a practical methodology for assessing the exposure and risk of engineering nanoparticles (ENM) in food, drink, and feed applications. This guidance identifies several exposure mechanisms and stipulates that any use of ENM in food or feed applications should follow appropriate scientific risk assessment based on ENM characterization, identification, physical–chemical properties, toxicity, and effective exposure levels to the ENM. The report also elaborates that toxicity data for non-nanoscale material is not enough for the full analysis but indicates that the presence of nanoform of a material should not stipulate presumed toxicity.

Characterization of ENM was viewed as one of the key steps in the risk analysis. The report listed all the characteristics of an ENM required to detect, identify, and quantify that includes chemical composition, particle size, physical form and morphology, particle and mass concentration, specific surface area, surface chemistry and charge, solubility and partition properties, pH and viscosity for liquid dispersions, density and dustiness for dry powders, chemical reactivity/catalytic activity, and photocatalytic activity. It is interesting to note that EFSA requires characterization of the size of the ENM by two independent analytical techniques, with one being electron microscopy. The report also acknowledges that the detection and characterization of ENM prior to its use in real food may be more straightforward than that in a final product because of the presence of complex matrices and usually low concentrations of ENM. The report states that appropriate in vitro and in vivo studies on the ENM should be undertaken to identify hazards and obtain dose–response data to characterize the hazard. Despite the lack of oral exposure data, the report insists on the use of ENM's absorption, distribution, metabolism, and excretion (ADME) data for in vivo testing. It also recognizes that the exposure quantification could be measured in several ways as mass, number of nanoparticles, or their effective surface area. Six cases for toxicity evaluation were outlined by the report and suggestion to go further when migration from food contact and transformation during digestion happens. Test on ENM's effects on in vitro and in vivo experiments are requested. These tests must focus on gene mutation, structural and numerical chromosome aberrations, cytotoxicity, oxidative stress, potential for inflammation, and immunotoxicity of ENMs. In vivo studies are required to follow the in vitro results. ADME information was quoted in the report as necessary data set to describe toxicokinetics of ENMs. For ADME studies, two different dose levels must be used so that this information may aid in dose setting in other toxicity studies.

With regard to the exposure levels of ENMs, the report recommends the use of existing methodology for non-nanoforms. The risk is therefore evaluated accordingly.

1.6 Future Trends and Conclusions

Nanotechnology has been applied by the packaging industry since a few years. According to recent reports (iRAP Inc. and BBC Research), the total nano-enabled food and beverage packaging market in the year 2008 was US$4.13 bln and forecasted to grow to US$7.3 bln by 2014. In food products, bakery and meat products have attracted the most nano-packaging applications, and in beverages, carbonated drinks and bottled water dominate; however, only a few of these systems have been developed and are being applied now. Among the regions, Asia/Pacific, in particular Japan, is the market leader in active nano-enabled packaging, with 45% of the current market, valued at US$1.86 bln in 2008 and projected to grow to US$3.43 bln by 2014. It is worth noticing that China, Taiwan, and South Korea have already taken a firm foothold in this market and are poised to be challenging competitors in this market. In the United States and Australia, improved and active packaging is already being successfully applied to extend shelf life while maintaining nutritional quality and ensuring microbiological safety.

The application will certainly increase when the cost of material and processing will decrease, when benefits of the novelty together with the assurances of no risks are not only demonstrated but also properly communicated. To date, with the exception of some materials such as nanoclays, the costs of material and processing are generally too great, compared to the advantages. Consequently, most packagings incorporating nanoparticles are currently receiving attention only at the research. This great opportunity for advancement will continue to be overlooked by the packaging industry until the cost becomes more affordable and consumers clearly accept the novelty.

Once perfected, insured for the health and environment, and produced at a competitive ratio cost/performances, the new material with improved mechanical, barrier, and antimicrobial properties and also able to trace and monitor the condition of food during transport and storage will be very attractive for use in several applications for delivering safe food to everyone. Moreover, in today's global market, it is necessary to harmonize worldwide the regulation requirements to improve both the level of food safety and the transparent communication to consumers.

In the next decade (Helmut Kaiser Consultancy), it is predicted that nanotechnology will change 25% of the food packaging business. The rocketing market growth comes mainly from the rapid multiplication of the applications employing nanotechnology. The new technologies hold great promise for contributing to meet the needs of safe food for everyone.

In conclusion, although there is a great deal of future promise for potential applications as high-performance materials, the industrial applications are coming slowly. In general, it appears that there is a reluctance to embrace this new technology due to costs of the material and variability in the quality and drawbacks in the production methodology (pre-polymerization and post-polymerization methods), together with the consumers' attitude to be sceptical in accepting new technology involving food. In addition to dreading the health and environmental consequences of nanofood consumption, consumers are also concerned about ethical or moral issues, including labeling and the right to choose, improvement of human abilities, engineering of living matter, privacy, equitability of socioeconomic distribution, and naturalness of the food supply. The processing conditions' optimization becomes a crucial point in the production. To use different material with different properties, it is necessary to have equipment conversion that accepts new material through recalibration. This is certainly a big investment for converters that must not be wasted by the fact that the new technologies are not adequately introduced to the public and accepted. At the same time, industry must provide to the public transparent evidence about its activities and governments must be alert for making the industry to respect the regulatory procedures.

The development of the new packaging nanotechnologies to improve food safety and food supply and to contribute to human health and wellness must be searched through the ethical and responsible development of nanotechnologies in accordance with precautionary, transdisciplinary, and life-cycle approaches. In particular, following steps are needed:

- Implementation of methodologies that take up PNFP beyond using just particle size and reactivity
- Promotion of mutually beneficial communication and science society dialogue between food technologists and consumers and other stakeholders all over the world with the continuous assessment of risks and benefit

- Dissemination of updated information on the possible impacts on human health and environment derived from the use, re-use, recycling and/or final treatment, and disposal of the new PNFP
- Development of harmonized standardization measures to ensure regulatory compliance of characterization of safe nanostructured materials is ensured for a faster penetration of the markets
- Increase of educational opportunities for young researchers in Europe based upon growing commercial success and advantages of the new nanomaterials

References

1. S.C. Prescott and B. Proctor, *Food Technology*, McGraw-Hill, New York (1937), p. 630.
2. J.D. Floros et al., *Comp. Rev. Food Sci. Food Safety*, 0 (2010) 1.
3. D.R. Paul and L.M. Robeson, *Polymer*, 49 (2008) 3187.
4. S. Sinha Ray and M. Bousmina, *Polymer Nanocomposites and Their Applications*, American Scientific Publishers, Los Angeles, CA (2008). ISBN: 158883-099-3.
5. M.C. Roco, C.A. Mirkin, and M.C. Hersam, *Nanotechnology Research Directions for Societal Needs in 2020: Retrospective and Outlook*, the World Technology Evaluation Center (WTEC) and Springer, Boston, MA, 2010. Further information at http://www.wtec.org/nano2. http://www.wtec.org/nano2/docs/Nano2-Brochure-Final-04-14-11.pdf
6. FAO 2009WSFS 2009/2, World summit on food security Rome, November 16–18 (2009), Declaration of the world summit on food security.
7. http://www.sustainablepackaging.org/content/?type=5&id=definition-of-sustainable-packaging
8. C. Silvestre, S. Cimmino, and D. Duraccio, *Prog. Polym. Sci.*, 36 (2011) 1766.
9. A. Arora and G.W. Padua, *J. Food Sci.*, 75 (2010) R43, January/February.
10. Q. Chaudhry, M. Scotter, J. Blackburn, B. Ross, A. Boxall, L. Castle, R. Aitken, and R. Watkins, *Food Addit. Contam.*, 25 (2008) 241.
11. Henriette M.C. De Azeredo, *Food Res. Int.*, 42 (2009) 1240.
12. C.I. Moraru, C.P. Panchapakesan, Q. Huang, P. Takhistov, S. Liu, and J.L. Kokini, *Food Technol.*, 57(2003) 24.
13. P. Sanguansri and M.A. Augustin, *Trends Food Sci. Tech.*, 17 (2006) 547.
14. G.L. Robertson, *Food Packaging: Principles and Practice*, 2nd edn., CRC Press, Boca Raton, FL (2006). ISBN 978-0-8493-3775-8.
15. K. Marsh and B. Bugusu, *J. Food Sci.*, 72 (2007) R39.
16. W. Soroka, *Fundamentals of Packaging Technology*, 4th edn., DEStech Publications, Inc., Lancaster, PA (2009).
17. A.K. Mohanty, M. Misra, and H.S. Nalwa, ed., *Packaging Nanotechnology*, Stavenson Ranch, CA, American Scientific Publisher (2009). ISBN 1588831051.
18. A.L. Brody, *Food Technol.*, 61 (2007) 80.

19. J.M. Lagarón, ed., *Multifunctional and Nanoreinforced Polymers for Food Packaging*, Woodhead Publishing Limited (2011). ISBN 1 84569 738 3, ISBN-13: 978 1 84569 738 9.
20. J. Jordan, K.I. Jacob, R. Tanenbaum, M. Sharaf, and I. Jasiuk, *Mater. Sci. Eng. A*, 393 (2005) 1.
21. S.E.M. Selke, J.D. Culter, and R.J. Hernandez., eds., *Plastics Packaging: Properties, Processing, Applications, and Regulations*, 2nd edn., John D. Culter Publications, Inc., Carl Hanser Verlag, Munich, Germany (2004). ISBN 3-446-21404-6, Gardner.
22. P. Hook and J.E. Heimlich, A history of packaging, Ohio State University Fact Sheet. http://ohioline.osu.edu/cd-fact/0133.html (accessed on October 26, 2005).
23. D.K.R. Robinson and M.J. Morrison, Nanotechnologies for food packaging: Reporting the science and technology research trends: Report for the ObservatoryNANO, August 2010. www.observatorynano.eu
24. K. Noller, M. Schmid, C. Schönweitz, D. Guerin, and C. Stinga, FlexPakRenew Workshop, Lyon, France, May 10, 2011, no. 13. http://www.flexpakrenew.eu/docs/filesProject/1/C1A9C54C-188B-310B-B8DF99255638B647.pdf (accessed on June 24, 2011).
25. C. Silvestre, M.L. Di Lorenzo, and E. Di Pace. In Vasile C., ed., *Handbook of Polyolefins: Second Edition, Revised and Expanded*, Marcel Dekker, New York (2000), Chapter 9, pp. 223–248. ISBN 0-8247-8603-3.
26. C. Silvestre, S. Cimmino, and E. Di Pace, Morphology of polyolefins. In Vasile C., ed., *Handbook of Polyolefins: Second Edition, Revised and Expanded*, Marcel Dekker, New York (2000), Chapter 7, pp. 175–205. ISBN 0-8247-8603-3 (2000).
27. A.L. Brody, B. Bugusu, J.H. Han, C.K. Sand, and T.H. Mchugh, *J. Food Sci.*, 73 (2008) R107.
28. D. Duraccio, C. Silvestre, M. Pezzuto, S. Cimmino, and A. Marra, Chapter 6, this book.
29. A. Emamifar, Application of antimicrobial polymer nanocomposites in food packaging. In A. Hashim, ed., *Advances in Nanocomposite Technology*, Publisher InTech (2011), Chapter 13. ISBN978-953-307-347-7.
30. V. Vladimirov, C. Betchev, A. Vassiliou, G. Papageorgiou, and D. Bikiaris, *Compos. Sci. Technol.*, 66 (2006) 2935.
31. S. Tang, P. Zou, H. Xiong, and H. Tang, *Carbohydr. Polym.*, 72 (2008) 521.
32. X. Jia, Y. Li, Q. Cheng, S. Zhang, and B. Zhang, *Eur. Polym. J.*, 43 (2007) 1123.
33. W. Chen, X. Tao, P. Xue, and X. Cheng, *Appl. Surf. Sci.*, 252 (2005) 1404.
34. Y. Bin, M. Mine, A. Koganemaru, X. Jiang, and M. Matsuo, *Polymer*, 47 (2006) 1308.
35. H. Zeng, C. Gao, Y. Wang, P.C.P. Watts, H. Kong, X. Cui, and D. Yan, *Polymer*, 47 (2006) 113.
36. S. Morlat-Therias, E. Fanton, J.L. Gardette, S. Peeterbroeck, M. Alexandre, and P. Dubois, *Polym. Degrad. Stab.*, 92 (2007) 1873.
37. J.Y. Kim, S.I. Han, and S.H. Kim, *Polym. Eng. Sci.*, 47 (2007) 1715.
38. K. Prashantha, J. Soulestin, M.F. Lacrampe, P. Krawczak, G. Dupin, and M. Claes, *Compos. Sci. Technol.*, 69 (2009) 1756.
39. J.Y. Kim, S.I. Han, D.K. Kim, and S.H. Kim, *Compos. Part A Appl. Sci. Manuf.*, 40 (2009) 45.
40. T. Ramanathan, A.A. Abdala, S. Stankovich, D.A. Dikin, M. HerreraAlonso, R.D. Piner, D.H. Adamson et al., *Nat. Nanotechnol.*, 3 (2008) 327.

41. K. Wakabayashi, C. Pierre, D.A. Dikin, R.S. Ruoff, T. Ramanathan, L.C. Brinson, and J.M. Torkelson, *Macromolecules*, 41 (2008) 1905.
42. C. Borriello, A. De Maria, N. Jovic, A. Montone, M. Schwarz, and M.V. Antisari, *Mater. Manuf. Process.*, 24 (2009) 1053.
43. Y. Chen, X. Cao, P.R. Chang, and M.A. Huneault, *Carbohydr. Polym.*, 73 (2008) 8.
44. E. Kristo and C.G. Biliaderis, *Carbohydr. Polym.*, 68 (2007) 146.
45. H.M.C. Azeredo, L.H.C. Mattoso, D. Wood, T.G. Williams, R.J. Avena-Bustillos, and T.H. McHugh, *J. Food Sci.*, 74 (2009) N31.
46. H.M.C. Azeredo, L.H.C. Mattoso, R.J. Avena-Bustillos, G.C. Filho, M.L. Munford, D. Wood, and T.H. McHugh, *J. Food Sci.*, 75 (2010) N1.
47. C. Bilbao-Sáinz, R.J. Avena-Bustillos, D.F. Wood, T.G. Williams, and T.H. McHugh, *J. Agric. Food Chem.*, 58 (2010) 3753.
48. A. Dufresne, J.Y. Cavaillé, and W. Helbert, *Macromolecules*, 29 (1996) 7624.
49. W. Helbert, J.Y. Cavaillé, and A. Dufresne, *Polym. Compos.*, 17 (1996) 604.
50. P. Podsiadlo, S.Y. Choi, B. Shim, J. Lee, M. Cuddihy, and N.A. Kotov, *Biomacromolecules*, 6 (2005) 2914.
51. M.A.S.A. Samir, F. Alloin, J.Y. Sanchez, and A. Dufresne, *Polymer*, 45 (2004) 4149.
52. K. Oksman, A.P. Mathew, D. Bondeson, and I. Kvien, *Compos. Sci. Technol.*, 66 (2006) 2776.
53. X. Cao, Y. Chen, P.R. Chang, M. Stumborg, and M.A. Huneault, *J. Appl. Polym. Sci.*, 109 (2008) 3804.
54. D. Yang, Y. Hu, L. Song, S. Nie, S. He, and Y. Cai, *Polym. Degrad. Stab.*, 93 (2008) 2014.
55. F. Zhang, H. Zhang, and Z. Su, *Polym. Bull.*, 60 (2008) 251.
56. X.D. Ma, X.F. Qian, J. Yin, and Z.K. Zhu, *J. Mater. Chem.*, 12 (2002) 663.
57. V.E. Yudin, J.U. Otaigbe, S. Gladchenko, B.G. Olson, S. Nazarenko, E.N. Korytkova, and V.V. Gusarov, *Polymer*, 48 (2007) 1306.
58. S.S. Ray and M. Okamoto, *Prog. Polym. Sci.*, 28 (2003) 1539.
59. H.M. Park, X. Li, C.Z. Jin, C.Y. Park, W.J. Cho, and C.S. Ha, *Macromol. Mater. Eng.*, 287 (2002) 553.
60. L. Cabedo, J.L. Feijoo, M.P. Villanueva, J.M. Lagarón, and E. Giménez. *Macromol. Symp.*, 233 (2006) 191.
61. G. Gorrasi, M. Tortora, V. Vittoria, E. Pollet, M. Alexandre, and P. Dubois, *J. Polym. Sci. Part B Polym. Phys.*, 42 (2004) 1466.
62. S.S. Ray and M. Bousmina, *Prog. Mater. Sci.*, 50 (2005) 962.
63. N. Ogata, G. Jimenez, H. Kawai, and T. Ogihara, *J. Polym. Sci. Part B Polym. Phys.*, 35 (1997) 389.
64. M.A. Osman, J.E.P. Rupp, and U.W. Suter, *Polymer*, 46 (2005) 1653.
65. Y. Kojima, A. Usuki, M. Kawasumi, A. Okada, Y. Fukushima, T. Kurauchi, and O. Kamigaito, *J. Mater. Res.*, 8 (1993) 1185.
66. Y. Kojima, A. Usuki, M. Kawasumi, A. Okada, T. Kurauchi, and O. Kamigaito, *J. Polym. Sci. Part A Polym. Chem.*, 31 (1993) 1755.
67. C.E. Powell and G.W. Beall, *Curr. Opin. Solid State Mater. Sci.*, 10 (2006) 73.
68. S.Y. Gu, J. Ren, and B. Dong, *J. Polym. Sci. Part B Polym. Phys.*, 45 (2007) 3189.
69. S. Cimmino, C. Silvestre, D. Duraccio, and M. Pezzuto, *J. Appl. Polym. Sci.*, 119 (2011) 1135.
70. S. Zhang, T.R. Hull, A.R. Horrocks, G. Smart, B.K. Kandola, J. Ebdon, P. Joseph, and B. Hunt, *Polym. Degrad. Stab.*, 92 (2007) 727.
71. A.R. Horrocks, B.K. Kandola, G. Smart, S. Zhang, and T.R. Hull, *J. Appl. Polym. Sci.*, 106 (2007) 1707.

72. B.K. Kandola, G. Smart, A.R. Horrocks, P. Joseph, S. Zhang, T.R. Hull, J. Ebdon, B. Hunt, and A. Cook, *J. Appl. Polym. Sci.*, 108 (2008) 816.
73. G. Smart, B.K. Kandola, A.R. Horrocks, S. Nazaré, and D. Marney, *Polym. Adv. Technol.*, 19 (2008) 658.
74. D. Porter, E. Metcalfe, and M.J.K. Thomas, *Fire Mater.*, 24 (2000) 45.
75. J.W. Gilman, C.L. Jackson, A.B. Morgan, R. Harris, E. Manias, E.P. Giannelis, M. Wuthenow, D. Hilton, and S.H. Phillips, *Chem. Mater.*, 12 (2000) 1866.
76. A.B. Morgan, L.L. Chu, and J.D. Harris, *Fire Mater.*, 29 (2005) 213.
77. K. Galic, M. Scetar, and M. Kurek, *Trends Food Sci. Technol.*, 22 (2011) 127.
78. J.M. Lagaron, E. Gimenez, M.D. Sánchez-García, M.J. Ocio, and A. Fendler, Novel nanocomposites to enhance quality and safety of packaged foods. In: *Food Contact Polymers*, Rapra Technologies, Shawbury, U.K. (2007), paper 19.
79. K. Amar Mohanty, M. Misra, and Hari S. Nalwa, ed., *Packaging Nanotechnology*, American Scientific Publisher, Stevenson Ranch, CA (2009). IISBN: 1-58883-105-1.
80. M.E. Doyle, *2006 Nanotechnology: A Brief Literature Review*. Food Research Institute Briefings, June 2006. http://www.wisc.edu/fri/briefs/FRIBrief_Nanotech_Lit_ Rev.pdf (accessed on August 4, 2008).
81. A.L. Brody, *Food Technol.*, 60 (2003) 92.
82. C.L. Wu, M.Q. Zhang, M.Z. Rong, and K. Friedrich, *Compos. Sci. Technol.*, 62 (2002) 1327.
83. J.M. Lagaron, R. Català, and R. Gavara, *Mater. Sci. Technol.*, 20 (2004) 1.
84. D.A.P. De Abreu, P.P. Losada, I. Angulo, and J.M. Cruz, *Eur. Polym. J.*, 43 (2007) 2229.
85. S. Cimmino, D. Duraccio, C. Silvestre, and M. Pezzuto Isotactic, *Appl. Surf. Sci.*, 256 (2009) S40.
86. R. Kotsilkova, C. Silvestre, and S. Cimmino, Thermoset nanocomposites for engineering applications. In: Kotsilkova R., ed., *Thermoset Nanocomposites for Engineering Applications*, Rapra Technology, Shawbury, U.K. (2007), Chapter 3. ISBN 978-1-84735-062-623.
87. Z. Mogri and D.R. Paul, *Polymer*, 42 (2001) 2531.
88. A. Hiltner, R.Y.F. Liu, Y.S. Hu, and E. Baer, *J. Polym. Sci. Part B Polym. Phys.*, 43 (2005) 1047.
89. H. Wang, J.K. Keum, A. Hiltner, E. Baer, B. Freeman, A. Rozanski, and A. Galeski, *Science*, 323 (2009) 757.
90. M.A. Osman, J.E.P. Rupp, and U.W. Suter, *J. Mater. Chem.*, 15 (2005) 1298.
91. K. Yano, A. Usuki, A. Okada, T. Kurauchi, and O. Kamigaito, *J. Polym. Sci., Part A Polym. Chem.*, 31 (1993) 2493.
92. J.H. Chang, K.M. Park, D. Cho, H.S. Yang, and K.J. Ihn, *Polym. Eng. Sci.*, 41 (2001) 1514.
93. S. Nazarenko, P. Meneghetti, P. Julmon, B.G. Olson, and S. Qutubuddin, *J. Polym. Sci. Part B Polym. Phys.*, 45 (2007) 1733.
94. D. Pereira, P.P. Losada, I. Angulo, W. Greaves, and J.M. Cruz, *Polym. Compos.*, 30 (2009) 436.
95. W.J. Choi, H.-J. Kim, K.H. Yoon, O.H. Kwon, and C.I. Hwang, *J. Appl. Polym. Sci.*, 100 (2006) 4875.
96. M.D. Sanchez-Garcia, E. Gimenez, and J.M. Lagaron, *J. Plast. Film Sheet.*, 23 (2007) 133.
97. M.A. Osman, V. Mittal, M. Morbidelli, and U.W. Suter, *Macromolecules* 36 (2003) 9851.

98. J.M. Yeh, S.J. Liou, M.C. Lai, Y.-W. Chang, C.-Y. Huang, C.-P. Chen, J.-H. Jaw, T.Y. Tsai, and Y.-H. Yu, *J. Appl. Polym. Sci.*, 94 (2004) 1936.

99. C. Thellen, C. Orroth, D. Froio, D. Ziegler, J. Lucciarini, R. Farrell, N.A. D'Souza, and J.A. Ratto, *Polymer* 46 (2005) 11716.

100. P.B. Messersmith and E.P. Giannelis, *J. Polym. Sci., Part A Polym. Chem.*, 33 (1995) 1047.

101. K.E. Strawhecker and E. Manias, *Chem. Mater.*, 12 (2000) 2943.

102. J.C. Grunlan, A. Grigorian, C.B. Hamilton, and A.R. Mehrabi, *J. Appl. Polym. Sci.*, 93 (2004) 1102.

103. M. Sirousazar, M. Yari, B.F. Achachlouei, J. Arsalani, and Y. Mansoori, *e-Polymer* (2007) #027, ISSN 1618–7229.

104. M. Pezzuto, Polyolefins based hybrid nanocomposites, PhD Thesis, Chapter 6, University of Federico II, Naples (2011).

105. C. Lotti, C.S. Isaac, M.C. Branciforti, R.M.V. Alves, S. Liberman, and R.E.S. Bretas, *Eur. Polym. J.*, 44 (2008) 1346.

106. S. Dadbin, M. Noferesti, and M. Frounchi, *Macromol. Symp.*, 274 (2008) 22.

107. H.M. Park, W.-K. Lee, C.-Y. Park, W.-J. Cho, and C.-S. Ha, *J. Mater. Sci.*, 38 (2003) 909.

108. H. Nandivada, B.G. Pumplin, J. Lahann, A. Ramamoorthy, and N.A. Kotov, *Science*, 318 (2007) 80.

109. Y.C. Li, J. Schultz, S. Mannen, C. Delhom, B. Condon, S. Chang, M. Zammarano, and J.C. Grunlan, *ACS Nano*, 3 (2010) 3325.

110. D. Restuccia, U.G. Spizzirri, O.I. Parisi, G. Cirillo, M. Curcio, F. Iemma, F. Puoci, G. Vinci, and N. Picci, *Food Control*, 21 (2010) 4125.

111. M. Ozdemir and J.D. Floros, *Food Sci. Nutr.*, 44 (2004) 185.

112. M.D. Cobb and J. Macoubrie, *Nanopart. Res.*, 6 (2004) 395.

113. Q. Li, S. Mahendra, D.Y. Lyon, L. Brunet, M.V. Liga, D. Li, and P.J.J. Alvarez, *Water Res.*, 42 (2008) 4591.

114 G. Gaskell, T.T. Eyck, J. Jackson, and G. Veltri, *Public Underst. Sci.*, 14 (2005) 81.

115. M. Siegrist, M.E. Cousin, H. Kastelnhoz, and A. Wiek, *Appetite*, 49 (2007) 459.

116 N. Sozer and J.L. Kokini, *Trends Biotechnol.*, 27 (2009) 82.

117. D.A. Pereira de Abreu, D. Antonio et al., *Packag. Technol. Sci.*, 23 (2010) 59.

118. M. Cushen, J. Kerry, M. Morris, M. Cruz-Romero E. Cummins et al., *Trends Food Sci. Technol.*, 24 (2012) 30.

119. K. Powers, S. Brown, V. Krishna, S. Wasdo, B. Moudgil, and S. Roberts, *Toxicol. Sci.*, 90 (2006) 296.

120 M.R. Wiesner and J.Y. Bottero, *Environmental Nanotechnology: Applications and Impacts of Nanomaterials*, McGraw-Hill Professional, New York (2007).

121. P. Simon, Q. Chaudhry, and D. Bakos, *J. Food Nutr. Res.*, 47 (2008) 105.

122. C.F. Chau, S.H. Wu, and G.C. Yen, *Trends Food Sci. Technol.*, 18 (2007) 269.

123. K. Lyons and J. Whelan, *NanoEthics*, 4 (2010) 53.

124. FDA Regulation of Nanotechnology, Mark N. Duvall, February (2012).

125. http://www.whitehouse.gov/sites/default/files/omb/inforeg/foragencies/Principles-for-Regulation-and-Oversight-of-Emerging-Technologies-new.pdf

126. http://www.whitehouse.gov/sites/default/files/omb/inforeg/foragencies/nanotechnology-regulation-and-oversight-principles

127. *EFSA J.*, 958 (2009) 1.

128. *EFSA J.*, 9 (2011) 2140.

2

Ethics, Communication, and Safety in the Use of PNFP

Yasemin J. Erden

CONTENTS

2.1 Introduction

2.1.1 Why Ethics? Why PNFP?

If it is not already apparent why a chapter on ethics would be important to the discussion of developments in polymer nanomaterials for food packaging (PNFP), then this chapter will show how and why it ought to be. To do this will involve discussion of both specific and general ethical issues arising from the development of PNFP, and engage with pertinent environmental, health, and safety (EHS) issues by way of example and analysis. There will also be discussion regarding the role of communication and stakeholder dialogue for ensuring that the testing, production, and marketing of PNFP follow models such as the precautionary principle. The primary aim of the chapter is to introduce "ethical design" as a concept that should play a key role in the development of PNFP. Ethical design will be offered as a paradigm for putting key ethical principles, such as the precautionary principle, into practice and includes issues about accountability and regulation.

Before that, however, there are some other questions that require our attention, the first of which being why PNFP? While the answer to this question
might seem more obvious to the reader than the question that began this
chapter, I will show that some of these typical answers rely on assumption and narrow thinking. For example, we might ask, what problems do
PNFP solve? This first question begets a second: are there other (potentially
more ethical) ways to achieve the solution of these perceived problems?
For example, while polymers may solve issues associated with packaging
weight, thereby lessening the environmental impact of transportation of
consumables, or by limiting food spoilage, this does not mean that other
hazards or ethical issues are likewise avoided. For instance, how does this
solve the problem of overconsumption and perceptions around disposabil-
ity, more broadly conceived? Might these sorts of problems be better solved
through the promotion of more limited consumption models, for example?
Furthermore, improvements to the weight of packaging may not mean that
this packaging is better for the environment at the disposal stage, nor that
it reduces the volume of packaging waste in general, or fit with other environmental concerns. For instance, PNFP might release harmful toxins into
the environment whether during production, or at postproduction stages,
including after disposal.

This chapter will engage with these and associated issues and in so doing
offer the view that ethical questions of the sort raised before, alongside others considered in the next section, need to be at the forefront of design from
the conceptual stages onward. Ethical design, which puts into practice key
ideas like the precautionary principle, presents a model of accountability,
responsibility, and ethical consideration commensurate with our current
understanding of all sorts of risks and hazards, whether they are known,
unknown, or unknowable. In this discussion, it is in fact better to think in
terms of hazard, rather than in more traditional terms of risk. The key difference between the two is that hazard involves looking at where something
may be potentially dangerous, but if handled well, need pose no risk. Where
circumstances change, however, it may then become a risk. On this view,
risk can be understood as hazard *times* exposure. Within ethical design, as
I explain later, focus should be on hazard assessment, particularly where
nanomaterials are involved. This is because high levels of uncertainty about
nanomaterials means that risk assessment in traditional terms becomes much
more difficult. Despite which, case studies from the past will be invoked to
show why there can be no justification in claiming ignorance because of epis-
temological uncertainty.

To do this, I offer later the first in a series of examples that will provide
some answers to the first question that was posed earlier: why ethics? To
put it another way, why should ethical thinking and planning be applied
throughout all stages of the design and development of PNFP? After this,
section two of this chapter will look at some of the benefits and setbacks
associated with PNFP, and in so doing introduce some examples from history

that show where ethical planning, early on, can be used to prevent future hazard. Section three will introduce the concept of ethical design and offer a paradigm by which hazards of the sort noted earlier might be avoided. The chapter ends with a section on legislation and discussion about the future for PNFP. Before this, however, we will begin by considering some of the broader issues associated with packaging.

2.1.2 Case Study: Packaging, Environment, and Public Perception

Packaging is everywhere. It covers, protects, and can offer direct advertising to consumers, and in this way, influence consumer behavior. According to a report by the Advisory Committee on Packaging, at least 10 million tons of packaging are used each year.[1] Consumables, technology, toys, health supplements, almost everything comes with packaging, and not all of this is going anywhere fast. In fact, since a substantial quantity of modern packaging is biopersistent, much of this is likely to remain long beyond the individual persistence of the product for which the packaging was produced. Excepting those instances, where the product itself is also biopersistent. These sorts of environmental issues have already begun to impact on regulation regarding the production, use, and disposal of packaging materials. Indeed, public awareness in Europe regarding environmental issues about packaging is currently high, although the question that this claim generates, concerns the difference between perception and understanding. For example, local and central governmental campaigns for recycling in the United Kingdom, including drives to limit the availability of single use carrier bags, have ensured that the issue of packaging waste is at the forefront of consumer concerns.[2] As a result, initiatives to encourage the recycling and reuse of packaging are so familiar that the RecycleNow brand claims their logo is recognizable to at least 65% of people in England.[3]

The same might also be said of the *chasing arrows* logo, based on the Mobius triangle, which is now a universal symbol associated with the recyclability, or otherwise, of a given material. The logo was created in 1970 by Gary Anderson (then a student of the University of California) and submitted to a design contest held by the Container Corporation of American (CCA). After an unsuccessful bid to trademark the logo by the CCA, the logo "fell into public domain."[4] It has since been used internationally to identify the specific material of a product, especially within the plastics industry. The likelihood of finding this logo on plastic material is especially high because in 1988 The Society of the Plastics Industry (SPI) "instituted a voluntary labeling system," consisting of a code number "placed inside the symbol to specify the primary resin used in the product (SPI 1988)."[4] This code has proved so successful that it is now commonplace on a range of materials. What is of particular interest within this story is the gulf that seems to have arisen between what the logo represents and what it is perceived to represent. As Siddique explains, "many people across the globe associate the logo with recycling while the products on which they appear do not have to be recyclable."[4] In fact, its association with the "reduce, reuse, recycle" campaign in the United Kingdom has only further cemented its identification with recycling. The problems with this are manifold. In particular, it leads to uncertainty about what can and cannot be recycled, and as such it increases the likelihood of the contamination of a recycling batch (potentially making the batch unusable). In his American study, Siddique found that "the perception of all plastics as recyclable is widespread," for which he offers a number of likely causes, some of which he claims may stem from "actions taken by plastic manufacturers and the plastic industry as a whole."[5] Within these he includes "misperceptions" arising from the widespread use of the Mobius logo, with 67% of respondents in his study "thinking the chasing arrows guarantees recyclability."[6]

What this example shows is that when it comes to packaging, consumer perception and decisions have an important role to play, even if these perceptions are misguided. If, for instance, consumer perception about PNFP proves to be negative, this could impact dramatically on consumer acceptance of PNFP, and subsequently on legislation. As such, there are a number of issues to consider: what packaging does and does not do, what it should and should not do, and understanding, informing, and engaging with public perception about these matters. To begin this discussion we need to return to our earlier question about why PNFP, and before that, why packaging in general?

2.2 Benefits and Setbacks of PNFP

2.2.1 Benefits

As noted earlier, food packaging serves a number of functions. Some of these primary functions are to contain consumables safely, securely, and in so doing protect the contents and the area within which the product is being stored, handled, and distributed. In protecting the contents, packaging typically has a number of roles to fulfill. To prolong the shelf life and ensure quality, to ensure that contaminants, pathogens, gases (including oxygen), and pests cannot get in, and that important qualities of the product are not lost, including moisture and gases. Once the product is in retail stores, packaging often serves as the first point of contact between products and the consumer, and as such it simultaneously allows consumers to identify the product and to understand its contents and how to use them. It may also influence purchase decisions. Once purchased, packaging needs to be accessible (easy to open, close, and sometimes reuse during its purposeful lifetime), and should be easily disposable, recyclable, or reusable. Food packaging should not contribute negatively to the products it contains, or impact upon shelf life. Nor should it adversely affect the properties of the content with regard to health and safety.

Plastics are useful for packaging for a number of reasons, not least because they weigh less and are more durable and dependable than more traditional materials, such as tin or glass. These factors are important for the environment, both in terms of energy saved during shipping and storage and the fact that many plastics are now also recyclable. There are often economic benefits, since plastics are typically relatively inexpensive. Furthermore, plastics are largely efficient. Plastic can be bent and can remain flexible, which means it is less likely to be damaged. Even where damage occurs, it is less likely that the contents will be affected. This is true for shipping and storage, as well as in consumer use later on. In addition, plastic offers a broad range of malleable properties, including a number of impermeable barrier properties with regard to water and gases. PNFP can significantly enhance a number of these properties.

Since physicochemical properties can change at the nanoscale, nanomaterials will contain properties that behave differently from normal materials. The small size offers a large surface area (thereby offering more of the same), such that just small quantities of particles can offer large reactivity and functionality. The use of these properties for PNFP is manifold, as Hunt[7] shows,

> Modern food processing can change the biochemical and other properties of food, and nanotechnology now means that there is an even greater potential for molecular changes. Essential biomolecules such as amino acids, sugars and DNA are in the nanoscale and new processing methods will bring a range of other ingredients and substances into the nanoscale for manipulation. Nano-filtration, surface engineering and encapsulation will be important in processing nanostructured emulsions; biopolymeric nanoparticles and even nano-laminates for edible films and coatings are under development by food technologists. Nanocomposites of polymer-clay, which enhance the mechanical, thermal, moisture-stability, barrier and wear-resistance properties for packaging applications, are already becoming widespread and other nano-composites are under development e.g. using nano-silver. Nanotechnology sensors for detecting pathogen and contamination, for tracking and traceability are already a promising development.

In a similar vein, Moore[8] notes, "Incorporation of nanoparticles into a polymer matrix will improve many properties including: mechanical bulk properties; surface properties; dimensional stability; thermal stability; surface appearance; decreased permeability to gases, water, etc." All of which, he adds, can offer improvements to the *usability, durability,* and *branding* of products, as well as in "adding value through the use of nanocomposites and 'smart' developments e.g., nanobarcodes."[8] Furthermore, he suggests, nanocomposites can assist in the reduction of packaging waste in the following ways: through the use of "monolayer films rather than multilayer which are not recyclable; enhanced material performance—less packaging required to obtain desired performance; use of naturally occurring polymers with enhanced properties; for flexible packaging applications."[8]

With the enhanced barrier properties that nanomaterials can offer, the potential for spoilage may be limited and the life of products prolonged. Recent research developments indicate even greater future potential in this area. This includes spoilage warning indicators on packaging, such as have been developed by InMat®. These can serve as a way to reassure consumers about the safety and quality of the food contained within PNFP. Alongside barrier properties is the potential for antimicrobial properties. These active antimicrobials (such as metal oxide nanoparticles in silver or gold, among others) are certainly attractive. As Kuorwel et al. explain, "Active packaging (AP) is a system in which the product, the package, and the environment interact in a positive way to extend shelf life or improve microbial safety or sensory properties whilst maintaining the quality of food products."[9] Research into the application of biopolymeric materials is, they say, being driven by consumer demand for "preservative-free, high-quality food products, packaged in materials that create less environmental impact."[9] To this they add that "In combination with antimicrobial (AM) packaging systems, biopolymer materials with AM properties are emerging as one of the more promising forms of active packaging systems."[9] However, this packaging, they add, is not yet suitable for all foods types:

> Developing commercial biodegradable films with improved physical and mechanical properties is still a challenge due to their hydrophilic nature that limits their application for packaging of food products with a high water activity. The biodegradable and bio-compostable materials are also, many times, more expensive, and more difficult to process, a fact that further increases their cost compared to synthetic polymers.[9]

As such, costs and benefits need to be carefully weighed and evaluated:

> When considering the cost of a package, the total "cradle to grave" economic approach should be evaluated. Thus, the economic evaluation should include not only the cost of the packaging material and of processing the material into a package but also the cost of disposing of the final package namely, recycling, and/or incineration, and/or land filling. This is very important especially for the last option, taking into consideration the decreasing number of land filling sites and the diminishing space for garbage disposal in the developed countries. If such considerations are taken into account, the difference between the cost of biodegradable/bio-compostable and synthetic polymers becomes much smaller.[9]

With these innovations we might, it is hoped, reduce the amount of food wasted in the United Kingdom. Current estimates put this at roughly 7 million tons of food waste per year.[10] Other current and in-development applications of nanotechnology (by which is meant the manipulation of matter at the nanoscale within a range of disciplines) in packaging include the use of recyclable materials such as EcoSynthetix®, which produces sustainable and renewable materials by transforming biomaterials. The benefits seemingly cannot be denied, and yet this is by no means where the story ends.

2.2.2 Problem with Polymers

The first problem to highlight is that polymers themselves do not biodegrade. This is not to say that they cannot be recycled, but rather that once disposed of, they persist. This generates a number of important EHS issues, not least of which is the problem of chemical leaching. Furthermore, any given controls that are introduced to prevent hazards and exposure can, and too often do, fail. This means that even where precaution is taken, this cannot be taken for granted. One pertinent example concerns Polyvinyl Chloride (PVC), one of the most common plastics, which is said to cause more hazardous byproducts than any other product. As Boyd and Tomlin explain, PVC is not only one of the most common plastics; it is also one of the most toxic:

> Two of the main ingredients in PVC—ethylene dichloride and vinyl chloride monomer—are known carcinogens. By-products of PVC production and disposal include dioxins, furans, PCBs, and hexachlorobenzene—all extremely toxic substances. Phthalates and heavy metals are added to PVC in the manufacturing process [...]. The chemicals produced and released in the manufacture of PVC cause cancer, disrupt the hormone system, impair normal child developments, suppress the immune system, and lead to birth defects and brain damage.[11]

This means that even if the polymer itself is benign, it is likely that the modifiers will not be. In fact, according to one account, additives or modifiers can equal up to 60% of the product.[12] These include plasticizers, such as phthalates, to improve flexibility; heat stabilizers, such as lead and cadmium, to prevent thermal degradation; compatibilizers, such as acrylonitrile-butadiene rubber (ABS) to allow two or more materials to coexist in a polymer; dyes to alter the color; fillers to improve strength; light stabilizers to reduce degradation from light; antioxidants to prevent oxidation; and flame retardants to prevent the spread of fire.[13]

With the increasing use of nanotechnology and nanomaterials in the development and production of packaging with novel properties there arises an exponential increase in ethical and EHS issues. An example from the medical industry, which already makes broad use of nanomaterials and particles, might help explain this point. Superparamagnetic iron oxide (SPIO) nanoparticles have been used in magnetic resonance imaging (MRI) scanning for a number of years now. They are injected into the body and an image taken of their aggregation. Yet, and despite their now common use, it remains unclear what the destiny is of the SPIO nanoparticles following the scan. Questions that still require answers include whether there is the possibility of translocation, what potential there may be for bioaccumulation of nanoparticles in different parts of the body, and broader questions about toxicity over both the short and the long-term future.[14] The potential for translocation of nanoparticles is a major issue that will need to be addressed. On this the European Commission Directorate General for Health and Consumers notes the following:

> Nanoparticle translocation can occur to a greater extent and to different sites than occurs with larger particles. There can therefore be a systemic distribution and accumulation of such particles. There is evidence that nanoparticles can translocate from their portal of entry and can reach other parts of the body, including the blood and the brain, although again very few studies have been performed and the extent and significance of this translocation is unclear. It is uncertain whether nanoparticles can reach the foetus. Obviously, in medical applications involving parenteral [piercing the skin] administration of nanoparticles, systemic distribution is probable.[15]

It is likely that these sorts of questions will need to be asked about many other nanomaterials, and in particular those that are most likely to be taken into the body, whether through medical procedures, or through ingestion such as via leaching from PNFP.

The American Chemistry Council, in their website entry on "Food Containers and Packaging," admit that "Virtually all food packaging materials contain substances that can migrate into the food they contact, including the plasticizers used in plastic food wrap."[16] It is to some extent less certain what particles will migrate, and what short or long-term effect there may be (whether toxic or otherwise), however this lack of certainty is not a reason to neglect the potential for harm. This is particularly true for nanomaterials, where uncertainty is at the heart of the matter. The difficulties associated with predicting migration effects from packaging, for example, are revealed in a study by Vitrac et al.:

> To prevent the risks of excessive contamination of food, the European regulation defined positive lists of substances authorized for the formulation of materials intended to come into contact with food, with specific migration limits (SML) for the substances that are the most biologically active. A SML is derived from a Tolerable Daily Intake (TDI) value by assuming that a 60 kg body weight consumer eats daily 1 kg of food packed in 6 dm2 of the considered material. This assumption results often in over-estimating the actual exposure of consumers, while the exposure of some groups (e.g. infants and young children) may be underestimated. In addition, the current approach does not take into account

the relative weight of each type of packaging material on the market and the type of food usually conditioned in it. More accurate assessments of food contamination and consumer exposure are major issues for the coming regulation on food contact materials (EC-DG SANCO-D3 2002a). Yet unregulated areas (e.g. multilayer, active packaging, coatings, polymerization aids, solvents, lubricants, contaminants connected to recycling) are likely to correspond to lesser uses than plastics and hence to lower median exposures (EC-DG SANCO-D3 2003). The new framework 2004/1935/EC directive (EC 2004) on food contact materials entails the use of risk assessment procedures by the Industry to assess the safety of their packed food products. However, for both point risk estimates and large sanitary surveys, the main bolts are respectively i) the great analytical difficulty to detect traces of contaminants in real foods or in simulants and ii) the impossibility of analyzing food samples on very large markets as the European Union.[17]

As Hunt and Riediker explain, there are a number of inherent problems associated with the issue of *uncertainty* with regard to nanomaterials.[18] In a survey of nanosafety experts (using a modified Delphi Method to interview scientists involved in different areas of nanotechnology research) on the issues of safety and risk they found that the level of uncertainty regarding nanoparticles is still very high, and that there is more than one kind of uncertainty involved. Of this are highlighted two sorts in particular, the first of which is perhaps expected for a technology in its infancy, and that is a lack of data.[18] The importance of this is not to be underestimated. The use of PNFP is still relatively new, and it cannot be stressed enough that we simply have not had enough time to show how bioaccumulation and biopersistence of nanoparticles will play out over the coming years and decades. The second sort of uncertainty is more fundamental, and is at the very core of why nanomaterials are employed in PNFP in the first place, namely the novel properties of these particles at the nanoscale. More specifically, there are intrinsic uncertainties that arise as a consequence of the complexity of living systems in their responses to nanoscale entities. Another way to think about this is to recognize that there remains a general uncertainty about the overall functioning of our bodies as a complex organism. The assumption that changes occurring within a body do so discretely is continuously challenged, and yet the presumption of this persists through the ways in which we both treat and perceive illness, health, and our physical relationship with the world.

Other issues about uncertainty arising from the Hunt and Riediker survey include questions about "adequacy of data," which involves, for instance, the difficulty of translation from in vitro to in vivo; and (as mentioned earlier) the lack of long-term studies. Added to which are questions about the "adequacy of methods." This includes inadequacy in protocols, definitions, standardization, reference materials, and comparability of experimental outcomes. Finally there is what is known as "Nanotoxicological Complexity," or more specifically the bionano interface, and nonlinear systems, emergent properties, and the need for computerized mass-data stochastic approaches.[18] We can add to this list uncertainty about bioaccumulation, as already touched upon earlier. There is evidence to show that these concerns are not without precedent.

It has been known for a number of years that certain chemicals can be hazardous to health. Before I consider some specific examples, it is worth looking at the bigger picture. In 1991, the Wingspread Conference, with the heading *Chemically-Induced Alterations in Sexual Development: The Wildlife/ Human Connection* offered the following consensus:

> We are certain of the following: a large number of man-made chemicals that have been released into the environment, as well as a few natural ones, have the potential to disrupt the endocrine system of animals, including humans. Among these are the persistent, bioaccumulative, organohalogen compounds that include some pesticides (fungicides, herbicides, and insecticides) and industrial chemicals, other synthetic products, and some metals.[19]

Among those chemicals listed (and the list extends to well over 20 chemicals and associated compounds) are those found in everyday items, such as non-biodegradable detergents, modified polystyrene, and PVCs (as already discussed earlier), soy products and laboratory animal and pet food products. Following regulatory consensus, an endocrine disruptor is understood as "an exogenous substance or mixture that alters function(s) of the endocrine system and consequently causes adverse health effects in an intact organism, or its progeny, or (sub)populations."[20] Whereas, a *potential* endocrine disruptor "is an exogenous substance or mixture that possesses properties that might lead to endocrine disruption in an intact organism, or its progeny, or (sub)populations."[20] Endocrine disruptors work in a number of ways, but the most important point for our purpose here is to recognize that "endocrine disruption is inherently described as a mode of action," and that these modes of action "are among several that may potentially lead to adverse effects on reproduction, growth, and development."[20] With this in mind, it is now worth considering some specific examples.

2.2.3 Case Study: BPA

Bisphenol-A (BPA) is a chemical used in the production of polycarbonate plastic and epoxy resins, as well as in the synthesis of polysulfones, polyether ketones, and as an antioxidant in some plasticizers. Polycarbonate plastic, which degrades, is currently found in a variety of products and packaging including baby and water bottles, medical devices, dental fillings, sealants, lenses, CDs, DVDs, and household electronics. Epoxy resins containing BPA are used as the inside coating of almost all food and beverage cans. Repeated studies have shown that BPA can leach from the plastic lining of canned foods and even polycarbonate products, especially those that are cleaned with harsh detergents or are used to contain acidic or high-temperature liquids, such as baby bottles. Yet even without these provisos, there may be evidence of leaching. In a seminal study by Carwile et al. it was shown that just "1 week of polycarbonate bottle use increased urinary BPA concentrations by two thirds."[21] To this they add that "Regular consumption of cold beverages from polycarbonate bottles is associated with a substantial increase in urinary BPA concentrations irrespective of exposure to BPA from other sources."[21] In fact, it is claimed that food packaging as a source of chemical contaminants, such as BPA, is a "potentially relevant route of human exposure to endocrine disrupting chemicals (EDC)."[22]

Debates about the likely toxicity of BPA for humans remain unresolved, but the question of why there is a lack of resolution is a complex one. In fact, suspicions about the hazardous nature of BPA have been around since the 1930s. As vom Saal explains, "BPA was reported by Dodds and Lawson in 1936 to be a full estrogen agonist, before chemical engineers determined in the 1950s that this hormonally active drug could be polymerized to produce polycarbonate plastic."[23–25] In recent times concern has been growing, while opinion and therefore reaction has been divided, however the reason for division is controversial. To get a sense of the division, let us consider some recent history in BPA regulation.

In March 2010, the Danish government released the following statement:

> On the basis of a new assessment by the National Food Institute at the Technical University of Denmark (DTU Food), the Danish government has, together with the Danish People's Party, decided to invoke the principle of precaution and introduce a temporary national ban on bisphenol A in materials in contact with food for children aged 0–3 years (infant feeding bottles, feeding cups, and packaging for baby food).[26]

The ban has been introduced despite what they term *uncertainty* about its potential for harm:

> Danish experts say there is no clear evidence that bisphenol A has harmful effects on the behaviour observed. However, the experts find that the new studies raise uncertainties about whether even small amounts of bisphenol A have an impact on the learning capacity of new-born rats. In my opinion these uncertainties must benefit the consumers, so we will utilize the precautionary principle to introduce a national ban on bisphenol A in materials in contact with food for children aged 0–3 years.[26]

The ban, they say, will remain in effect until such time as "new studies document that low doses of bisphenol A do not have an impact on development of the nervous system or on the behavior of rats." As of October 2011 it had not yet been revoked. The French National Assembly followed suit in May 2010.[27]

Then, in November 2010, the EU announced that it was to introduce a ban, to come into force in 2011, on BPA in polycarbonate baby bottles. This they say was over fears that the chemical "could be hazardous to the health of young children." At the same time, however, the UK Food Standards Agency (FSA) released the following statement: "The Agency's current position is that exposure to BPA from food contact materials does not represent a risk to consumers based on current scientific evidence that has been reviewed by independent experts."[28] As this brief overview shows there is a clear division between those who think BPA is safe, and those who do not, and it might appear from the reports cited before that the division is fairly evenly split. Yet before the announcements noted earlier (by the EU and FSA), and as early as June 2010, there was a clear warning about the dangers of BPA. This came in the form of a letter from more than 20 leading scientists (alongside scores of NGOs) to the European Food Safety Authority (EFSA) calling for the reduction in "levels of BPA exposure, particularly in groups at highest risk, namely young infants and pregnant mothers." Within this letter they admonish the EFSA for the quantity, and even quality, of the research through which the safety of BPA had frequently been determined. During the previous 15 years, they claim, "several hundred peer reviewed scientific papers, have been published that have highlighted potential adverse health effects associated with BPA exposures." Of which, they add, "Only a tiny minority of studies have articulated that BPA exposure is completely safe, and many of these research papers have been criticised in academic commentaries and responses as having serious flaws, but it is these few flawed studies that EFSA previously relied on to declare BPA safe."[29] In fact, as early as 2006, vom Saal and Welshons had published results of their rather extensive research review on BPA, with the following damning remarks that seem to lend support to the claims made earlier:

> There are 109 published studies as of July 2005 that report significant effects of low doses of BPA in experimental animals, with many adverse effects occurring at blood levels in animals within and below average blood levels in humans; 40 studies report effects below the current reference dose of 50 µg/kg/day that is still assumed to be safe by the US-FDA and US-EPA in complete disregard of the published findings. The extensive list of significant findings from government-funded studies is compared to the 11 published studies that were funded by the chemical industry, 100% of which conclude that BPA causes no significant effects.[30]

Yet at the time of writing, it seems there remains little consensus regarding the status of BPA. In September 2011, the French Agency for Food Health Safety (ANSES) published a report in which are highlighted "apparent health risks from exposure to bisphenol A (BPA),"[31] and then in October 2011 the French Health Minister gave his backing "to a law that would outlaw the use of bisphenol A in all food packaging from 2014—as well as proposing that the chemical should be banned in all packing aimed at children from 2013,"[32] As a result of the September report the EFSA have promised to "liaise" with ANSES. EFSA told one website, "EFSA and ANSES liaise very closely. EFSA has received ANSES reports on BPA and will be analysing the data contained in those reports in coming days. EFSA is always ready to consider any new relevant findings which become available."[31]

Despite this assurance, however, the last time EFSA updated their advice on BPA was back in September 2010, in which they stated:

> Following a detailed and comprehensive review of recent scientific lit-erature and studies on the toxicity of bisphenol A at low doses, scientists on the European Food Safety Authority's (EFSA) CEF Panel conclude they could not identify any new evidence which would lead them to revise the current Tolerable Daily Intake for BPA of 0.05 mg/kg body weight set by EFSA in its 2006 opinion and re-confirmed in its 2008 opin-ion. The Panel also state that the data currently available do not provide convincing evidence of neurobehavioural toxicity of BPA.[33]

My objective in this discussion is not to offer a solution to the controversy. My main concern is to offer an example whereby government, and indeed industry-led change—since these early warnings did not in themselves lead to significant reduction in the use of BPA until relatively recently—may not be forthcoming when ethical and EHS issues are at stake. The exception to which is often where public concern is brought to the fore. In October 2011, for example, the American Chemical Council (ACC) called on federal authorities to ban BPA, because they said it would "make clear to consumers that the substance is no longer used in the manufacture of [baby bottles and cups]." This is despite the fact that they nevertheless "remained convinced that inclusion of BPA in any food contact material was safe."[34]

It may in fact be difficult to ever establish without doubt that BPA is as dangerous as some research has suggested, but this should not matter. In a report by Aschberger et al. from 2010, it was noted that "some epidemiologi-cal studies have linked potential health effects such as diabetes, obesity, liver enzymes abnormalities, and cardiovascular diseases with the exposure to BPA."[35] To this the authors add "In this case, interpretation is made difficult by the fact that some confounding factors might not have been accounted for."[35] Proving causality is indeed complicated, and it is not always easy to attribute causes, nor identify those that can be said to be either necessary or sufficient in and of themselves. Individual causes may often be contribu-tory, and it is difficult to show that one cause is more directly relevant than any others. In fact, proof of causation involves many factors, some of which can be difficult to correlate, including temporality, strong association, dose-response, consistency, and experimental variation. That said, there might not be the same difficulties in establishing biological plausibility, which in turn may prompt a precautionary approach. In this way serious consideration ought to be given to the issue of what *might* happen; and would this be serious and irreversible? Indeed, the BPA situation as detailed earlier offers a pertinent reason for the adoption of the precautionary principle. For while there may be more or less uncertainty about the issue on the one hand, there is enough evidence to suggest the *potential* for hazard or harm to our current and future health, on the other. The precautionary principle would suggest that the lack of *certainty* would not in itself be sufficient reason to deny the plausibility of *potential for harm*.

2.2.4 Other Early Warnings

BPA is not the only example of early warnings that have sometimes been ignored or given insufficient priority. Some past examples where polymer production and distribution have caused seriously adverse EHS effects include Phthalates, Polychlorinated Biphenyls (PCBs), and Brominated flame retardants. Before I move on to the next section, and to show that the aforementioned situation with BPA is not an isolated case, I will briefly explain the history of early warnings associated with PCBs specifically. As Koppe and Keys explain, an awareness of the dangers to health of PCBs was present as early as 1899, when chloracne was identified in workers from the chlorinated organic industry. Then, in 1968, 1800 people were poisoned by PCB-contaminated rice oil in Japan. Just over 10 years later, in 1979, 2000 people were poisoned by polluted rice oil in Taiwan, and during the 1990s there was evidence to show that PCBs were "associated with IQ and brain effects in children exposed in utero to mothers' PCP-contaminated diets."[36] Despite the overwhelming evidence that is shown by the aforementioned fact, the EU directive to eliminate PCBs did not come until 1996, "with phase-out by 2010."[36] This is a full century after the earliest indications that the chemical might be hazardous. This is an excellent example where the burden of proof remains to show that the chemical is indeed toxic, rather than adopting the precautionary principle, which would instead have ensured such early warnings are taken seriously. It seems fair to suggest that the burden of proof should in fact be to show that these substances are safe, and not the other way around.

It is reasonable to accept that since PNFP offer benefits (as already detailed earlier), it would be wrong to dismiss such developments entirely and without due consideration. That said, the evidence also shows that the process by which benefit versus risk, or more appropriately hazard (as discussed earlier in this chapter), is measured needs reconfiguration. In a report by Arnall in 2003, and published by Greenpeace, the following environmental concerns about nanoparticles were noted:

> The potential impact of nanostructured particles and devices on the environment is perhaps the most high profile of contemporary concerns. Quantum dots, nanoparticles, and other throwaway nanodevices may constitute whole new classes of nonbiodegradable pollutants that

scientists have very little understanding of. Essentially, most nanoparticles produced today are mini-versions of particles that have been produced for a long time. Thus, the larger (micro) versions have undergone testing, while their smaller (nano) counterparts have not (ETC Group, 2002a). For example, Vicki Colvin, Executive Director of Rice University's Centre for Biological and Environmental Nanotechnology (CBEN) has recently postulated that nanomaterials provide a large and active surface for adsorbing smaller contaminants, such as cadmium and organics. Thus, like naturally occurring colloids, they could provide an avenue for rapid and long-range transport of waste in underground water.[37]

It is for these reasons, as well as all the reasons noted earlier (and likely many others yet to follow) that ethical design needs serious consideration. In this way a proper balance between benefit and hazard might be struck.

2.3 Ethical Design

As we have seen, the potential for novel properties found at the nanoscale presents a fertile area for exploration and exploitation by science and industry. It offers a way in which to tackle both age-old and contemporary problems. Within the former we might include pollution and the scarcity of resources. With regards to the latter, it may provide a way in which to preempt and respond to new and emerging problems, such as an aging population and higher consumer expectations. It is already recognized that, on the one hand, nanotechnology offers ways in which to tackle perceived problems and bring a broad range of benefits. On the other hand, and as already discussed at some length, there is also the likelihood for risks and hazards, whether known and perceived, currently unknown, or even unknowable. There is however sometimes a reluctance to recognize those assumptions upon which traditional ideas about *benefit* and *risk* rely. For example, with regard to PNFP, one perceived benefit (as noted earlier) is the potential for the reduction or better decomposability of packaging waste. This, as a solution to an ever-growing problem of packaging waste might seem ideal. What is not always considered, however, is the supposition that drives this. Namely, that the creation of more packaging is an adequate solution to the problem of too much packaging waste. In fact, the assumption does not allow for the possibility that the creation of more packaging *in general* (and the expectations that this packaging brings), irrespective of its composition, may itself be part of the problem.

Keeping ethical, social, and legal issues at the fore might also mean considering alternatives to the manufacture of PNFP. These might involve reduction, whether of materials used or of products consumed; refinement, for example of existing materials; or creative recycling, whereby we reuse what already exists in novel ways. The benefits of these sorts of alternatives should not be discounted in favor of new PNFP designs without full consideration. That said, even alternative solutions that appear promising ought also to be judged according to ethical design requirements. For example a new form of Bioplastic biodegradable packaging, developed in 2010 and made from fruit skins[38] is, according to researchers, durable and economic to produce. What is not so clear however are the sorts of environmental impacts there may be with regard to the energy required to produce these materials. This is not to say that this new form of packaging is in some way more harmful than traditional plastics, in terms of energy required in production, but only to repeat that apparent environmental promise should not be determined solely by means of limited factors.

Holistic approaches to packaging ought to take into account more than just the final waste product and its necessary disposal. It ought also to include production costs to the environment, as well as broader implications about food waste and airmiles. For example, might improved packaging somehow discourage mounting concerns about food miles? This last question brings us back to the issue of perception, and highlights a phenomenon well known to advertisers. This is where the perception or emphasis of positive attributes of a thing can mean that other equally important attributes are neglected. One such example concerns so-called "health foods," that although low in calories or fat, are yet high in salt and/or sugar. Seat belts in motor vehicles are another example. Contrary to popular belief, it is not in fact a given that the addition of seatbelts to motor vehicles has led to improved safety when driving. According to a seminal study by Adams in 1985 the reason for this is "risk compensation," whereby the act of wearing a seat belt results in a tendency of the driver to think they are now safer.[39] One of the consequences of this perceived safety is that many drivers will then drive at higher speeds. This is despite the fact that there is in fact a higher likelihood of serious accidents occurring at higher speeds, while any injuries sustained as a result are likely to be more serious.

This last example is offered as a way to show that by focusing on one particular area where risk has been minimized, a tendency to ignore other important hazards can result. In terms of PNFP, thinking ethically about the use of nanomaterials, requires that we think about a wider range of impacts than a simple or limited risk/benefit analysis can provide. Just as seat belts are not the only factor for safety when driving, neither should the environmental impacts of improved packaging weight, for instance, be the primary determining factor for deciding the value of new packaging design. In particular, ethical design needs to incorporate views about the need for nanomaterials specifically, and new and emerging technologies more broadly. These needs are not *ipso facto*, and the question of alternatives ought always to be a feature. As already noted, a broader understanding of what constitutes risk and hazard also needs to be adopted.

So far I have presented the need for ethical design by giving specific examples of hazard, and by making general points about risk and hazard for EHS, but this does not clarify how we might go about establishing what constitutes risk, hazard, or harm. As Wickson et al. explain, "While everyone may agree that scientists, policy makers and citizens should work to ensure that nanotechnology does not harm 'nature' or 'the environment,' there are very different ideas about what these concepts mean, what constitutes harm, and the reasons why we might wish to avoid it."[40] Added to which is the relative difficulty in establishing such terms for nanomaterials in particular, as the following discussion shows:

> Nanomaterials have very large surface areas and singular properties that do not necessarily follow the general physical laws for macro-size materials, e.g., gravity, reactivity, etc. Often, their high surface area leads to greater or faster reactivity. Smaller size can lead to higher probability of movement and entry into sites such as pores that cannot be reached by larger particles. Movement may affect nanoparticle transport within a package material or from a package material to food contents, etc.
> Studies on particulates such as carbon black and titanium dioxide suggest that properties of nanoparticles are profoundly different from those of conventionally sized particles of the same molecule. Materials that are "safe" in macro and micro sizes might be more or less hazardous in nano size. Research is steadily accumulating to better understand potential effects of nanoscale materials on biological systems.[41]

As Hunt and Riediker explain regarding the life-cycle of nanomaterials:

> On the whole, those working in occupational and public health or in human and environmental toxicology have the impression that they do not sufficiently understand the impact of manufactured nanomaterials on living systems. This is perceived as a source of concern.[19]

There is, they conclude, "a substantial need for long-term documentation and traceability of MNM application and use," as well as a need to raise awareness of the kinds of complexity involved across the life-cycle of nanomaterials.[19] On this issue, Smolander and Chaudhry note the following:

> The available information is currently very sparse in terms of experimental data on the toxicology of ENPs that are (or can be) used to develop new food contact materials to provide a basis for adequate risk assessment. The risk is, however, dependent on both hazard and exposure—where the absence of one means no risk. Therefore, in the absence of Hazard information, the focus is on the estimation of potential exposure.

They further add that while there has been some migration testing already, for example, in the safety of nanotechnology–derived food contact materials, "more testing on other types of materials is needed to build a broader picture in regard to the migration patterns for other nanocomposites that may be used in food packaging." Finally they state that focus from detailed toxicological studies needs to be on "the health consequences of the long-term exposure to potentially lower levels of ENPs via food and drinks."[42]

In a similar vein, Hunt et al. point out the following:

> One recognized concern is that agglomerates thought to be harmless may break up under certain specific conditions that occur in food, in feed, and in the gastro intestinal tract or in biological tissues. Another concern is evidence indicating that ENM can react with proteins, lipids, carbohydrates, nucleic acids, ions, and other biological substances. The nano-specific properties and characteristics of an ENM (as opposed to its bulk form) are therefore likely to affect its toxico-kinetic behavior and the toxicity profile. The risk assessment should therefore be performed on a case-by-case, or product-by-product basis, based on the proper identification and comprehensive characterization of the nanoparticles or nanomaterials concerned.[43]

Some of the implications of this for the ethical design of PNFP are that there will need to be early and full hazard assessments for both the polymer nanomaterials and modifiers. From the aforementioned assessment it seems clear that the use of chlorine and BPA, among other hazardous chemicals should be avoided in the production of PNFP. This would also include a reassessment of standard polymers that have already triggered early warning signs of toxicity or other EHS hazards, including where such use in the production of nanocomposites is being proposed. This includes, for example, nanomaterials in the production of PVC and polycarbonate. It might also mean that development ought to move away from fossil oil-based polymer production and toward bio-based polymers as vehicles for nanomaterials, and that these should be of a monolayered rather than multilayered design so as to ensure they are recyclable. This list is by no means exhaustive and so the difficulty remains regarding how to ensure emerging hazards are taken seriously. The by no means unrelated question of who funds preproduction EHS research is also one that will need serious consideration.

A study by Owen and Goldberg from 2010 gives reason to be concerned in these respects. In their research they examined ten "risk registers" that were completed by researchers applying for funding from the Engineering and Physical Sciences Research Council. These risk registers required applicants to identify potential environmental, health, societal, as well as ethical concerns that may result from the innovation process', as well as to offer an "appraisal of risk," some identification of the level of associated uncertainty, and the details about who in the team would be "accountable for managing any identified risks."[44] The results are perhaps not surprising:

> All risk registers were completed conservatively, with the overwhelming majority of impacts identified by the applicants being narrowly focused on health impacts associated with exposure to nanomaterials during nanoparticle synthesis, manipulation, and prototype device fabrication. In the majority of cases, the risks associated with these were judged by the applicants as being low or sometimes moderate in nature, with low uncertainty.[44]

In fact, "Few potential impacts on the wider natural environment (e.g., associated with eventual device use or end of life) were identified in the risk registers and no future societal impacts were identified at all."[44] Owen and Goldberg suggest some of the reasons for these perceptions to be the "often unpredictable nature of innovation, its interaction with society" as well as "the need to put in place processes to understand this and feed back into decision making."[44] We might also add to this list the effects of viewing technologies as essentially value-neutral (more on this next).

Owen and Goldberg state that during telephone interviews about the process, many of the researchers considered the risk register activity worthwhile and shared the view that scientists "should reflect on the wider implications of their proposed research."[44] One participant did not agree however, and offered the following contention: "these issues would be best considered by those applying the technology later in development rather than by the scientists themselves."[44] It is of course difficult to judge how many scientists might share this latter view, particularly when talking amongst themselves instead of during interviews conducted by external researchers. Indeed, while it may be true to say of those researchers interviewed by Owen and Goldberg, or of any who agree with the positive appraisal of the register, that they recognise the *theoretical value* of considering the wider implications of their research, this is little indication of what they will then do in practice.

The view of the lone scientist noted here (that EHS and ethical issues should be considered only later on in development), has serious implications, particularly because the results of such research can quickly become embedded in practice. As Feng explains, "over time some technological practices become so entrenched in society that it becomes difficult to do things differently," whereas "early on in the design process technologies are often malleable enough to be produced and implemented in a number of ways. Hence the need for ethical discussion to take place early on in the design of technologies."[45] As the evidence offered in the preceding sections should show, once a technology becomes entrenched, and sometimes this can happen relatively quickly, it cannot always be removed or changed easily. While new baby bottles may no longer be produced using BPA, this does not mean that all old baby bottles will therefore be discarded. Such things can even be passed down through the generations, whether for sentimental or for socio-economic reasons, and often without awareness of possible hazards. It is difficult, even impossible to assess the impact this may have for future generations.

Furthermore, ethical design is important because where the drive is primarily to solve technical problems, for example by amending particles at the molecular level in order to improve durability, this does not promote the need for consideration to be given to the EHS impact of these new materials. In the matter of durability, for instance, a balance needs to be struck between the need to produce something that is stable enough that it will last for the desired length of its useful life, but also something that is not so stable it will not break down over time once at the disposal stage. The study by Owen and Goldberg seems to support this view, since those proposals that showed "a strong commitment to responsible science and innovation" were simultaneously proposals that also "promoted continuous reflexivity, embedding suites of multidisciplinary approaches around the innovation research core to support decisions modulating the trajectory of the innovation research in real–time."[46] This approach is one also highlighted by Feng, who claims that:

> Technologies rarely satisfy social and ethical concerns because of some innate goodness on their part; rather, they satisfy such concerns because people bring them to the attention of engineers, executives, government officials, the media, and the public. In the absence of social advocates, economic interests tend to dominate the design process.[45]

On this view, where priority is given to economic interests this is not a *fait accompli*, but is instead because of very "human choices,"[45] and these choices are not value-neutral nor "autonomous":

> The implication of the value–neutral myth is that ethics is relevant only in terms of an object's *use*, not its design. Hence, social concerns such as privacy, democracy, or social justice need not be considered during the design of technology, but can instead be left until after the technology has been built. In practice, when such concerns are set aside, other values (e.g., economic ones) tend to dominate the design process, sometimes resulting in technologies that are at odds with... more "social" values.[47]

This rather limited view, itself a barrier to ethical design, implies "social and ethical goals are irrelevant to the design of new technologies,"[45] and can obscure the importance of other values and goals, whereas,

> the determination of the "best" design results from a complex set of factors including cultural assumptions, economic interests, political considerations, organizational constraints, and so on. In a very deep sense, the design of technology has to be seen as a value-laden activity.[45]

As noted earlier, some of the benefits that nanomaterials bring include those directly affecting EHS, while at the same time the risks or hazards they present may do likewise. If a product's design generates as many problems as it solves, or has the potential to do so, then this needs to be considered early on. The question is then how to assess and balance these in relation to other needs, alongside emerging regulation. In other words, how should ethical design be implemented?

In its simplest form, ethical design can be described as involving a number of key factors. The primary task would be to adopt broader testing aims, such as, for example, by looking for similarities between polymers and human hormones, as well as thinking about broader outcomes relative to the design of a product (including EHS considerations). One of the ways to achieve this might be through upstreaming early on, so for instance involving stake-holders at both the design and production stages. This may also help limit the chance of consumer rejection at point of sale, and so could bring economic benefits. One way to achieve this is through the process of subsidiarity, whereby decisions are influenced by decision-making at consumer level (although here again there is the danger of purely economic decisions playing a primary role). Another way would be to extend the amount of collaborative work, so for instance, including production process engineers in the laboratory design process, and thus limiting the potential for harmful modifier additions post nanopolymer design stage.

This latter point is particularly important in terms of EHS, and there are plenty of examples of what can happen when there is limited communication across stages, some of which have already been considered (e.g., BPA). Typical to these examples is the fact that unexpected hazards might have been easily avoidable. For example, McDonough and Braungar describe how some mass-produced polyester clothing or water bottles in fact contain *antimony*, which is "a toxic heavy metal known to cause cancer under certain circumstances."[46] The reason it is there is because it is used in the production process. As they explain, antimony is typically used as "a catalyst in the polymerization process and is not necessary for polyester production."[46] The problem with this, as discussed by Billatos and Basaly is that heavy metals cannot be destroyed by incineration. They are therefore "accumulated both in the environment and within human bodies."[47]

At this juncture it would be pertinent to consider the similarities between the term *ethical design*, as promoted earlier, and other labels that appear to denote a similar ethos. This includes *green-*, *eco-* or *environmental*-design. The problem with these other terms, however, is that (whether true or not) they suggest an emphasis on external or environmental issues, and as the examples about the recycling logos show at the beginning of this chapter, perception is a powerful tool. *Sustainable*-design might rectify this image in some ways, but again the emphasis remains one in which we think specifically about the external, in this case, *sustainability*. To a layperson it might be difficult to see how a term such as sustainability would also encompass issues about health, or social and political concerns. For this reason in particular ethical design is preferred as an umbrella term, since it promotes a broader picture of the related and integrated issues that require consideration. As the following discussion shows:

> Ethical and sustainable packaging is an umbrella term which covers a number of different types of packaging, including recycled, biodegradable, and reduced/lightweight packaging. It also includes the issue of reducing environmental impacts of packaging in terms of both production and waste. Ethical packaging is constantly evolving as companies begin to find ethical sustainability is more of a threat towards their corporate social responsibility (CSR).[48]

On sustainable design Otto says that it should deliver "the best (social, environmental and economic) performance or result for the least (social, environmental and economic) cost," involving three "so-called pillars" at its crux: social, environmental and economic.[49] These, in turn involve *people, planet, profit*, and are not discrete categories, or "separate, static entities," but rather elements of the same picture that need to work in harmony. She then offers the metaphor of juggling balls and suggests the trick "is to keep them working together in a single, smooth process." Yet, she acknowledges, "we often don't juggle too well."[49] This account is one that would fit neatly under the ethical design umbrella, with some of the qualifications as noted earlier.

In practice, as already noted, ethical design would incorporate subtle appreciations for broader ethical and EHS issues, including recognition of the complexity and interconnectedness of EHS, ethics, and socio-political situations. With regard to PNFP, this would mean that the design of polymers would be based on an understanding of potential EHS and economic harms (e.g., rejection by public), *before* the production process, and entry onto the market. This would mean that there is not simply a *reaction* when something goes wrong; but rather that priority is given to maximizing benefits to human welfare and EHS more broadly understood. Ethical design would require everyone involved in the design and production of PNFP to be proactive, and in so doing position themselves such that they might preempt problems *before* they arise. As already noted earlier, this could include multidisciplinary communication and feedback from all levels, including stakeholders and those involved in the production process.

2.4 Regulation and the Future of PNFP

The issue of regulation with regard to nanotechnology broadly conceived is a tricky one. Some of the major problems inherent to regulation in these fields arise because of quite how broad the spectrum is that encompasses the term *nano*. Included within this are troubles arising from, for instance, distinctions that need to be drawn between natural nanomaterials and engineered nanomaterials. Further difficulties arise with both the characterization and definition of any nanomaterials as, in fact, *nano*. New attempts to characterize nanomaterials by the European Union have resulted in the following definition:

> The International Organization for Standardization defines the term "nanomaterial" as material with any external dimensions in the nanoscale or having internal structure or surface structure in the "nanoscale." The term "nanoscale" is defined as size range from approximately 1 nm to 100 nm.[50]

Yet it remains to be seen whether this definition is adequate, or even accepted within the scientific community. For instance, Maynard calls for the nondefinition of nano, and claims that "Basing regulations on a term with no scientific justification will do more harm than good."[51] Aside from this there are issues around characterizing nanoparticles based only on size. Particularly as size is not necessarily the most important factor for nanomaterials, while reactivity and shape are significant defining characteristics.

Despite these difficulties, both characterization and standardization are likely to play an increasingly important role in the development of new legislation, or in the expansion of existing legislation, and there may be much to be gained from this. The focus of characterization is, for example, to improve usability and durability, in order to produce functional and economical products, as well as to provide the groundwork for the development of proportional legislation. Some of the most pressing questions arising from these processes, however, are as follows: First, which central principle determines how this process occurs? And second, what are the priorities? The problem of priority is, of course, key, for it is often claimed that this is where serious tensions arise. McDonough and Braungar claim that "At its deepest foundation, the industrial infrastructure we have today is linear: it is focused on making a product and getting it to a consumer quickly and cheaply without considering much else."[50]

In fact, these sorts of issues make it difficult to establish how and to what extent nanotechnology is already covered by existing legislation, and whether, and in what ways, new legislation ought to be enacted. The focus of this chapter has not been, nor could it be, to offer a solution to these issues as such, but rather to show how these problems might be incorporated within the ethical design process as outlined earlier. On this, Hunt states, "There can be no doubt that in the next decade all EU companies involved in nanotechnology will have to make such assurances, act on them, and support rather than resist regulation meant to protect the public from chemical hazards."[7]

On legislation and proportionality, Kosta and Bowman offer the following comment. They say that existing legislation, along with "engineering based solutions" should suffice, so long as researchers and manufacturers can be *encouraged* to consider integrating precautionary systems within their designs during the early stages of the products' development.[52] While they are specifically talking about privacy with regard to information and computing technology (ICT) and nanotechnology combined implants, the point stands. One way we might achieve this would be to give some priority to stakeholder engagement on issues rather than simply prioritizing top-down approaches. Stakeholder, as a term, is used here in its broadest possible sense and as Hunt et al. explain, this includes "almost all of those outside the academic research field."[43] The reason for this, they say, is because

> We all crave a healthy future for ourselves and our kin. If nanomaterial-enhanced products become ubiquitous, we all have a stake in the safety aspects of these goods. Furthermore, business needs safe products to be competitive and regulators are there to ensure benefits to society.[43]

Legislation has an important role to play in this process, both for consumers, but also for manufacturers. In fact it is not a given that manufacturers are against legislation *per se*. Instead emphasis may be on how such decisions are made. For example, in May 2010, a number of prominent electronics companies, including Acer, Dell, Hewlett Packard, and Sony Ericsson, alongside environmental organizations, released a joint call for tighter EU restrictions on the use of "hazardous substances in electronic products in 2015," the reason being "to avoid more global dioxin formation."[33] A representative from Acer commented,

> The transition away from environmentally sensitive substances, such as brominated flame retardants and PVC is well under way at Acer. However we do not have the leverage to move the entire supply-chain on our own. Legislators can help in this process. [...] By introducing restrictions, and thereby ensuring that the entire supply-chain is on board, costs are kept down and availability of safer alternative material is promoted.[33]

In his discussion of research challenges for nanomaterials, Riediker claims that

> Many industries assess the risks associated with their products, e.g., the medical device, food, pharmaceutical, cosmetics, and chemical industries. It is reasonable to assume that those industries that are developing nanotechnology-enabled products are conducting safety evaluations, despite the fact that current regulations do not oblige them to do so for materials of equivalent chemical composition to ones that are already approved in bulk scale.[53]

To this he adds,

> Whilst regulators may not require this information under current regulations, there is an exception in the case of nanomaterials that are embedded

in food packaging materials, where it is necessary to show that the material is inert and that additives to do not migrate out of the material in unacceptable quantities (Framework Regulation (EC) 1935/2004 (L338/4)).[53]

The new Framework Regulation on materials and articles intended for food contact was published in the Official Journal in November 2004 and came into force in December 2004 (replacing Framework Directive 89/109/EEC and Directive 80/590/EEC).[54] The Framework Regulation (EC) 1935/2004 (L338/4) states that food contact materials should be safe, and that they must not transfer their components into the food in quantities that could endanger human health, change the composition of the food in an unacceptable way, or deteriorate the taste and odor of foodstuffs. In addition, the Regulation also includes important provisions regarding labeling. For instance, If an article is intended for food contact it must be labeled as such, or bear the glass-and-fork symbol. The exceptions to this would be where food contact is obvious by the very nature of the article, including, for example, knives, forks, and wine glasses. Furthermore, labeling, advertising, and presentation of food contact materials must not mislead the consumer.

Whereas until this point specific measures existed only for ceramics, regenerated cellulose and plastics, the 2004 Regulation establishes 17 groups of different materials and articles for which specific measures can now be determined. These include active and intelligent materials and articles; adhesives; ceramics; cork; rubbers; glass; ion-exchange resins; metals and alloys; paper and board; plastics; printing inks; regenerated cellulose; silicones; textiles; varnishes and coatings; waxes; and wood. The following comment is made on active and intelligent packaging in particular:

> The Regulation includes definitions on active and intelligent packaging. If the materials release substances in the food that change the food composition or properties, then these substances must comply with food legislation e.g., food additives. These materials and articles cannot be used to mask spoilage of food and mislead the consumer.[54]

It is clear that steps are being taken to ensure that EHS issues are addressed for the use of nanomaterials in food packaging, yet these do not in themselves address broader potential ethical issues arising in the design of PNFP, as the examples already noted in this chapter indicate. Nor is it clear how legislation alone will be able to account for the sheer variety of issues arising from a field as broad as nanotechnology, and it remains to be seen whether the aforementioned legislation will go far enough. Few deny that the use of nanomaterials involves risk and hazard, but the difficulty is in understanding what those risks and hazards may be, and trying to account for them before it is too late to prevent harm occurring. The application of ethical design ideas presented earlier is offered as one way to achieve this.

Acknowledgments

I am very grateful to the support, guidance, and information given to me by Prof. Geoffrey Hunt in the preparation of this chapter.

References

1. http://www.incpen.org/docs/PackaginginPerspective.pdf (accessed on September 21, 2011).
2. http://www.carrierbagchargewales.gov.uk/retailers/?lang=en (accessed on October 1, 2011).
3. http://www.recyclenowpartners.org.uk/retail_high_street/brand_benefits.html (accessed on September 21, 2011).
4. D. Siddique, Plastic recycling and public perception. *Plastic Recycling* (2005) 2. http://www.cbc.ca/documentaries/doczone/2009/foreverplastic/report.pdf (accessed on September 3, 2011).
5. D. Siddique, Plastic recycling and public perception. *Plastic Recycling* (2005) 10. http://www.cbc.ca/documentaries/doczone/2009/foreverplastic/report.pdf (accessed on September 3, 2011).
6. S. Barr, *Household Waste in Social Perspective: Values, Attitudes, Situation and Behaviour*, Hampshire, U.K.: Ashgate (2002).
7. G. Hunt, Nano chemicals and the REACH controversy: Ethics and politics (2005).
8. Cited in Current & future opportunities for nanomaterials in packaging nanomaterials. http://www.nanocentral.eu/wp-content/uploads/Graham-Moore1.pdf (accessed on September 10, 2011).
9. K. K. Kuorwel, M. J. Cran, K. Sonneveld, J. Miltz, and S. W. Bigger, Antimicrobial activity of biodegradable polysaccharide and protein-based films containing active agents, *Journal of Food Science*, 76(3) (2011) R90–R102.
10. Cited in http://www.defra.gov.uk/environment/waste/ (accessed on September 26, 2011).
11. D. R. Boyd and B. Tomlin, *Dodging the Toxic Bullet: How to Protect Yourself from Everyday Environmental Health Hazards*, Vancouver, British Columbia, Canada: Greystone Books (2010) p. 135.
12. http://online.sfsu.edu/~jge/html/functional_additives.html (accessed on October 1, 2011).
13. Cited in http://www.zerowaste.org/publications/06m_plastics_101.pdf (accessed on October 1, 2011).
14. N. Singh, G. J. S. Jenkins, R. Asadi, and S. H. Doak, Potential toxicity of superparamagnetic iron oxide nanoparticles (SPION), *Nano Reviews*, 1 (2010) 1–15.
15. http://ec.europa.eu/health/opinions2/en/nanotechnologies/l-3/9-conclusion.htm (accessed on October 1, 2011).
16. http://phthalates.americanchemistry.com/Phthalates-Basics/Food-Containers-and-Packaging (accessed on September 21, 2011).

17. O. Vitrac, B. Challe, J.-C. Leblanc, and A. Feigenbaum, Contamination of packaged food by substances migrating from a direct-contact plastic layer: assessment using a generic quantitative household scale methodology, *Food Additives and Contaminants*, 24 (1) (2007) 75.

18. G. Hunt and M. Riediker, Building expert consensus on uncertainty and complexity in nanomaterial safety, *Nanotechnology Perceptions*, 7(2) (2011) 82–98.

19. http://www.endocrinedisruption.com/files/wingspread_consensus_statement.pdf (accessed on September 20, 2011).

20. G. H. Degen and J. W. Owens, Xenoestrogens and xenoantiandrogens. In: H. Greim and R. Snyder (eds.), *Toxicology and Risk Assessment: A Comprehensive Introduction*, Chichester, U.K.: John Wiley & Sons (2008) 583.

21. J. L. Carwile, H. T. Luu, L. S. Bassett, D. A. Driscoll, C. Yuan, J. Y. Chang, X. Ye, A. M. Calafat, and K. B. Michels, Polycarbonate bottle use and urinary Bisphenol A concentrations, *Environmental Health Perspectives*, 117(9) (2009) 1368.

22. J. Muncke, Endocrine disrupting chemicals and other substances of concern in food contact materials: An updated review of exposure, effect and risk assessment, *The Journal of Steroid Biochemistry and Molecular Biology*, 127(1–2) (2011) 118.

23. F. S. vom Saal, Bisphenol A eliminates brain and behavior sex dimorphisms in mice: How low can you go? *Endocrinology*, 147(8) (2006) 3679.

24. E. C. Dodds and W. Lawson, Synthetic estrogenic agents without the phenanthrene nucleus, *Nature*, 137 (1936) 996.

25. E. C. Dodds and W. Lawson, Molecular structure in relation to oestrogenic activity. *Compounds without a Phenanthrene Nucleus, Proceedings of the Royal Society*, London, U.K., 125 (1938) 222.

26. http://www.fvm.dk/Default.aspx?ID=18488&PID=169747&NewsID=6014 (accessed on March 1, 2011).

27. http://www.foodproductiondaily.com/Quality-Safety/French-deputies-adopt-BPA-baby-bottle-ban (accessed on March 1, 2011).

28. http://www.foodproductiondaily.com/Quality-Safety/EU-to-ban-bisphenol-A-in-baby-bottles (accessed on February 25, 2011).

29. http://www.env-health.org/spip.php?article582 (accessed on February 25, 2011).

30. F. S. vom Saal and W. V. Welshons, Large effects from small exposures. II. The importance of positive controls in low-dose research on Bisphenol A, *Environmental Research*, 100 (2006) 50.

31. http://www.foodproductiondaily.com/Quality-Safety/ANSES-highlight-BPA-health-risks. (accessed on October 10, 2011).

32. http://www.chemsec.org/news/news-2010/april-june/553-leading-electronics-companies-and-environmental-organisations-urge-eu-to-restrict-more-hazardous-substances-in-electronic-products-in-2015-to-avoid-more-global-dioxin-formation (accessed on October 1, 2011).

33. http://www.efsa.europa.eu/en/press/news/cef100930.htm (accessed on October 10, 2011).

34. http://www.foodproductiondaily.com/Packaging/US-chemical-industry-urges-feds-to-ban-bisphenol-A-in-baby-bottles. (accessed on October 11, 2011).

35. K. Aschberger, P. Castello, E. Hoekstra, S. Karakitsios, S. Munn, S. Pakalin, and D. Sarigiannis, Bisphenol A and baby bottles: Challenges and perspectives, European Commission, *Joint Research Centre, JRC Scientific and Technical*

Reports, EUR 24389 EN 2010; http://publications.jrc.ec.europa.eu/repository/bitstream/111111111/14221/1/eur%2024389_bpa%20%20baby%20bottles_chall%20%20persp%20(2).pdf (2010) p. 9.

36. J. G. Koppe and J. Keys, PCBs and the precautionary principle. In: P. Harremoës et al. (eds.) *The Precautionary Principle in the 20th Century*, London, Sterling, U.K.: Earthscan Publications (2002), 64.

37. A. H. Arnall, Future technologies, today's choices: Nanotechnology, artificial intelligence and robotics; A technical, political and institutional map of emerging technologies. A report for the Greenpeace Environmental Trust (2003) 36.

38. http://www.foodproductiondaily.com/Packaging/Bioplastic-packaging-made-from-fruit-skins (accessed on September 1, 2011).

39. J. G. U. Adams, Smeed's Law, seat belts and the emperor's new clothes. In: L. Evans and R. C. Schwing (eds.), *Human Behaviour and Traffic Safety*, New York: Plenum Press (1985).

40. F. Wickson, K. Grieger, and A. Baun, Nature and nanotechnology: Science, ideology and policy, *International Journal of Emerging Technologies and Society*, 8(1) (2010) 7.

41. A. L. Brody, Case studies on nanotechnologies for food packaging, *Food Technology*, 61(7) (2007) 102.

42. M. Smolander and Q. Chaudhry, Nanotechnologies in food packaging. In: Q. Chaudhry, L. Castle, and R. Watkins (eds.), *Nanotechnologies in Food*, London, U.K.: RSC Publishing (2010) 86.

43. G. Hunt, K. Schmid, M. Riediker, and D. Hart, Stakeholders and their interests: Wrapped up in nano: How to inform the public about nano-enhanced food contact materials. *Deliverable 4.1c under the European Commission's Seventh Framework Programme*, for Project NanoImpactNet (2010) 4.

44. R. Owen and N. Goldberg, Responsible innovation: A pilot study with the U.K. *Engineering and Physical Sciences Research Council, Risk Analysis*, 30(11) (2010) 1699, 1701–1703.

45. P. Feng, Rethinking technology, revitalizing ethics: Overcoming barriers to ethical design, *Science and Engineering Ethics*, 6(2) (2000) 208–210, 212–213.

46. W. McDonough and M. Braungar, *Cradle to Cradle: Remaking the Way we Make Things*, New York: North Point Press (2002) p. 26.

47. S. B. Billatos and N. A. Basaly, Green technology and design for the environment, London, U.K.: Taylor & Francis (1997) 115.

48. N. Horton, Trends in ethical and sustainable packaging, *Business Insights* (2008) 17.

49. B. K. Otto, Sustainability (2007) pp. 1–2. http://www.designcouncil.org.uk/en/About-Design/Business-Essentials/Sustainability (accessed on September 26, 2011).

50. http://ec.europa.eu/environment/chemicals/nanotech/pdf/commission_recommendation.pdf (accessed on October 20, 2011).

51. A. D. Maynard, Don't define nanomaterials, *Nature*, 475 (2011) 31.

52. E. Kosta and D. M. Bowman, Treating or tracking? Regulatory challenges of nano-enabled ICT implants, *Law and Policy*, 33(2) (2011) 256.

53. M. Riediker, Chances and risks of nanomaterials for health and environment. In: A. Schmid et al. (eds.), *Nano-net: 4th International ICST Conference, Nano-Net 2009*, Lucerne, Switzerland, October 18–20, 2009. Proceedings. Berlin, Germany, Springer-Verlag (2009), p. 128.

54. http://ec.europa.eu/food/food/chemicalsafety/foodcontact/framework_en.htm (accessed on September 27, 2011).

3

Evolution of Rheology, Structure, and Properties around the Rheological Flocculation and Percolation Thresholds in Polymer Nanocomposites

Rumiana Kotsilkova, Evgeni Ivanov, Ekaterina Krusteva, Clara Silvestre, Sossio Cimmino, and Donatella Duraccio

CONTENTS

3.1 Introduction

There is a high level of interest in using nanoscale reinforcing fillers, such as nanolayers, nanoparticles, and recently carbon nanotubes (CNT), for producing polymeric nanocomposite materials with exceptional properties.[1,2] The immense internal interfacial area and the nanoscopic dimensions between the particles lead to the formation of a hybrid structure, which fundamentally differentiates polymer nanocomposites from traditionally filled plastics. A unique feature of polymer nanocomposites is that the improved performance is reached at low filler content, thus resulting in lightweight materials.

Technology of polymeric nanocomposites is concerned with nanoparticles dispersed in polymer matrix; thus, nanocomposites combine two concepts, the composites and the nanometer-scale materials. The aim is to gain control on structure at atomic, molecular, and supramolecular level and to maintain the stability of interfaces in order to efficiently manufacture these materials. Because of the small nanoscale size of the filler and the chemical processes at the nanoparticle–matrix interface, nanocomposites exhibit novel combinations of properties that ensue from the nanoscale structure. Therefore, polymer nanocomposites provide opportunities to outperform conventional reinforced plastics, enhancing the promise of nanoengineered materials applications.[2–4]

There are references in literature for the high potential of carbon nanotubes to improve the properties of polymers, compared to the conventionally filled polymers.[1–3,5] Carbon nanotubes exhibit exceptionally high aspect ratios in combination with unique mechanical, thermal, and electronic properties, which make them a potential candidate for the reinforcement of polymers. However, the key point is to transfer the extraordinary properties of carbon nanotubes to the polymer in the composite. This may be achieved by their uniform dispersion, orientation in the polymeric matrix, and control on interfacial interactions between carbon nanotube surfaces and the polymer matrix.[6]

Polypropylene (PP) nanocomposites with carbon nanotubes are studied extensively as a new way for obtaining high-performance PP engineering plastics for variety of applications.[3,4,7–10] Authors reported on the effects of nanotube filling for the structure and nucleation mechanism, the mechanical property improvement and reinforcement, as well as enhancement of the thermal stability and flammability of PP. Different approaches for creation of nanocomposites producing different strength of interface interaction are referred in the literature. Three methods are currently used for preparation of PP nanocomposites, such as (i) melt mechanical mixing of carbon nanotubes with polymer,[11,12] (ii) solution mixing of suspension of nanotubes in dissolved polymer,[9] and (iii) in situ polymerization of nanotube–monomer mixture.[10] In the present study, we apply another preparation procedure for the nanocomposites by melt extruder mixing of PP with a commercial masterbatch containing carbon nanotubes.

The processing and application of nanocomposites requires information on their rheological responses. The quantitative characterization of carbon nanotube dispersion in polymer is a difficult task, so different approaches are applied in this field including direct microscopic observation and indirect estimative methods. One of the indirect methods is the dynamic rheological measurement, which demonstrates different behavior of storage modulus and dynamic viscosity, according to the dispersion state of nanofiller in the matrix polymer. A few studies reported on the dispersion and rheological aspects of carbon nanotubes in polymer melts.[2,7,13–17] In our previous study,[18] we provided rheological results for the effect of the melt processing conditions on the degree of dispersion of multiwall carbon nanotubes in iPP. The present study is stressed on characterization of the rheological parameters of carbon nanotube/PP composites, as they are of major significance for many industrial applications, as far as they determine the technology of producing numerous products important for industry.

The concentration dependence of viscosity reveals some specific features. Formation of percolating network of nanofiller in polymer[19] (e.g., geometric percolation threshold) and its synergistic effects with electrical properties,[15] mechanical properties,[20–28] and crystallization behavior[7,29–35] of polymer nanocomposites is widely reported. Passing through the geometric percolation threshold, which depends on the amount of filler and on their aspect ratio, many of the properties increase suddenly.[24–28] This is an interesting issue in terms of industrial applications, because it facilitates processing and contributes to lower final price. While the geometric percolation has been widely studied and modeled,[36] few reports concern the rheological percolation of thermoplastic polymer nanocomposites, that is, the evolution of rheological properties around the geometric percolation threshold.[2,15,37–39] First observations were on clay–polymer nanocomposites,[4,38,39] where the effect of nanofiller volume fraction on the rheological behavior is discussed as a change from a liquid-like to pseudosolid-like flow behavior above a critical volume fraction. This rheological threshold was presumably related with the mesoscale structure formed by the percolation of nanofiller, and further discussed with respect to the relationship with nanocomposite properties.

The carbon nanotube/PP composites are recently studied for nanoreinforced, active, and smart food packaging material applications.[40–43] Improved tensile strength/modulus is found by addition of carbon nanotubes to PP.[41,42] Carbon nanotubes have also been reported to have antibacterial properties, as direct contact with aggregates of carbon nanotubes may kill *E. coli*,[43] which makes carbon nanotubes attractive in active food packaging material applications. Another possible application of polymer/carbon nanotube composites is as nanosensors, which, when embedded into packaging, may be able to respond to environmental changes during storage. The recent findings open new opportunities for carbon nanotube/polymer composites in smart food packaging applications. Nevertheless, there are important safety concerns about nanotechnology applications to food contact materials, which require further investigations.[44]

Surface mechanical testing of polymeric materials, including indentation and hardness tests, and tests of scratch resistance have been used with limited success and primarily for qualitative comparisons and quality control. Scratch performance of polymers is determined both by the scratch stress field associated with the indenter geometry and by material properties. The geometry of the indenter has significant effects on scratch resistance of the polymer.[45] More recent efforts have been aimed at measuring quantitative material properties and understanding relationships between surface properties and performance characteristics. In most of these studies, single-probe testing devices, including depth-sensing indentation and scratch systems, are used. With the development of nanophased materials in the recent years, attempts were made to develop nanoparticulate-filled polymer composites for improving the scratch performance of the matrices.[46,47] In general, for low-modulus polymers, an increase in modulus and a decrease in friction coefficient can reduce the scratch depth and size of the plastic zone on the scratch surface.[48] Typically, two approaches are applied to improve the tribological properties and scratch resistance of a given plastic material, namely, (i) modifying the polymer molecular structure, for example, crystallinity, and (ii) producing polymer composites with various fillers and additives.

In this chapter, we study nanocomposites prepared by melt mixing of iPP with a commercial masterbatch of PP containing multiwall carbon nanotubes. At first, we reveal the influence of technological conditions on system homogenization, as well as determine two rheological thresholds of the dispersion structure depending on nanofiller loading. Then, we investigate the evolution of the structure and macro-, micro-, and nanomechanical properties around the rheological thresholds, thus explaining the performance by rheology–structure–mechanical property relationships. The effects of carbon nanotube filler amount in PP are discussed, with respect to the ability of decreasing the friction coefficient and enhancing the hardness and rigid filler effectiveness. The study aims to improve the knowledge on the design of carbon nanotube/PP nanocomposites with a possible application in eco-sustainable food packaging technology.

3.2 Basic Principles of Rheology for Nanocomposite Characterization

In our previous studies,[2,28,29] we have proposed a rheological approach to gain control on nanocomposite processing, which comprises (i) methods for characterization of the degree of nanofiller dispersion and (ii) characterization of the supramolecular structure of nanocomposites formed by interconnectivity of nanoparticles and polymer–nanofiller interactions. This rheological approach is proposed as a highly useful analytical tool in the development and optimization of nanocomposites, as well as for routine and nonexpensive control of nanocomposite preparation technology, in order to identify interesting samples at an early stage of their preparation and to establish control on the nanodispersion structure.

3.2.1 Rheology Control on the Degree of Dispersion

This rheological method allows control on the degree of nanofiller dispersion in a polymer matrix by using experimental data from the low amplitude oscillatory shear flow and the steady-state shear flow combined with rheological models.[2] Both the shear thinning exponent (n) and the terminal regime exponent of storage modulus (m) can be used to compare the degree of nanofiller dispersion in different nanocomposite samples at a fixed filler concentration. In general, n = 1 indicates a Newtonian flow system (typical for monomers and low-viscosity polymer samples), as well as m = 2 is the value of the terminal flow behavior of fully relaxed polymers. If the filled samples behave as the matrix polymer, essentially Newtonian (n ≈ 1), or present terminal behavior (m ≈ 2), they are usually not nanocomposites and such behavior indicates the presence of micron-sized aggregates. In contrast, nanocomposite samples demonstrate considerable shear thinning (n ~ 0) and solid-like behavior (m ~ 0) at a relatively small filler volume fraction, usually with a morphology of smooth, finely dispersed nanoscale filler. Additionally, samples with moderate values of n ≤ 0.5 and m ≤ 1 behave as not perfectly dispersed nanocomposites.

3.2.2 Rheological Flocculation and Percolation Thresholds of Nanodispersions

Authors determine mostly the rheological percolation threshold and relate it with the increase of some physical and mechanical properties of nanocomposites.[37–39] In our previous study,[2] we found that the rheological response and properties of polymer nanocomposites are very sensitive to both the short-range and long-range interconnectivity of nanofiller particles. Hence, we assumed two critical concentrations accounting for structural transitions of nanodispersions with increasing nanofiller content, named rheological flocculation threshold, (ϕc), and rheological percolation threshold, (ϕp), as follows:

1. The rheological flocculation threshold (ϕc) expresses the critical nanofiller concentration, where the short-range connectivity of the nanofiller particles becomes significant resulting in flocculation. The fractal floccules consist of near uniform nanofiller agglomerates penetrated with a matrix polymer. This threshold is usually displayed by deviation of the viscosity function vs. volume fraction from the Einstein relation.

2. The rheological percolation threshold (ϕp) represents the long-range connectivity of the fractal flocs within the polymer matrix, resulting in the formation of a three-dimensional (network) supramolecular structure. This threshold is usually displayed by the sudden rise of the viscosity vs. volume fraction function. The percolation threshold may be determined also based on the scaling concept.[49,50]

In our previous studies,[2,28,29] we found that the values of the two rheological thresholds depend significantly on the size and shape of nanofiller particles, as well as on the viscosity of the polymer matrix. The possible implication of the relatively small values of both (ϕc) and (ϕp) for the nanolayers and nanoparticles, as compared to that of micron-sized fibers and particles, is associated with the high surface area of the nanofiller particles and the strong interactions at the nanofiller–polymer interfaces. Relationships between the rheological response and some properties around the two rheological thresholds of flocculation and percolation are reported for the polymer nanocomposites.[2] Particularly, for the nanocomposites of iPP with multiwall carbon nanotubes,[28] correlations were found between the rheological flocculation threshold and the improvement of the thermal and macromechanical properties, which were explained with the structural transitions observed at the rheological flocculation thresholds.

3.3 Experimental

3.3.1 Materials and Preparation

A commercial masterbatch of 20 wt.% multiwall carbon nanotubes in PP was used, as obtained in pallet from Hyperion Catalysis Int. The multiwall carbon nanotubes (MWCNTs) are commercially manufactured from high purity, low-molecular-weight hydrocarbons in a continuous, gas phase, catalyzed reaction. Typical outside diameter range of the tubes is from 10 to 15 nm, the lengths are between 1 and 10 μm, and their density is approximately 1.75 g/cm^3. The masterbatch is produced by initially dispersing intertwined agglomerates of nanotubes into the PP. Isotactic polypropylene (iPP) "Buplen"6231 (Lukoil Neftochim) with density, $\rho = 901$ kg/m^3, and shear viscosity, $\eta_o = 200$ Pa.s (at 190°C), was used as a matrix polymer.

Nanocomposites were produced by direct melt compounding in Brabender model DSE 35/17D twin-screw extruder according to a two-step process, as published elsewhere.[18] The masterbatch was diluted to different carbon nanotube concentrations in the range of 0.1–3 wt.% with virgin iPP at melt temperature of 200°C and a screw speed of 30 rpm. The extruded composites were cooled and pelletized. In order to improve the carbon nanotube dispersion, the compositions were extruded in three runs and further calendared as sheets with thickness of about 1.5 mm under the same temperature conditions. Test specimens taken after each extrusion run were punched from the sheets for further characterization.

3.3.2 Methods of Characterization

Rheological properties of MWCNT/iPP melts were investigated in the oscillatory shear flow mode and the measurements were carried out using a cone-plate rotational viscometer at a melt temperature of 190°C. Dynamic viscosity, storage, and loss moduli were measured in the angular frequency range of 0.06–100 rad/s within the linear viscoelastic range at low strain amplitude of 0.02. The specimens for the viscometric tests were discs with diameter of 14 mm and thickness of 1.5 mm.

Thermal properties and crystallization processes during cooling were studied by differential scanning calorimetric (DSC) analysis performed with a Mettler Toledo 822 differential scanning calorimeter. The calorimetric properties were investigated by heating from −80°C to 200°C at a rate of 10°C/min (first run); after cooling at 30°C/min, the sample was heated again from −80°C to 200°C at a rate of 10°C/min (second run). The melting temperature (T_m) was obtained from the maxima of the endothermic peaks of the second run curves. The glass transition temperature (T_g) was determined from the peak of the first derivative of the second run curves. The crystallization behavior was studied by a standard procedure, as samples of about 8 mg were heated from room temperature to 200°C and held there for 10 min to eliminate any thermal history of the material. Subsequently, the samples were cooled to −50°C at 10°C/min.

Wide angle X-ray diffraction (WAXD) analysis, carried out on a Philips XPW1730 power 16diffractometer (CuNi-filtered radiation) equipped with a rotated sample holder device, was applied to study the effect of nanotubes on the crystal structure and crystallinity of iPP. The percentage crystallinity ($X_c\%$) was calculated by using the following procedure: the baseline was drawn between two points, chosen so that all diffraction patterns had minima at these points; the amorphous peak was chosen by drawing a line connecting the two extreme minimum points of the baseline and the minima of the crystalline peaks; and the ratio of the area under the crystalline peaks and the total area, multiplied by 100, was taken as the percentage of crystallinity.

Optical measurements were carried out with an Axioskop polarizing microscope (Zeiss, Thornwood, NY) equipped with a THMS 600 hot stage (Linkam Scientific Instruments Ltd., UK), Linkam TMS91 temperature control unit, and a video camera model JVC TK-1085E. Samples for optical microscopy were prepared by squeezing a small quantity of the material onto a glass cover slip.

Microstructure and morphology of the bulk nanocomposites was studied by scanning electron microscopy (SEM), using JEM 2CX (Joel), with a resolution of 40A and accelerated voltage of 100 kV. Samples were cut in liquid nitrogen and covered with vacuum-evaporated carbon and gold. The hierarchy of the nanocomposite structure was estimated at different magnifications within the range from 100 to 0.2 μm.

Nanoindentations were carried out using a Nanoindentation Tester (UNMT) with atomic force microscope (AFM) and PRO500 3D Profilometer, produced by the Center for Tribology (CETR), the United States. The sample surface was polished by means of Leica RM2245 microtome with stereomicroscope Leica A60S and a diamond knife from Leica Microsystems, Germany. The hardness and elastic modulus were calculated from the recorded load–displacement curves. Calculation methods to determine hardness and modulus are typically based on the methods of Oliver and Pharr.[51,52] The indentation impressions were then imaged using optical microscope and AFM. Indenter type Berkovich Diamond with Tip Radius 70 nm was used for indentations in force control mode of 5 mN. A series of 48 (4 × 12; spacing between indents 80 μm) indentations was performed for each sample. A typical indentation experiment consists of the subsequent steps: (i) approaching the surface, (ii) loading to the peak load of 5 mN for 15 s, (iii) holding the indenter at peak load for 10 s, (iv) unloading from maximum force of 5 mN to 10% for 15 s, (v) holding at 10% of max force for 15 s, and (vi) final complete unloading for 1 s (load function 15s-10s-15s trapezoid). The hold step was included to avoid the influence of creep on the unloading characteristics since the unloading curve was used to obtain the elastic modulus of the material.

The scratch test was performed on the Tribology Tester UMT-2 (CETR). In this test, polymeric materials are scratched applying the force control method with a constant force of 2 N, using a single microcutting blade of composite diamond with a tip radius of 0.8 mm at a speed of 0.083 mm/s, over a distance of 10 mm and time of 120 s. The distance between the plaque surface and the bottom of the groove is calculated from the carriage displacement and defined as the "scratch depth." The friction coefficient is recorded and calculated by the ratio between tangential force and normal load.

The tensile experiments were carried out with the "Tiratest 2300" universal testing machine. The moving speed of the crosshead was 10 mm/min for the tensile measurements and the ambient temperature was 20°C. Standard test specimens with size of 100 × 8 × 1 mm were used. The values of the tensile characteristics were calculated as the average over six samples for each composition.

3.4 Results and Discussion

3.4.1 Rheology

3.4.1.1 Effect of Processing on Carbon Nanotube Dispersion

The effect of processing (homogenizing) was investigated by testing samples that were extruded once, twice, and three times (1st, 2nd, and 3rd runs, respectively). As an example, Figure 3.1a–d presents the three steps of

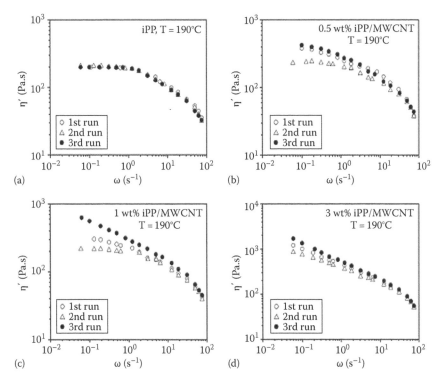

FIGURE 3.1
(a–d) Dynamic viscosity, η′ vs. angular frequency, ω of the example systems. (a) pristine iPP and MWCNT/iPP nanocomposites with (b) 0.5 wt.%, (c) 1 wt.%, and (d) 3 wt.% MWCNTs, as varying the number of extrusion runs.

homogenization for the dynamic viscosity, η′, as a function of the angular frequency, ω, thus comparing the pristine PP and the nanocomposites containing 0.5, 1, and 3 wt.% MWCNTs.

The rheological results do not provide indications for degradation of the pristine polymer during the three extrusion runs, as seen from the good fit of the viscosity curves of iPP, obtained after each of the three steps of homogenization. For the three MWCNT/iPP composites after the 3rd extrusion run, the dynamic viscosity at low frequencies rises significantly, about 2–3 times, if compared to the second run, which is associated with a better dispersion stage. The results obtained for the first extrusion run are not stable, while the second extrusion run produces low values of the viscosity, these leading to a conclusion that the carbon nanotubes are not well dispersed and some orientation in the flow field may appear during calendaring after the first 2 runs of homogenization Therefore, the 3rd extrusion run is accepted further on as the best processing condition and those samples were punched from the sheets for the characterization of properties.

3.4.1.2 Rheological Flocculation and Percolation Thresholds

Figure 3.2a and b shows the viscous and viscoelastic properties of the MWCNT/iPP melts by plotting the dynamic viscosity, η', and the storage modulus, G', as a function of angular frequency, ω, at a temperature of 190°C, by varying the carbon nanotube content from 0.1 to 3 wt.%.

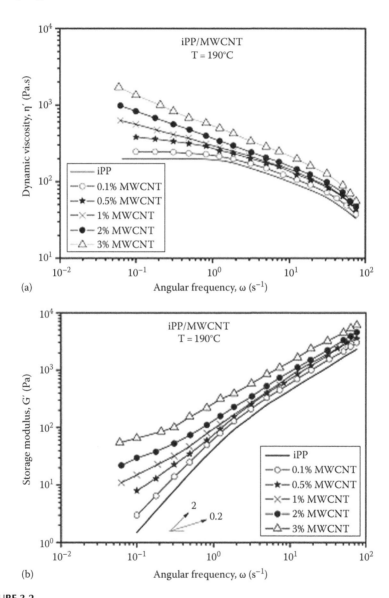

FIGURE 3.2

(a, b) Dynamic viscosity η' and storage modulus G' vs. angular frequency ω of MWCNT/iPP melts at 190°C, as varying the nanotube content from 0.1 to 3 wt.%.

The values of both functions strongly increase by increasing the nano-filler content, and this effect is most pronounced at low frequencies below 1 s^{-1}. The character of the viscosity curves η' (ω) of the composites at very low nanotube content resembles that of the pristine polymer having a Newtonian plateau at low shear rates. However, the Newtonian plateau disappears gradually above 0.5 wt.%, indicating a transition to plastic behavior. The frequency dependence of the storage modulus G' (ω) of the iPP melt shows typical homopolymer-like terminal behavior with the scaling properties of G' ∞ ω^2, indicating that polymer chains are fully relaxed at low frequencies. However, nonterminal behavior appears even at very small nanotube content around 0.5 wt.%, and the slope of the low-frequency storage modulus decreases with increasing filler content having a tendency to a plateau. Hence, the storage modulus of the MWCNT/iPP melts exhibits scaling behavior with the low frequencies, for example, G' ∞ $\omega^{1.3}$ at 0.5 wt.% MWCNTs, while G' ∞ $\omega^{0.2}$ at 3 wt.% MWCNTs.

The nonterminal behavior is indicative for a transition from liquid-like to solid-like behavior. Similar phenomenon has been observed by other researchers studying rheology of polymer/carbon nanotube dispersions.[7,13,49] Both the remarkable shear thinning and the nonterminal behavior of MWCNT/iPP melts observed with increasing nanotube loading may be attributed to nanotube connectivity and surface interactions, which restrain the long-term motion of the polymer chains and result in percolation. At high frequencies around 1 s^{-1} and above, the effect of filler loading on the viscosity is weaker because of nanotube entanglement within the flow field. These rheological results give the opportunity of determining the structural transitions of the melt dispersions with increasing nanotube content.

Figure 3.3a shows the evolution of the low-frequency storage modulus G$_0'$ (at $\omega = 0.1$ s^{-1}) as a function of nanotube weight fraction. At low nanotube content, the modulus depends weakly on the weight fraction. However, above a certain value of nanotube loading, the modulus at low frequency increases more steeply. Such behavior is expected from the percolation models for the low-frequency plateau moduli and is associated with the concentration of percolation. Hence, an approximate method is proposed here, which determines the rheological percolation threshold from the G$_0'$(φ) function, calculating a value of $\varphi_p = 2$ wt.% for the MWCNT/iPP composites.

Further information on the microstructure of melt dispersions is obtained by analyzing the high-frequency viscosity data. While the low-frequency modulus gives information about the aggregates and the percolation structure formed by aggregates, the high-frequency dynamic viscosity is always dominated by the polymeric matrix contribution. Figure 3.3b shows the evolution of the high-frequency dynamic viscosity η' (at $\omega = 10$ s^{-1}) against the weight fraction φ of MWCNTs. Two important results can be inferred from these data. The concentration dependence of viscosity shows the expected linear increase at low concentrations, corresponding to the noninteracting Einstein-like dispersion viscosity. At concentrations around and above 0.5 wt.%, the deviations

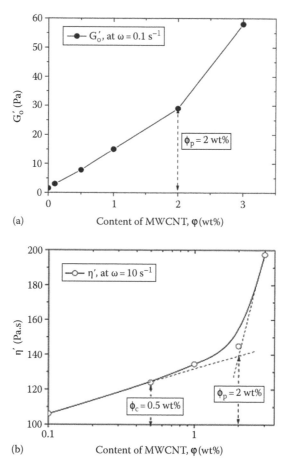

FIGURE 3.3
(a,b) Log–log plots of the low-frequency storage modulus, G_0' (at $\omega = 0.1$ s^{-1}), and high-frequency dynamic viscosity, η' (at $\omega = 10$ s^{-1}), against nanotube concentration for the MWCNT/iPP composites. Arrows show the rheological flocculation and percolation thresholds $\phi_c = 0.5$ wt.% and $\phi_p = 2$ wt.%.

from the linear regime become noticeable indicating the beginning of nanotube interactions and eventually entanglements in the dispersed composite. This concentration region is indicative for short-range connectivity of carbon nanotubes and formation of fractal flocs. Therefore, this nanotube content is called rheological flocculation threshold, having value of $\phi_c = 0.5$ wt.% for the MWCNT/iPP composites studied. Additionally, the rheological percolation threshold could be determined also from the η' (ϕ) function, by the sharp increase of its slope at approximately $\phi_p = 2$ wt.%. This behavior accounts for the formation of long-range connectivity of carbon nanotubes, that is, infinite cluster (network). The value of ϕ_p may be also determined with high correctness from the plot of the η' (ϕ) function (Figure 3.3b).

The rheological percolation threshold represents the physical gelation characterizing the critical gel point of the MWCNT/iPP composites at φ_p = 2 wt.%. Researchers reported similar gel points of about 2 wt.%[53] and 1.6 wt.%[14] for the CNT/PA composites and the CNT/PC composites, respectively, which are comparable with those found for polymer gels. Moreover, for these composites, authors found the gel points to coincide with the percolation threshold of the electrical conductivity and the high strength in CNT/polymer composite applications.[53,54]

Some authors suggest[49] that at nanotube loading above overlap concentration the tubes form an elastic network in the matrix. Physical junctions of this network are strong and stable enough to provide a rubber-like elastic response with very slow relaxation. The long-range connectivity in the CNT/polymer material, according to other researchers,[53,14] originates from a combination of entanglement of CNTs[55-57] and interactions between CNTs and polymer chains,[52] although this interaction is noncovalent. For this kind of CNT/polymer gels, it can be expected that the gel point will strongly depend on the aspect ratios of CNTs, dispersion of CNTs within polymer matrix and the interactions between CNTs (functionalized or not) and the polymer matrix (polar or nonpolar). A research based on molecular simulation suggests[58] that at the percolation threshold, an interconnecting structure of nanotubes is formed, which hinders straightforward relaxation of the polymer chains, thus increasing the storage modulus in several decades.

As shown in Figure 3.2b, the storage modulus of the MWCNT/iPP composites studied increases nearly two decades by increasing the filler content from 0.1 to 3 wt.%, compared to neat iPP. The results allow the assumption that the percolation threshold of these nanocomposites may be associated with the formation of a physical gel of interacting floccules that build a relatively weak elastic network in the matrix.

3.4.2 Structure

3.4.2.1 Crystallization Behavior

We study the evolution of the structure and crystallization kinetics of the MWCNT/iPP composites around the rheological flocculation and percolation thresholds, φ_c = 0.5 wt.% and φ_p = 2 wt.%, respectively.

DSC nonisothermal analysis is performed to study the calorimetric properties of MWCNT/iPP composites by varying the nanotube content. Figure 3.4a and b shows (a) the thermoanalytical curves, DSC (second run), and (b) the first derivative of the DSC thermograms of iPP and its composites with 1, 2, and 3 wt.% MWCNT. Data for all investigated compositions, such as the melting temperature (T_m) and the glass transition (T_g), obtained from the DSC curves and determined by the peaks of the first derivative of the DSC curves, respectively, are summarized in Table 3.1.

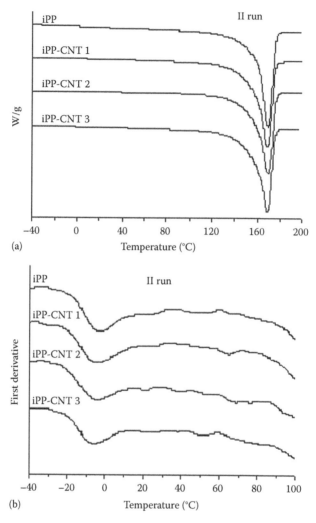

FIGURE 3.4

(a,b) DSC thermograms (a) and the first derivative of DSC thermograms (b), as taken from the second run of iPP and its composites, containing 1 wt.%, 2 wt.%, and 3 wt.% MWCNTs.

The results in Figure 3.4 and Table 3.1 show that the two characteristic temperatures are changed insufficiently by increasing the MWCNT content. Therefore, the addition of MWCNT nanotubes within the concentration range of 0.1%–3 wt.% does not influence the calorimetric properties of iPP. The results indicate weak interface interaction between the iPP matrix and the carbon nanotube surfaces.

The effect of the MWCNT masterbatch on the crystallization behavior of iPP is studied by analyzing the dynamic thermograms of the nonisothermal

TABLE 3.1

DSC and WAXD Results

Samples	Nanotube Content, (wt.%)	T_g (°C)	T_m (°C)	T_{onset} (°C)	T_c (°C)	$T_{m'}$ (°C)	X_c (%)
iPP	0	−4	169	129	122	167	54
iPP-CNT0.1	0.1	−5	170	132	126	168	52
iPP-CNT0.5	0.5	−5	170	135	127	168	51
iPP-CNT1	1	−6	169	140	128	168	50
iPP-CNT2	2	−5	169	140	128	168	51
iPP-CNT3	3	−6	169	140	129	168	51
PP masterbatch	20	—	—	142	132	169	—

T_g and T_m (data from Figure 3.4); T_{onset}, T_c, and $T_{m'}$ (data from Figure 3.5); and total crystallinity, X_c (%) (data from Figure 3.6) of iPP and MWCNT/ iPP composites.

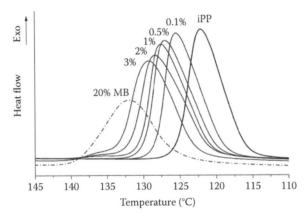

FIGURE 3.5
Nonisothermal crystallization DSC curves at 10°C/min of iPP and MWCNT/iPP composites with various nanotube contents from 0.1 to 3 wt.% and the masterbatch (20 wt.% mb).

DSC analysis. Figure 3.5 shows example exothermic heat flow curves obtained for neat iPP and its composites containing 0.1 to 3 wt.% MWCNTs and the masterbatch containing 20 wt.% of nanotubes. The crystallization onset temperature (T_{onset}), the crystallization peak (T_c), and the apparent melting temperatures ($T_{m'}$) of crystallized samples as a function of MWCNT content are reported in Table 3.1. The addition of nanotubes to iPP produces a strong rise in the peak temperatures of T_{onset} and T_c, indicating an obviously enhanced overall crystallization rate due to the nucleation effects of carbon nanotubes.

A more interesting finding is as follows. For nanocomposites with MWCNT content below the rheological flocculation threshold of $\varphi_c = 0.5\%$, the relative shift of T_{onset} and T_c is quite evident. Further increase of MWCNT concentration produces much smaller but continuous increase in the crystallization temperatures, accompanied with a significant broadening of the heat flow peak. The aforementioned results indicate that below the rheological flocculation threshold, at $\varphi < \varphi_c$, the crystallization rate of MWCNT/iPP composites increases faster with increasing nanotube concentration, thus indicating a significant nucleation effect of carbon nanofiller added in very small quantities to iPP. Obviously, at $\varphi > \varphi_c$, the crystallization rate of nanocomposites is slower, accounting for suppressed nucleation.

These results confirm our previous studies,[29] reporting the significant nucleation effect of carbon nanoparticles added in very small amounts to iPP. Authors proposed that the nanotube entanglement, producing the short-range interconnectivity, may restrain diffusion of polymer chains in the undercooled melt during crystallization, leading to changes in crystallization kinetics of the polymer matrix.[33] Furthermore, no significant changes in the melting point after crystallization ($T_{m'}$) of iPP were detected in the nanocomposites.

3.4.2.2 Crystalline Morphology

The x-ray powder diffraction profiles of iPP and MWCNT/iPP are shown in Figure 3.6. The iPP crystallized in the α-form as indicated by the peak at $2\theta = 17°–19°$ in the diffraction pattern. The α-form is that generally present in iPP obtained from cooling or quenching from the melt. Similar WAXD patterns are obtained for all MWCNT/iPP samples within the range of 0.1–3 wt.% carbon nanotubes, indicating that, for the preparation conditions used, iPP crystallizes always in the α-form independently of the nanotube content. Small amounts of

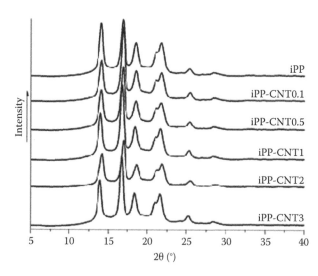

FIGURE 3.6
WAXD patterns of iPP and composites by varying the carbon nanotube content from 0.1 to 3 wt.%.

0.1 wt.% nanotubes cause a slight decrease of the crystallinity value, whereas further addition does not produce any further decrease in crystallinity. The percentage crystallinity (X_c%) is calculated from the data in Figure 3.6, using the procedure described in Section 3.3, as well as the results shown in Table 3.1. The nanotube loading has no significant influence on the total percentage crystallinity. The results confirmed those obtained by the thermoanalytical DSC characterization.

Figure 3.7a–d shows the optical micrographs of the crystalline morphology at the end of the dynamic crystallization process. For the neat iPP, large spherulites of average diameter ~80 µm are visible in Figure 3.7a.

The addition of a small amount of 0.1 wt.% MWCNT, Figure 3.7b, reduces significantly the dimension of iPP spherulites (to average diameter ~10 µm), which confirms that very low content of nanotubes, below the flocculation threshold, $\varphi \leq \varphi_c$, enhances significantly the nucleation

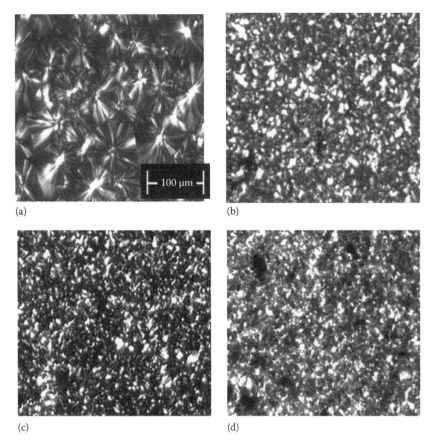

(a)

(b)

(c)

(d)

FIGURE 3.7
Optical micrographs of (a) iPP, (b) 0.1 wt.% MWCNT/iPP, (c) 0.5 wt.% MWCNT/iPP, and (d) 2 wt.% MWCNT/iPP at the end of the dynamic crystallization at 10°C/min. Magnification of 100 µm.

process of iPP. Further increase in the nanotube content exerts insignificant effect on the size reduction of spherulites. Globally, relatively well-homogenized MWCNT/iPP composites are observed at large scales below and around the rheological flocculation threshold, $\varphi_c = 0.5$ wt.%, as shown in Figure 3.7b and c. However, we found worsened homogeneity of dispersions when the nanotube content increased around and above the percolation threshold, $\varphi_p = 2$ wt.%. Particularly, in Figure 3.7d at 2 wt.% MWCNT content, aggregates of nanotubes with average size of 10–30 μm become visible as black spots in the optical micrographs.

3.4.2.3 Microstructure of Nanocomposites

Figure 3.8a and b shows the SEM micrographs presenting the microstructure and morphology at the surface of the fresh-cut nanocomposite samples. In Figure 3.8a, the quality of the dispersion of 1 wt.% MWCNT in iPP can be seen at large scales (at a magnification of 10 μm). On the micrograph, one can see a lot of long nanotubes, connecting two small aggregates of masterbatch. Figure 3.8b shows the microstructure of 0.5 wt.% MWCNT/iPP composite at high magnification of 0.2 μm, visualizing inside an aggregate of masterbatch. Obviously, such aggregate has floccule-like structure, which incorporates entangled and bundled nanotubes penetrated by the PP matrix.

Therefore, the SEM results confirm our previous findings from rheological investigations that by extruder mixing the carbon nanotube masterbatch disperses in the iPP matrix in single nanotubes and small floccule-like aggregates. Similar results were reported by Huang and Terentjev,[49] suggesting that the dispersion of carbon nanotubes in high-viscosity polymers is possible in nonparallel bundled nanotube aggregates.

3.4.3 Nano-, Micro-, and Macromechanical Properties of Nanocomposites

3.4.3.1 Nanoindentation

Nanomechanical properties are studied by instrumented nanoindentation. Representative load–displacement curves for the pristine iPP and MWCNT/iPP composites, obtained from the nanoindentation tests at loading of 5 mN, are shown in Figure 3.9a and b. The maximum load was held for 10 s for proving that the viscoelastic contribution was negligible and the Oliver–Pharr model was reliable. The addition of a small amount of 0.1% MWCNT displaced the curves to lower penetration depths, that is, the material had higher resistance to penetration. By adding 0.5–1 wt% of MWCNT, the behavior was very similar to that of the 0.1% MWCNT/iPP composite, while the sample containing 3 wt.% of nanotubes was the one that opposed the highest resistance to the penetration. This evolution evidences that increasing the

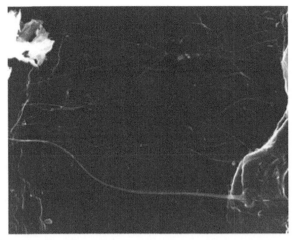

(a)   10 μm   ————

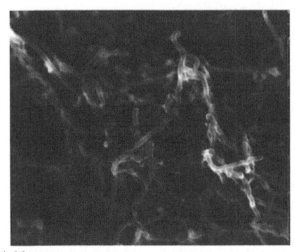

(b)   0.2 μm   ————————

FIGURE 3.8
SEM micrographs presenting the microstructure of MWCNT/iPP composites at different nanotube content: (a) 1 wt.% MWCNT, scale 10 μm; (b) 0.5 wt.% MWCNT, scale 0.2 μm.

reinforcement rate above 0.1 wt.% MWCNT loading has a hardening effect in the nanocomposites tested.

The average values of the apparent elastic modulus and hardness measured for the different carbon nanotube content, in Figure 3.10a and b, follow the same trend expected from the load–displacement curves. Among nanocomposites, the elastic modulus and hardness grew with the reinforcement rate. Interestingly, they were much higher than those of the neat polymer in the case of a very low addition of 0.1 wt.% MWCNT.

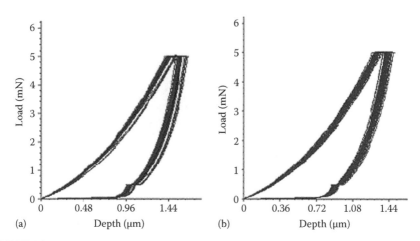

FIGURE 3.9
Representative load–displacement curves from nanoindentation tests of (a) pristine iPP and (b) 1 wt.% MWCNT/iPP loaded at 5 mN.

At nanofiller content around the rheological flocculation threshold, that is, between 0.5 wt.% and 1 wt.%, the values of the modulus and hardness show a tendency to a plateau. In contrast, they increase strongly around and above the rheological percolation threshold, $\varphi \geq \varphi_p = 2$ wt.%. All the values measured showed experimental dispersion, but it was very low and similar to the neat resin in the case of nanoreinforced materials below the percolation threshold. At higher nanofiller content, the experimental dispersion of the values increases slightly, this revealing nonhomogeneous filler dispersion.

Figure 3.11 shows the correlation between the percentage of the rise in the values of the apparent elastic modulus and hardness. In general, as the apparent elastic modulus increased, the hardness also did. As it can be observed, small addition of 0.1 wt.% MWCNT results in a significant improvement of the nanomechanical properties. For example, at nanotube content around the rheological flocculation threshold, $\varphi_c = 0.5$ wt.%, the apparent elastic modulus and the hardness of nanocomposites are higher with 14% and 23%, respectively, compared to neat iPP. This improvement becomes much higher by increasing the nanofiller loading around and above the rheological percolation threshold. Specifically, loading with 3 wt.% MWCNT enhances strongly the nanomechanical properties of iPP, leading to about 45% improvement of both characteristics.

Figure 3.12 compares the values of the apparent elastic modulus measured by instrumented indentation techniques (average of 48 tests) and the storage modulus from DMTA at 25°C by varying the nanotube content from 0.1 wt.% to 3 wt.%. Results for DMTA are taken from the Ref. [28]. The values of the apparent elastic modulus obtained from the nanoindentation tests are in all cases slightly higher than those obtained from DMTA tests. However, both techniques presented the same tendency of improvement of the modulus with increasing MWCNT content. These results provide the grounds to

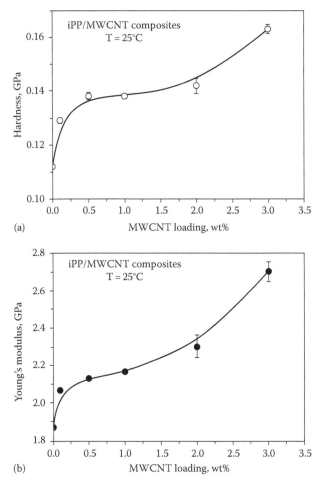

FIGURE 3.10
Average values of (a) hardness and (b) apparent elastic modulus vs. nanotube content from nanoindentation tests of MWCNT/iPP composites at maximum load of 5 mN.

propose MWCNTs as appropriate fillers for the nanoreinforcement of iPP, resulting in a significant improvement of the hardness and Young's modulus.

In order to explain the reinforcement of MWCNT/iPP composites at a nanoscale, it is important to evaluate the effects of both the dispersion structure and the crystalline morphology. We assume that the nanoreinforcement, visible at very low nanotube content around the rheological flocculation threshold, is probably due not only to nanotube entanglement but also to the significant nucleation effect of MWCNTs on the crystallization behavior of iPP. Also, the strong nanoreinforcement around and above the rheological percolation threshold probably consists in the formation of a new kind of physical gel in the MWCNT/iPP composites, which originates from a combination of several factors, such as (i) the formation of an

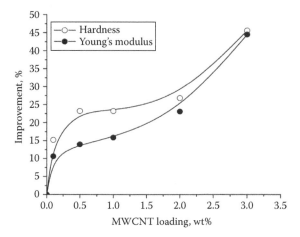

FIGURE 3.11
Percentage of improvement of the apparent elastic modulus and hardness vs. MWCNT loading, wt.%.

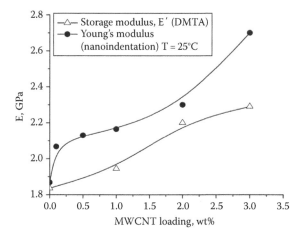

FIGURE 3.12
Elastic modulus of MWCNT/PP composites measured by DMTA and nanoindentation tests vs. carbon nanotube loading, measured at a temperature of 25°C. (Data for DMTA are taken from R. Kotsilkova, E. Ivanov, E. Krusteva, C. Silvestre, S. Cimmino, D. Duraccio, *J. Appl. Polymer Sci.*, 115 (2010) 3576.)

elastic network of overlapped and bundled nanotube aggregates penetrated by the PP matrix, (ii) interactions between CNTs and polymer chains, and (iii) physical interactions (crystallization behavior).

There is a competition between reinforcing effect of nanofiller and mobility of the polymer molecules. While the reinforcing increased the elastic modulus, larger mobility resulted in lower modulus.[15,59] This phenomenon probably would explain the results in Figure 3.10a and b around the rheological flocculation threshold, where the values of the elastic modulus and hardness increase insufficiently within 0.1 wt.%–1 wt.% nanotube loading. Obviously, the effects in the competition between reinforcing and molecular mobility are established with an equal weight in this range of nanotube loading. In contrast, at nanotube content around and above the percolation threshold, the reinforcing effect of nanofiller suppresses significantly the molecular dynamics of polymer segments, resulting in a strong increase of the elastic modulus and hardness.

To analyze this, the shape of the permanent prints made on the substrate is analyzed. Figure 3.13a and b presents the AFM image of example prints made on the samples of neat iPP and 1 wt.% MWCNT/iPP composite. These images provide information about the brittle–ductile behavior of materials because rigid ones show sink-in phenomenon in the indentation print and ductile ones show pile-up effect.[60]

Figure 3.13a and b shows that both the neat iPP and 1 wt.% MWCNT/ iPP composites demonstrate low plasticity behavior, in which pile-up effect was not observed in the line profile across the print. Similar results were obtained when the test was applied to all the nanocomposites containing 0.1 wt.%–3 wt.% MWCNT. The permanent prints observed suggested that the neat iPP and the nanocomposites have a rigid behavior and MWCNTs produce a reinforcement effect on iPP.

3.4.3.2 Scratch and Friction Behavior

With the development of nanophased materials in the recent years, attempts were made to develop nanoparticulate-filled polymer composites for improving the surface properties of the matrices.[61–64] It is well known that the hardness of PP is dependent on the degree of crystallinity and can also be affected by various factors such as type of fillers or additives used.[63] The parameters for describing the scratch resistance in this work are all associated with the surface properties of MWCNT/iPP systems and the modified polymer molecular structure of iPP nanocomposites prepared with various amount of MWCNTs. The scratch direction is chosen to be along the melt flow direction.

Typical surface scratch profiles are shown in Figure 3.14, which presents the depth (displacement)/time curves of iPP and nanocomposites by varying the nanotube content. Ridges of deformed material are produced on both sides of the scratch (groove). Extensive debris and cracking are observed at the bottom of the scratches, and this is mostly evident at high nanotube content of 3 wt.%. The increase of the nanotube content from 0.1 wt.% to 3 wt.% shifts the displacement/time curves to lower values, as well as a tendency to a plateau is observed.

The computed scratch depth is the average of at least three measurements along the scratch, and the results are shown in Figure 3.15. As can be seen from the plot, the unfilled iPP samples have a relatively deep scratch depth

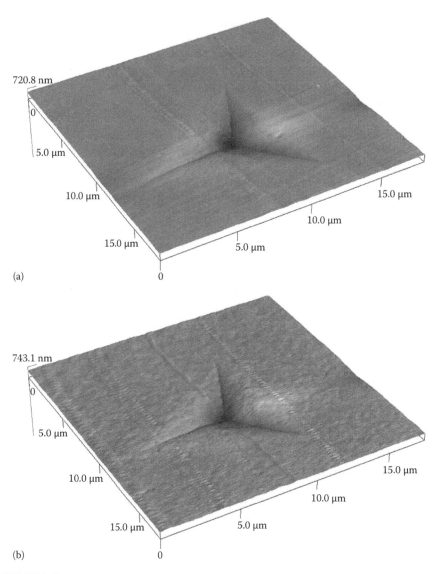

FIGURE 3.13
AFM image of prints made on (a) neat iPP and (b) 1 wt.% MWCNT/iPP composite.

at a time of 120 s. The scratch depths of the MWCNT/PP nanocomposites are significantly lower than those in pristine iPP. Interestingly, a very low nanotube content of 0.1 wt.% results in a sharp decrease (~26%) of the scratch depth. This effect is associated with the significant nucleation effect of the carbon nanotubes observed for very low nanotube amount, as shown in the data for the enhanced overall crystallization rate (Figure 3.5) and size reduction of spherulites (Figure 3.7).

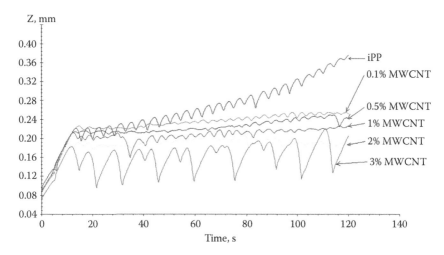

FIGURE 3.14
Carriage displacement (Z) vs. time for the iPP and nanocomposites by varying the nanotube content.

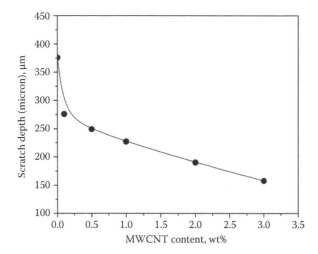

FIGURE 3.15
Scratch depth of 2 N load scratch vs. the MWCNT content.

Further increase of the MWCNT content, from 0.1% to 3 wt.%, results in a linear decrease of the scratch depth. Thus, the 3 wt.% MWCNT composites show 58% lower scratch depth values than those of neat iPP.

Figure 3.16 shows the variation of the coefficient of friction with the time of lateral displacement of the micro-cutting blade during the scratch test at a constant force of 2 N. Here the value of the coefficient of friction is defined as the ratio of the lateral force to the normal force. The coefficient of friction depends on several factors including the surface roughness and material properties of

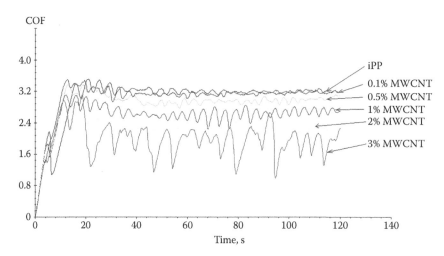

FIGURE 3.16
Surface profiles of the friction coefficient at scratch of iPP and nanocomposites by varying the MWCNT content.

the sample. Since the initial part of the data is influenced by surface morphology, the average was calculated using the data from the plateau region of the curves. Rather than just being a friction parameter, this value is a measure of the resistance to wear. A harder material would impose more resistance to scratch and consequently the blade will experience a larger lateral force.

Figure 3.17 plots the coefficient of friction at scratch vs. carbon nanotube content. It is seen that the addition of smaller amounts of MWCNTs can improve the friction-reduction of PP. Thus, the coefficient of friction decreases about 40% for the 3 wt.% MWCNT-reinforced composites, as compared to the neat

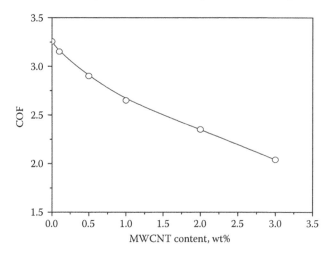

FIGURE 3.17
Coefficient of friction at scratch vs. carbon nanotube content.

iPP. The value of the coefficient of friction of the iPP and 3 wt.% MWCNT/iPP systems was found to be 3.3 and 2, respectively.

For packaging applications the surface properties, such as hardness and scratch resistance, are of special interest. The strong improvement of scratch resistance could be mostly related with the presence of carbon nanotubes and the nanocomposite concept. The amount of crystallinity is reduced slightly from 54% for the neat iPP to 51% for the 3 wt.% MWCNT/iPP composites (Table 3.1) and therefore the strong nucleation effect of MWCNTs and the size reduction of spherulites are the factors, which modify the polymer molecular structure and improve the surface properties of the nanocomposites.

3.4.3.3 Tensile Properties

Tensile strength as a function of carbon nanotube content of MWCNT/iPP composites is presented in Figure 3.18, where the effect of homogenization after the three extrusion runs is compared. Generally, by increasing the number of the extrusion runs, the tensile properties are improved, this related to a better dispersion state of carbon nanotubes in the matrix polymer. As seen, the difference in tensile strength between the samples processed by two and three runs is insignificant at low nanotube contents, as well as the statistic deviation of these data is small. However, deviations become relatively high by increasing the nanotube content, which indicates that the homogeneity of the composites is worsened.

An increase of about 13% of the tensile strength of iPP/MWCNT composites is observed at very low nanotube contents, below and around the rheological flocculation threshold, $\varphi \leq \varphi_c = 0.5$ wt.%, if compared to that

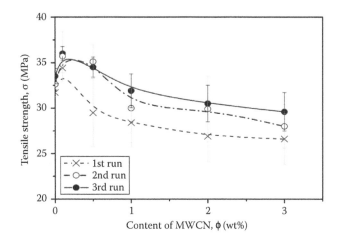

FIGURE 3.18
Tensile strength as a function of carbon nanotube content. The effect of homogenization on the tensile properties is compared within the range of 0.1–3 wt.% MWCNT, as varying the number of the extrusion runs.

of neat iPP. Higher nanofiller content results in a decrease of the tensile strength of the iPP/MWCNT composite applications. Similar finding was reported in our previous studies on the mechanical properties of polymer nanocomposites.[2,28]

These observations are very important for establishing a rheology–structure–mechanical property relationship of the MWCNT/iPP composites. Based on the results of rheology and structure, we assumed that the macroscale reinforcement, observed around the rheological flocculation threshold, originates mostly from the short-range connectivity of carbon nanotubes, as well as from the significant nucleation effect of carbon nanoparticles on iPP, while further increase of the MWCNT loading increases the nonhomogeneity of composite structure, resulting in a decrease of tensile properties.

3.5 Conclusions

In this chapter we prepare 0.1–3 wt.% MWCNT/iPP composites by melt compounding a commercial masterbatch (20% MWCNT in PP) with an iPP. The evolution of rheological properties, structure, crystalline behavior and morphology, nanohardness and Young's modulus, scratch, friction, and tensile strength against the nanotube content are investigated. The following conclusions are inferred from these investigations:

By extruder mixing in 3 runs, the carbon nanotube masterbatch disperses in iPP matrix in single nanotubes and small floccule-like bundled nanotube aggregates penetrated with a matrix polymer. By increasing the nanotube content, those aggregates interact forming an elastic network in the matrix PP. Using both low and high frequency rheological data, the microstructure of the melt dispersion is analyzed, and approximate methods are proposed for determining two rheological thresholds of flocculation, $\varphi_c = 0.5$ wt.%, and percolation, $\varphi_p = 2$ wt.%.

The low nanotube content, below and around the rheological flocculation threshold, $\varphi \leq \varphi_c$, has a significant nucleation effect on the crystallization kinetics of iPP. Hence, the iPP crystalline morphology is dramatically changed and the dimension of spherulites is reduced from ~ 80 μm in iPP to an average diameter of ~ 10 μm in the 0.1 wt.% MWCNT/iPP composites. Further increase of the nanotube content has an insignificant effect on the size reduction of spherulites, but the homogeneity of the dispersions is considerably worsened. However, the addition of MWCNT nanotubes up to 3 wt.% does not influence the glass transition (T_g) and the melting temperature (T_m) of iPP, thus accounting for a weak interface interaction between the iPP matrix and the carbon nanotube surfaces.

This evolution of the structure, crystallization behavior, and morphology around the rheological flocculation and percolation thresholds is found to be dominant influencing the nano-, micro-, and macromechanical properties of MWCNT/iPP composites. The nanohardness and apparent elastic modulus, measured by nanoindentation tests, increase at small additions of 0.1 wt.% MWCNT. Around the rheological flocculation threshold, $\varphi_c = 0.5$ wt.%, the apparent elastic modulus and the hardness of MWCNT/iPP composites are higher with 14% and 23%, respectively, than those of neat iPP. These improvements become much stronger around and above the rheological percolation threshold. Specifically, 3 wt.% MWCNT loading enhances both characteristics about 45%. The apparent elastic modulus obtained from the nanoindentation test has slightly higher values than that obtained from DMTA tests, but both techniques present the same tendency toward improvement of the modulus with increasing MWCNT content. The neat iPP and nanocomposites exhibit rigid behavior and the MWCNTs produce a reinforcement effect on iPP.

The surface properties of iPP, such as scratch resistance and friction reduction, are significantly improved by addition of MWCNT. Thus, the 3 wt.% MWCNT composites exhibit about 58% lower scratch depth and about 40% decrease in the coefficient of friction compared to the neat iPP. The strong improvement of the surface properties of nanocomposites are related to the presence of carbon nanotubes, the rigid filler effectiveness, and the strong nucleation effect of MWCNTs, which modify the polymer molecular structure of PP.

The tensile strength increases with about 13% at very low nanotube content, below and around the rheological flocculation threshold, $\varphi \leq \varphi_c = 0.5$ wt.%, if compared to that of the neat iPP. The higher nanofiller content results in a decrease of the tensile strength in the MWCNT/iPP composite applications, probably because of worse homogenization.

In general, the nano-, micro-, and macroscale reinforcement becomes visible at very low nanotube content around the rheological flocculation threshold, which is probably due not only to the nanotube entanglement and short-range connectivity but also to the significant effect of MWCNT on the crystallization morphology of iPP. The strong nano-reinforcement observed around and above the rheological percolation threshold is associated with the formation of a new kind of physical gel in the MWCNT/iPP composites, which originates from both the interactions between CNTs and polymer chains and the changes in the crystalline morphology.

Acknowledgments

This work was carried out with the support of the following projects: COST FA0904, FP7-INCO-BY NanoERA; and bilateral project between IMech, BAS and ICTP-CNR, Italy.

References

1. K. Friedrich (Ed.), *Polymer Composites—From Nano to Micro Scale*, Kluwer Academic Publisher, Dordrecht, the Netherlands (2005).
2. R. Kotsilkova (Ed.), *Thermosetting Nanocomoposites for Engineering Application*, Rapra Smiths Group, London, U.K. (2007).
3. K. Schulte, F.H. Gojny, B. Fiedler, G. Broza, and J.K.W. Sandler, Carbon nanotube reinforced polymers. A state of the art review. In: K. Friedrich (Ed.), *Polymer Composites—From Nano to Micro Scale*, Kluwer Academic Publisher, Dordrecht, the Netherlands (2005).
4. R. Kotsilkova, *Mech. Time-Depend. Mater.*, 6 (2002) 283.
5. E.T. Thostenson, Z. Ren, and T.W. Chou, *Compos. Sci. Tech.*, 61 (2001) 1899.
6. M.M. Hassan, Y. Zhou, and S. Jeelani, *Mater. Lett.*, 61 (2007) 1134.
7. D. Wu, Y. Sun, L. Wu, and M. Zhang, *J. Appl. Polym. Sci.*, 108 (2008) 1506.
8. R. Andrews, D. Jacques, M. Minot, and T. Rantell, *Macromol. Mater. Eng.*, 18 (2002) 395.
9. G.-W. Lee, S. Jagannathan, H.G. Chae, M.L. Minus, and S. Kumar, *Polymer*, 49 (2008) 1831.
10. A. Funck and W. Kaminsky, *Compos. Sci. Tech.*, 67 (2007) 906.
11. S.B. Kharchenko, J.F. Douglas, J. Obrzut, E.A. Grulke, and K.B. Migler, *Nat. Mater.*, 3 (2004) 64.
12. T. Kashiwagi, E. Grulke, J. Hilding, K. Groth, R. Harris, and K. Butler, *Polymer*, 45 (2004) 4227.
13. Q. Zhang, D.R. Lippits, S.N. Vaidya, and S. Rastogi, Dispersion and rheological aspects of SWNTs in ultra high molecular weight polypropylene: Nanotech 2006, *Technical Proceedings of 2006 NSTI Nanotechnology Conference and Trade Show*, 1, Chapter 2: Carbon nanotubes (2006), p. 206.
14. L. Chenyang, J. Zhang, J. He, and G. Hu, *Polymer*, 44 (2003) 7529.
15. A. Kanapitsas, P. Pissis, and R. Kotsilkova, *J. Non-Cryst. Solids*, 305 (2002) 204.
16. B. Cipiriano, T. Kashiwagi, S.R. Raghavan, Y. Yang, E. Grulke, K. Yamamoto, J. Shields, and J.F. Douglas, *Polymer*, 48 (2007) 6086.
17. S.B. Kharchenko, K.B. Migler, J.F. Douglas, J. Obrzut, and E.A. Grulke, Rheology, processing and electrical properties of multiwall carbon nanotube/polypropylene nanocomposites, *ANTEC*. May 16–20, 2004, Chicago, Illinois, Conference Proceeding, Vol. II, Soc. Plastic Engineering (2004) 1877–1881.
18. E. Ivanov, E. Krusteva, S. Djumalijski, R. Kotsilkova, R. Krastev, C. Silvestre, S. Cimmino, and D. Duraccio, Effect of Processing on Rheology and Structure of Polypropylene/Carbon Nanotube Composites. In: *Nanoscience and Nanotechnology*, Vol. 8 (E. Balabanova and I. Dragieva, Eds.), Prof. M. Drinov Acad. Publishing House, Sofia, (2008) 89–92.
19. D. Stauffer and A. Aharony. *Introduction to Percolation Theory*, Taylor & Francis, Washington, DC (1992).
20. V. Favier, J.Y. Cavaille, G.R. Canova, and S.C. Shrivastava, *Polymer Eng. Sci.*, 37 (1997) 1732.
21. J.Y. Cavaille, R. Vassoile, G. Thollet, L. Rios, and C. Pichot, *Colloid Polymer Sci.*, 269 (1991) 248.
22. C.M. Sinko, G.T. Carlson, and D.S. Gierer, *Int. J. Pharm.*, 114 (1995) 85.

23. P.A. Cirkel and T. Okada, *Macromolecules*, 33 (2000) 4921.
24. I. Balberg, N. Binenhaum, and N. Wagner, *Phys. Rev. Lett.*, 52 (1984) 1465.
25. N. Ueda and M. Taya, *J. Appl. Phys.*, 60 (1986) 459.
26. X.H. Yin, K. Kobayashi, K. Yoshino, H. Yamamoto, T. Natanuki, and I. Isa, *Synthetic Met.*, 69 (1995) 367.
27. I. Balberg and N. Binenhaum, *Phys. Rev. B*, 28 (1983) 3799.
28. R. Kotsilkova, E. Ivanov, E. Krusteva, C. Silvestre, S. Cimmino, and D. Duraccio, *J. Appl. Polymer Sci.*, 115 (2010) 3576.
29. C. Silvestre, S. Cimmino, M. Raimo, C. Carfagna, V. Vapuano, and R. Kotsilkova, *Macromol. Symp.*, 228 (2005) 99.
30. F. Du, R.C. Scogna, W. Zhou, S. Brand, J.E. Fischer, and K.I. Winey, *Macromolecules*, 37 (2004) 9048.
31. E. Assouline, A. Lustiger, A.H. Barber, C.A. Cooper, E. Klein, E. Wachtel, and H.D. Wagner, *J. Polym. Sci. Part B Polym. Phys.*, 41 (2003) 520.
32. L. Valentini, J. Biagiotti, J.M. Kenny, and S. Santucci, *J. Appl. Polym. Sci.*, 87 (2003) 708.
33. R. Arup, A.R. Bhattacharya, T.V. Sreekumar, T. Liu, S. Kumar, L.M. Ericson, R.H. Hauge, and R.E. Smalley, *Polymer*, 44 (2003) 2373.
34. D. Xu and ZH. Wang, *Polymer*, 49 (2008) 330.
35. C.H. Liu, J. Zhang, J. He, and G. Hu, *Polymer*, 44 (2003) 7529.
36. E.J. Garboczi, K.A. Snyder, and J.F. Douglas, *Phys. Rev. E*, 52 (1995) 819.
37. J. Vermant, S. Ceccia, M.K. Dolgovskij, P.L. Maffettone, and C.W. Macosco, *J. Rheol.*, 51 (2007) 429.
38. R. Krishnamoorti and A.S. Silva, *Rheological Properties of Polymer-Layered Silicate Nanocomposites.* In: *Polymer-Clay Nanocomposites*, (T.J. Pinnavaia and G.W. Beall, Eds.), John Wiley & Sons, NY (2000) 315–343.
39. L.A. Utracki. *Clay-Containing Polymer Nanocomposites, Smithers Rapra Technology*, Shawbury, U.K., Vol. 1 (2004) p. 456.
40. M.C. Henriette de Azeredo. Nanocomposites for food packaging applications, Review, *Food Res. Int.*, 42 (2009) 1240.
41. M.A. López Manchado, L. Valentini, J. Biagotti, and J.M. Kenny, *Carbon*, 43 (2005) 1499.
42. K. Prashantha, J. Soulestin, M.F. Lacrampe, P. Krawczak, G. Dupin, and M. Claes, *Compos. Sci. Tech.*, 69 (2009) 1756.
43. S. Kang, M. Pinault, L.D. Pfefferle, and M. Elimelech, *Langmuir*, 23 (2007) 8670.
44. H. Bouwmeester, S. Dekkers, M.Y. Noordam, W.I. Hagens, A.S. Bulder, C. De Heer, S.E.C.G. Ten Voorde, S.W.P. Wijnhoven, H.J.P. Marvin, and A.J.A.M. Sips, *Regul. Toxicol. Pharmacol.*, 53 (2009) 52.
45. B.J. Briscoe, L. Fiori, and E. Pelillo, *J. Phys. D: Appl. Phys.*, 31 (1998) 2395.
46. K. Friedrich, Z. Zhang, and A.K. Schlarb, *Compos. Sci. Tech.*, 65 (2005) 2329.
47. C.J. Schwartz and S. Bahadur, *Wear*, 237 (2000) 261.
48. C. Xiang, H.-J. Sue, and J. Chu, *ANTEC*, 99 (1999) 3463.
49. Y.Y. Huang and E.M. Terentjev, *Int. J. Mater. Form.*, 1 (2008) 63.
50. P.-G. De Gennes, *Scaling Concepts in Polymer Physics*, Cornell University Press, Ithaca, NY (1979).
51. W.C. Oliver and G.M. Pharr, *J. Mater. Res.*, 7 (1992) 1564.
52. W.C. Oliver and G.M. Pharr, *J. Mater. Res.*, 19 (2004) 3.
53. P. Pötschke, T.D. Fornes, and D.R. Paul, *Polymer*, 43 (2002) 3247.
54. M.H. Al-Saleh and U. Sundararaj, *Carbon*, 47 (2009) 2.

55. M.S.F. Shaffer and A.H. Windle, *Macromolecules*, 32 (1999) 6864.
56. M.S.F. Shaffer, X. Fan, and A.H. Windle, *Carbon*, 36 (1998) 1603.
57. K. Liao and S. Li, *Appl. Phys. Lett.*, 79 (2001) 4225.
58. J. Chen, H.Y. Liu, W.A. Weimer, M.D. Halls, D.H. Waldeck, and G.C. Walker, *J. Am. Chem. Soc.*, 124 (2002) 9034.
59. R. Kotsilkova, D. Fragiadakis, and P. Pissis, *J. Polym. Sci. Part B Polym. Phys.*, 43 (2005) 522.
60. H.A. Pour, M. Lieblich, A.J. Lopez, J. Rams, M.T. Salehi, and S.G. Shabestari, *Compos. Part A—Appl. Sci. Manuf.*, 38 (2007) 2536.
61. K. Friedrich, Z. Zhang, and P. Klein P. Wear of polymer composites. In: *Wear – Materials, Mechanisms and Practice*, (G.W. Stachowiak, Ed.), Tribology in Practice Series, Wiley (2005) 269–290.
62. Z. Zhang and K. Friedrich. Tribological characteristics of micro- and nanoparticle filled polymer composites. In: K. Friedrich, S. Fakirov, and Z. Zhang (Eds.), *Polymer Composites—From Nano- to Macro-Scale*. Berlin, Springer, Germany, Chapter 10 (2005) p. 169.
63. J. Chu, L. Rumao, and B. Colman, *Polym. Eng. Sci.*, 38 (1998) 1906.
64. Y. Hui Liu and J. Long Gao, *Adv. Mater. Res.*, 299 (2011) 798.

4

Characterization of Safe Nanostructured Polymeric Materials

Geoffrey R. Mitchell

CONTENTS

4.1 Introduction

The addition of small quantities of nanoparticles to conventional and sustainable thermoplastics leads to property enhancements with considerable potential in many areas of applications including food packaging,[1] lightweight composites, and high performance materials.[2] In the case of sustainable polymers,[3] the addition of nanoparticles may well sufficiently enhance properties such that the portfolio of possible applications is greatly increased. Most engineered nanoparticles are highly stable and these exist as nanoparticles prior to compounding with the polymer resin. They remain as nanoparticles during the active use of the packaging material as well as in the subsequent waste and recycling streams. It is also possible to construct the nanoparticles within the polymer films during processing from organic compounds selected to present minimal or no potential health hazards.[4]

In both cases, the characterization of the resultant nanostructured polymers presents a number of challenges. Foremost among these are the coupled challenges of the nanoscale of the particles and the low fraction present in the polymer matrix. Very low fractions of nanoparticles are only effective if the dispersion of the particles is good. This continues to be an issue in the process engineering but of course bad dispersion is much easier to see than good dispersion. In this chapter we show the merits of a combined scattering (neutron and x-ray) and microscopy (SEM, TEM, AFM) approach. We explore this methodology using rod-like, plate-like, and spheroidal particles including metallic particles, plate-like and rod-like clay dispersions, and nanoscale particles based on carbon such as nanotubes and graphene flakes. We will draw on a range of material systems, many explored in partnership with other members of Napolynet. The value of adding nanoscale particles is that the scale matches the scale of the structure in the polymer matrix. Although this can lead to difficulties in separating the effects in scattering experiments, the result in morphological studies means that both the nanoparticles and the polymer morphology are revealed.

4.2 Microscopy and Scattering

Microscopy techniques provide a powerful approach to the visualization of nanoparticles. A number of techniques (transmission electron microscopy TEM, scanning electron microscopy SEM, scanning probe microscopy SPM) have the necessary intrinsic resolution to image nanoscale particles and polymer nanoscale structure. Any microscopy study must take account of the fact that the high magnification required to image the nanoscale directly results in a small spatial area being sampled. In contrast, scattering techniques are usually applied to much larger areas typically with a scale of 100 µm–1 mm and hence the approach provides averages over many nanoparticles. However, the major limitation is that little information is available from scattering procedures on the dispersion of the nanoparticles or the spatial variation in structure. In recent years, considerable advances have been made with microbeam x-ray scattering techniques with a beam focus of less than 1 µm. However, such methodologies are only available at specialist synchrotron beam lines such as ID13 at the European Synchrotron Research Facility.[5]

The complementary advantages of microscopy and scattering are shown directly in Figure 4.1.

The methodologies used to produce these results will be discussed in later sections in this chapter. The right-hand image shows a TEM micrograph of self-assembled nanoparticle embedded in a crystallized polyethylene matrix.[6] As will be discussed later the nanoparticle has directed the crystallization of the polymer matrix, as can be seen been seen from the orientation of the lamellar crystals viewed edge on in the image with respect to the long axis of the nanoparticle. On the left-hand side is a small-angle

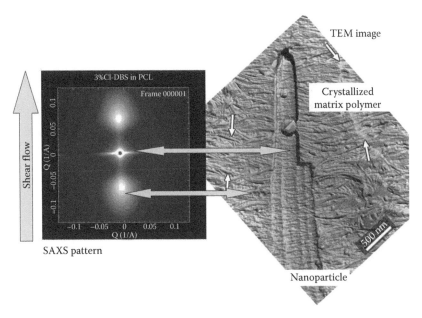

FIGURE 4.1

A small-angle x-ray scattering pattern obtained for a crystalline polymer containing 3% Cl-DBS after crystallization from a sheared melt alongside a TEM micrograph of a replica of an etched surface of a polymer treated in the same manner. (From Nogales, A. et al., *Macromol. Rapid Comm.*, 24, 496, 2003; Mitchell, G.R., *Macromol. Rapid Comm.*, 2012, to appear.)

x-ray scattering (SAXS) pattern obtained at the I22 SAXS beam line at the Diamond synchrotron facility.[7] The high level of azimuthal variation in intensity indicates directly the high level of anisotropy or preferred orientation present in the structure. The black circle in the center of the pattern is the beam stop which blocks the incoming direct beam from reaching the detector. The highly anisotropic scattering in the center of the pattern in the form of a horizontal streak arises from an extended object in which the length is considerably longer than the width. This illustrates the reciprocal nature of scattering. The intense spots above and below the center arise from stacks of lamellar crystals. The position, in terms of $|Q|$ where $|Q| = 4\pi \sin \theta/\lambda$, 2θ is the scattering angle and λ is the incident wavelength, provides quantitative information on the so-called long period which is the repeating distance involving the crystal thickness and the layer of amorphous material between the crystals. The arrows in the figure show how the features in the TEM image are connected to the SAXS pattern. Clearly the TEM image only contains one nanoparticle while the SAXS pattern was obtained from a volume $300 \times 300 \times 1000$ μm and hence the pattern is an average over $\sim 5 \times 10^{12}$ nanoparticles.

In general, microscopy at the nanoscale is restricted to static samples at fixed conditions, although there is growing use of scanning probe microscopy in the study of polymer crystallization particularly as the scan rates

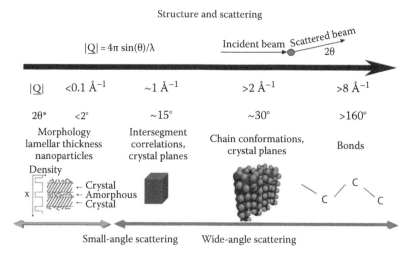

FIGURE 4.2

The structural scales which can be accessed using x-ray or neutron scattering technique. *Calculated using CuK radiation.

have increased.[8] In contrast, x-ray and neutron scattering techniques lend themselves to adaption with different sample environments and when coupled with intense x-ray sources available in the laboratory or at synchrotron facilities enable time-resolving techniques to follow the evolution of nanoscale structure.

Figure 4.2 shows a schematic of the structural scales available in scattering experiments and it is the breadth of this range which makes x-ray and neutron scattering procedures so powerful in the study of materials including the structure at the nanoscale. For microscopy, the technical challenges are centered on achieving a high level of magnification, while the reverse is the case with scattering. The challenge with small-angle x-ray and neutron scattering is measuring the signal at low Q (i.e., large real space distances). We can see the issues in the SAXS pattern in Figure 4.1. The dark circle is the beam stop. Clearly data cannot be recorded for Q values which are less than the edge of the beam stop. The size of the beam stop depends on the quality of the beam and the beam optics used to shape and define the beam. The design of the most recently built synchrotron facilities leads to high quality beams at the sample as can be seen in the pattern. On beam line I22 at Diamond with the longest sample-detector distance, Q_{min} is 0.002 Å^{-1}. To achieve the measurement of scattering at lower Q values corresponding to distances of ~200 nm requires highly specialized beam lines such as the newly announced ZOOM beam line at ISIS and the BW4 at DESY. This starts to overlap with the scales relevant to light scattering but this, of course, requires a level of transparency for the light scattering not present in many samples. In contrast, x-rays and especially neutrons do not have this limitation. In the case of polymers loaded with nanoparticles containing metallic

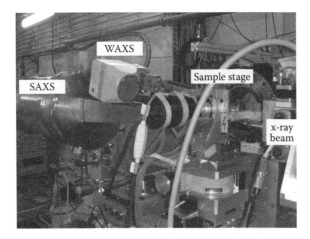

FIGURE 4.3
Beam line BM16 at the European Synchrotron Research Facility, Grenoble, France.

elements, the absorption may become significant in x-ray scattering procedures when the fraction increases above 0.01. Similar effects will be observed
in polymer heavily loaded with clay platelets. For neutron scattering absorption is much less substantial and in general will present no restrictions.

Figure 4.2 shows the two regimes of small-angle and wide-angle scattering. Generally, in order to obtain the highest quality scattering data, different
instrumentation is used. For time-resolving scattering there is tremendous
value in recording both SAXS and WAXS data simultaneously.

Figure 4.3 shows a photograph of the SAXS/WAXS setup on beam line
BM16 at the European Synchrotron Research Facility.[9] The SAXS 2-D detector
is out of the picture to the left at the end of a 5 m vacuum tube. Figure 4.4a
shows a SAXS pattern, obtained using the setup shown, for an isotactic
polypropylene (i-PP) based nanocomposite which was crystallized in situ
after being subjected to shear flow in the melt using a shear cell specifically
designed for this purpose.[10] This geometry gives a SAXS pattern which covers
the complete azimuthal range over a useful Q range. In contrast, the WAXS
detector which is offset from the beam line, only gives coverage of a limited
Q range and a particularly narrow azimuthal range. Figure 4.4b shows the
WAXS pattern recorded from the sample over the same time period as the
SAXS pattern, revealing a sequence of WAXS peaks which is characteristic of
the β phase of iPP.[11] As only a limited range of azimuthal WAXS data is accessible, interpretation of the pattern must take account of any anisotropy present especially when the level is high. Clearly, use of the WAXS data is simplest
when the reference patterns have been obtained ex situ using instrumentation which provides full coverage. Much of the data which is shown in this
chapter has been obtained using synchrotron beam lines but it is empha-
sized that very useful data can be recorded using laboratory sources when
evaluating static samples or when the time-resolution required is modest.

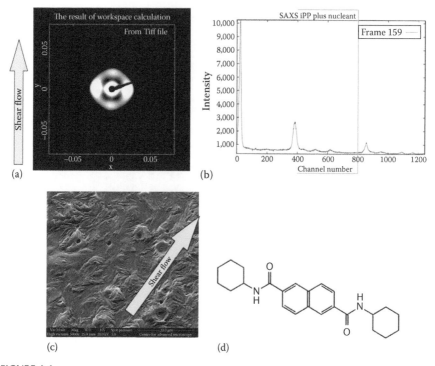

FIGURE 4.4

(a) SAXS of a sample iPP + 1% N'-Dicyclohexyl-2,6-Naphthalene dicarboxamide sheared in the melt at 210°C and then cooled at 10°C/min, (b) WAXS recorded from the same sample, (c) SEM micrograph of the etched surface of the sample at the point of the sample corresponding to the scattering experiments, and (d) chemical structure of % N'-Dicyclohexyl-2,6-Naphthalene dicarboxamide. (From Mitchell, G.R. and Felisari, L., submitted to *Polymer.*)

Microscopy techniques either provide an image of the surface topology of the sample (SEM, SPM) or generate contrast from the interior of the sample (TEM, SPM). The highest resolution is obtained in a SEM by forming an image using the low energy secondary electrons which originate in the very surface layer as shown schematically in Figure 4.5.

Using secondary electron imaging in a SEM equipped with a field emission electron source, a resolution ~1 nm can be obtained, as is the case for the FEI Quanta FEG 600 SEM in the Centre for Advanced Microscopy at the University of Reading shown in Figure 4.6.

Back scattered electrons are higher energy and the intensity at the imaging detector is in part dependent on the elemental composition in the sample. Using back scattered electrons can be a powerful tool in imaging nanoparticles in polymer matrix but the gain in contrast needs to be offset against the lower resolution (Figure 4.5). The incident electron beam also generates x-rays with energies which are characteristic of the elements present in the sample. These x-rays can be sorted in to a spectrum (Figure 4.6) and the peaks used to identify the composition of the sample. They can also be used

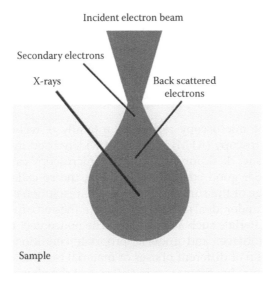

FIGURE 4.5
Schematic of the relationship between the incident electron beam in a SEM and the volumes from which the electrons and x-rays which are used to form images originate from.

FIGURE 4.6
The FEI Quanta 600 in the Centre for Advanced Microscopy at Reading which is equipped with field emission gun and x-ray spectroscopy. The left-hand screen shows an x-ray spectrum obtained from a nanoparticle sample shown in the right-hand screen.

to generate images or maps of the elemental distribution in the sample but the resolution is lower than that of the electron images, partly as the x-rays on average come from deeper in the sample where the activated volume is somewhat larger as shown schematically in Figure 4.6. More recent developments such as low vacuum and environmental SEM modes which allow uncoated samples to be examined also compromise the resolution which can be achieved.

Scanning probe microscopy represents a family of techniques in which atomic force microscopy (AFM) is the most widely used microscopy which is relevant to this work. Scanning tunneling microscopy can yield atomistic level images under good conditions. In AFM, the so-called contact mode provides an image of the surface topology. The resolution can approach the nanometer scale under ideal conditions but tip adhesion can limit the usefulness with soft materials such as polymers. The noncontact method involves an oscillating cantilever and under appropriate conditions, images of the spatial distribution of different phases or material can be formed where the contrast arises from the differences in stiffness of the phases.

Transmission electron microscopy provides high resolution images of samples using either absorption or diffraction contrast. As well as the imaging modes, the TEM also offers the opportunity for studying the structure through diffraction and x-ray spectroscopy. In some cases, for example with block copolymers and blends it is useful to stain the sample with a heavy metal salt.[12] The metal selectively binds to specific chemical species and provides an absorption contrast which reveals the distribution of that chemical species. TEM needs relatively thin sections typically <100 nm. This presents no difficulty in preparing nanoparticles for study, but may require the use of an ultramicrotome to produce thin sections from a bulk sample of a nanocomposite.

4.3 Nanoparticles

Nanoparticles are available in a variety of shapes and sizes, some in quantities suitable for manufacturing and some on a scale suitable for laboratory studies. The three basic shape classes of spheres, rods, and plates, with examples of materials, are shown in Figure 4.7.

As will be seen later, the shape has a critical impact on the resultant particle-matrix interactions. Moreover, the shape is intrinsically coupled to the function of the nanoparticle in the material system. For example, much of the reported work on polymers and nanoparticles is focused on the development of nanocomposites with enhanced mechanical properties in which case rod-like particles such as carbon nanotubes are particular suited. In contrast, the control of oxygen permeability in polymer films using increased

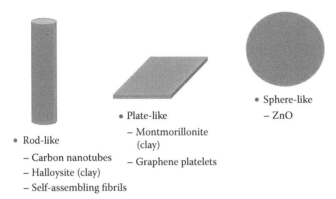

FIGURE 4.7
A schematic representation of the shapes of nanoparticles.

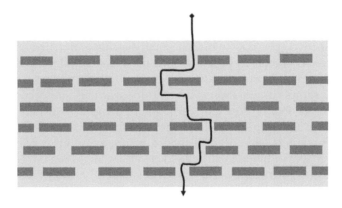

FIGURE 4.8
A schematic of a polymer film not to scale loaded with nanoclay. The line represents the tortuous path required for oxygen molecules or other gas molecules to pass through the film. The actual path would be more random due to the random processes associated with diffusion.

tortuosity is most effectively achieved using plate-like particles as shown schematically in Figure 4.8.

Within this simple shape classification there can be major differences in the nature of the nanoparticles and in particular in the pathways to the final polymer nanocomposites. This has an immediate impact on the approach used in the characterization of the particles.

4.3.1 Preformed Nanoparticles

The majority of nanoparticulates are preformed prior to mixing with the polymer. Of course, the effective dispersion of the nanoparticles in the polymer matrix remains a major challenge; nevertheless, the nanoparticles used in the nanocomposites can be usefully characterized prior to mixing. TEM provides

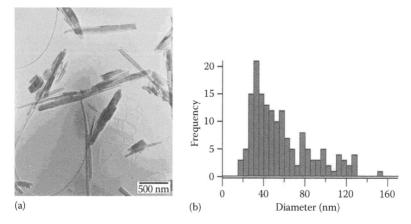

(a)   (b)

FIGURE 4.9
(a) TEM micrograph of halloysite nanotubes and (b) distribution of the recorded nanotube diameters measured from the TEM micrographs. (From Mitchell, G.R. et al., unpublished work.)

a highly effective approach to determining the shape and size of individual nanoparticles as shown in Figure 4.9a in the case of halloysite nanotubes.[13] Halloysite nanotubes are a naturally occurring mineral form.[14] For such systems, the sample preparation is straightforward. The nanoparticles are dispersed using alcohol on to a lacy carbon support grid and then inserted in to the TEM for imaging using simple absorption contrast. The dispersion of the particles on to the support grid should ensure a minimum of overlapping particles to enable individual particles to be observed. As can be seen from the TEM image in Figure 4.9a, the particles will generally exhibit a distribution of sizes, in the case of rods there are both the diameter and the length. Adequate sampling is the key to obtaining reliable and useful size distribution data as shown in Figure 4.9b which was obtained using several images similar to Figure 4.9b. The use of image analysis procedures may be helpful in measuring the particle size but success with such techniques will largely be determined by the quality of the sample preparation and the TEM image.

Once the nanoparticles have been dispersed in the polymer matrix, the task of imaging the particle and its distribution becomes considerably more difficult. The challenges center on the nanoscale of the particles and the low fraction present in the polymer matrix. There is the additional problem of the contrast between the nanoparticle and the polymer matrix. As we will see next, this is less of a challenge when the elemental composition of the nanoparticles is considerably different to the hydrocarbon basis of most polymers. It may be possible to use absorption contrast in the TEM or the backscattered electron imaging mode in SEM which are both sensitive to elemental composition.

The most straightforward techniques involve SEM, especially a high resolution system equipped with a field emission source or AFM. Both of these techniques are able to reveal the surface topology at high resolution.

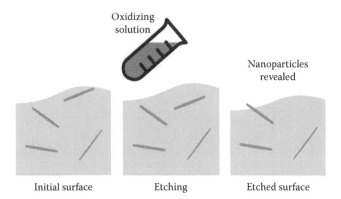

FIGURE 4.10

A schematic of the chemical etching technique use to reveal a surface with embedded nanoparticles and the morphology of the polymer matrix.

In the case of AFM, the tapping mode with phase imaging can be used to reveal differences in the material's dynamic response and hence image the distribution of different materials. However, the main challenge here is to reveal the nanoparticles which in general will be below the surface under the "encapsulating" polymer matrix. We have found that the most effective experimental methodology is to use a selective etching procedure which removes material from the surface and reveals both the nanoparticles and the morphology of the polymer matrix. Etching techniques and microscopy have a long history which started with metals. This concept was successfully extended to polymers by workers at Reading using oxidizing solutions based on permanganic acid.[15,16] The solution leads to fragmentation of the polymer chains and these are removed from the surface as shown in Figure 4.10.

The etchant may also fragment the nanoparticle. When the surrounding encapsulating polymer has been etched away the nanoparticle can also be washed from the surface. One of the particular advantages of this approach is that the etchant is also selective with respect to the crystalline and amorphous phases in semicrystalline polymers. As a consequence the crystalline and amorphous phases exhibit different differential etching rates which leads to a nanoscale structuring of the surface which reveals the semicrystalline morphology of the polymer matrix. In some cases, such as with polyurethane resin, the etching may reveal density variations. The etched surface can be examined in a SEM or using AFM techniques. In some cases, there may be advantages in using replication techniques to allow the study of the etched surface using the higher resolution of the TEM. In fact the image in Figure 4.1 was obtained in this manner using a two stage replication process.[6,17] Figure 4.11 shows a nanocomposite of iPP and multiwalled carbon nanotubes which has been etched in a similar manner, coated with gold and then examined in a SEM.[18] Bundles of CNTs are easily seen and these are uniformly distributed through the sample.

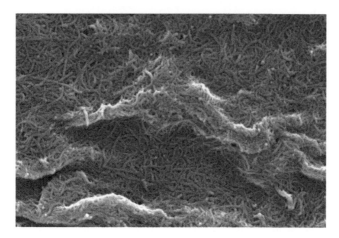

FIGURE 4.11
SEM micrograph of the etched surface of a sample of iPP containing multiwalled carbon nano-tubes. (From Olley, R.H. and Mitchell, G.R., unpublished data.)

4.3.2 Exfoliated Particles

Plate-like particles such as clay and graphene have an overwhelming propensity to stick together. As a consequence, a considerable proportion of the processing procedure is concerned with separating and stabilizing individual layers or packets of layers. This process is known as exfoliation. During the exfoliation process, platelets on the outer part of the packet are cleaved or separated off, exposing further platelets for separation. The process is generally helped through the use of compounds which can enter between the layers and bond to its surface. The gallery is the volume or space between clay platelets and its spacing depends on the extent of intercalation and the nature of the molecule or polymer occupying the space between the layers. Exfoliation is the breakup of this stack of platelets. Figure 4.12 shows a schematic of the various structures in clays.

X-ray scattering is a useful and quantitative tool for measuring the gallery spacing in nanoclays; in the case of montmorillonite the spacing in the unprocessed clay in ~1 nm. Figure 4.13 shows an x-ray scattering curve for the montmorillonite and the equivalent curve for a polymer nanocomposite.[19]

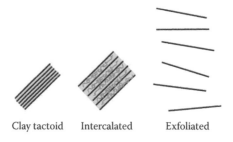

Clay tactoid   Intercalated   Exfoliated

FIGURE 4.12
Schematic of the different states of nanoclay in a polymer matrix.

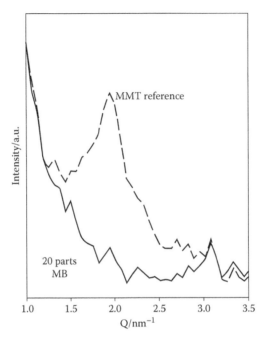

FIGURE 4.13
X-ray scattering curve of a sample montmorillonite (broken line) and a polyethylene nanocomposite containing montmorillonite. (From Green, C.D. et al., *IEEE Trans. Dielect. Elect. Insul.*, 15, 134, 2008. With permission.)

The absence of a well-defined maximum is generally taken to mean that the sample is completely exfoliated. In contrast, we examined the same material using SEM techniques after etching the sample using the procedures outlined earlier. These samples are based on a blend of branched polyethylene with 10% linear polyethylene. Figure 4.14a shows a SEM micrograph of the

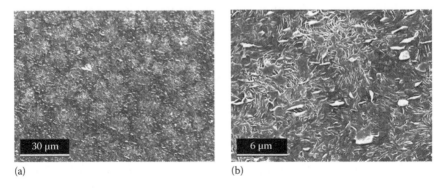

(a)   (b)

FIGURE 4.14
SEM micrograph of the same polyethylene nanocomposite containing montmorillonite as in Figure 4.13 at (a) low magnification (b) high magnification. (From Green, C.D. et al., *IEEE Trans. Dielect. Elect. Insul.*, 15, 134, 2008.)

etched surface and even at the moderate magnification used here we can see the montmorillonite tactoids can be clearly observed as bright elongated but small particles.[19] It is interesting to note, that these tactoids appear to have a common orientation which is close to the horizontal direction in the micrograph suggesting that the sample has retained a memory of the extrusion process used to prepare the blend, despite having been remelted and recrystallized. The etching process has also revealed the underlying spherultic morphology which almost fills the space as circular objects.

Figure 4.14b shows a sample which has been crystallized at a slightly lower temperature of 117°C and at this temperature banded spherulites form. The clay tactoids can be seen more distinctly at this higher magnification. The etching has revealed the polymer morphology and in particular the dominant crystalline lamellae can be seed edge on as bright lines. Since the process of image formation in the SEM is particularly dependent upon sample topography, it is difficult to determine the thickness of the montmorillonite aggregates in this sample. A comparison of the aggregates with the polyethylene lamellae in the SEM micrograph shows that these are of comparable thickness in the range of tens of nanometers thickness. There appears to be little interaction between the polymer matrix and the clay and there is no templating of the lamellar crystal growth by the clay. The comparison of the x-ray data and the SEM micrographs underlines the value of a combined approach. The x-ray diffraction is highly dependent on the presence of a highly ordered stack, whereas the SEM shows the envelope of the tactoid. Together they provide a comprehensive specification of the morphology of the clay in the polymer.

In situations where the clay has been exfoliated, the TEM provides a powerful tool with the required magnification and resolution. Of course this requires the preparation of a thin sample which has been taken from a larger processed sample. Although it is possible to deposit a sample directly on to a grid using, for example, spin coating or a drop of solution, this may exhibit a very different morphology to the bulk sample. Figure 4.15 shows a sample of iPP with montmorillonite which has been taken from a bulk sample using an ultramicrotome technique.[20] The contrast arises from the increased absorption of the clay as it contains higher atomic number elements. The image shows aggregates of montmorillonite which are ~100 nm in thickness.

4.3.3 Particles Prepared In Situ

It is self-evident that the formation of many nanoparticles such as metallic particles or nanotubes require detailed and specialist preparation before introducing them into a polymer matrix. We have explored the potential for preparing nanoparticles with specific shapes and sizes within the polymer matrix.[18] This requires the selection of suitable molecular units from which

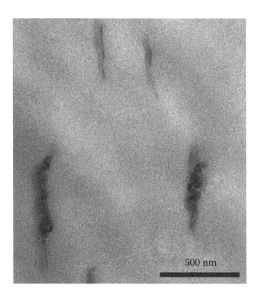

500 nm

FIGURE 4.15
TEM micrograph of a nanocomposite based on iPP and montmorillonite. (From Olley, R.H. and Mitchell, G.R., unpublished data.)

to organize the nanoparticles and the process will clearly depend on the interaction of the molecular units with the matrix. Reports in the literature show how the polymer matrix can be employed to control the particle size, shape, and habit.[22] The deployment of a polymer matrix certainly adds an additional control component to the already complex interactions between the solute and solvent at the face of the particle as it grows.[23] There is a rich variety of synthetic and natural polymers which have different architectures. These materials offer definable physical and chemical properties and this makes the potential of this approach particularly attractive.[24] However, the literature contains little on the potential for processing the particles within the polymeric matrix in order to further define the particle size or shape.

Within this framework, we have explored the preparation of nanoparticles using dibenzylidene sorbitol (DBS) which is a low molar mass compound as can be seen from the inset in Figure 4.16.

The polymer matrix in which we formed the nanoparticles was poly(ε-caprolactone) (PCL) but the process is not limited to this polymer. By forming the nanoparticles in situ we expect the major issue of dispersion of the nanoparticles which besets the mixing of preformed nanoparticles is largely eliminated and indeed this is found to be the case. Moreover, DBS is intrinsically multifunctional and offers significant potential for preparing derivatives with specific chemical or physical properties.

Figure 4.16 shows part of the phase behavior for the DBS/PCL system. This is based on PCL with Mw ~80,000 Da, in which the mixtures were obtained by solution blending using butanone as the cosolvent. The phase

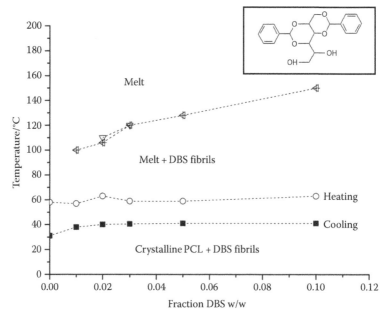

FIGURE 4.16
Phase diagram of PCL and DBS obtained using SAXS and SANS techniques DSC and optical measurements. The inset shows the chemical structure of DBS. (From Wangsoub, S. et al., *Macromol. Chem. Phys.*, 206, 1826, 2005. With permission.)

diagram was obtained using a combination of in situ SAXS and SANS coupled with calorimetry.[21] The upper line is the liquidus line which separates a high temperature homogenous solution of the DBS in PCL from a two phase state of liquid PCL and crystalline DBS. The lowest line is the crystallization temperature of the polymer matrix which is dependent on the concentration of the DBS at the lowest concentrations but then reaches a plateau when the concentration of the DBS is greater than 2%. We have estimated a solubility of ~1% DBS in PCL at 80°C. In situ SAXS and WAXS techniques were employed to characterize the nanoparticles which are created on cooling from the high temperature homogenous solution in to the two phase region.

To enable the combined effects of temperature and flow on the mixture and the nanoparticles which form to be studied we used a specially designed parallel-plate shear-flow system, shown schematically in Figure 4.17[10] in conjunction with the intense x-ray beams available at a synchrotron radiation source.

Essentially the sample is held between two thin mica disks supported on slotted plates within a small oven equipped with electrical heaters and a refrigerated gas cooling system. This temperature control system provided defined heating rates up to 20°C min[-1] and cooling rates up to 10°C min[-1]. This flow system allows x-ray and neutron scattering experiments to be

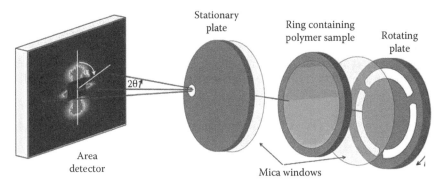

FIGURE 4.17
A schematic of the shear cell developed to enable time-resolved x-ray and neutron scattering measurements of samples subjected to a defined temperature and shear profile. A photograph of the shear cell mounted on a beamline is shown in Figure 4.3.

performed during and following the imposition of controlled shear flow. The incident x-ray beam lies normal to the plane of the sample and to the flow direction. The shear cell has been used at a number of synchrotron beam lime with a quasi-monochromatic incident beam with a size at the sample ~0.3 mm diameter. The beam size defines the range of shear rates effectively sampled within the parallel-plate cell which is ~2% of the stated value. In these experiments, the SAXS data were recorded using a 2-D detector and examples are shown in Figure 4.18.

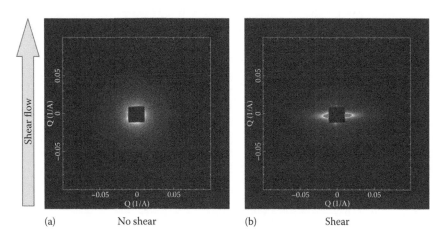

FIGURE 4.18
SAXS patterns of a sample of a PCL 3% w/w DBS held in the melt in the shear cell shown in Figures 4.17 under static conditions (a) and under a shear flow of 10 s^{-1}. (b) The dark square/circle in the center of the pattern is the beam stop. (From Wangsoub, S. et al., *Macromol. Chem. Phys.*, 206, 1826, 2005. With permission.)

These patterns are corrected for variations in the incident beam intensity and for the interception of the x-ray beam by the rotating spokes shown in Figure 4.17.

The so-called invariant, Ω, is a very useful quantity to evaluate and monitor the structural variations and this is given by[25]

$$\Omega = \int_{0}^{\pi/2} \int_{Q=0}^{Q_{max}} |\underline{Q}|^2 I(|\underline{Q}|, \alpha) \sin \alpha \, dQ d\alpha \qquad (4.1)$$

This is directly related to the average of the square of the electron density differences; if the density of the crystals or nanoparticles is constant this is proportional to the volume fraction of crystals. Figure 4.19 shows the value of the invariant as the sample is cooled from the homogenous melt in to the two phase region and finally to room temperature.[26] The first increase in the value of Ω arises from the formation of the nanoparticles. The second increase at ~30°C arises from the crystallization of the matrix. Figure 4.18a shows the SAXS pattern for a sample cooled from a temperature in the homogenous solution in to the biphasic region at 80°C which shows largely isotropic scattering, typical of a particulate system. The SAXS pattern (Figure 4.18b) become more distinct following the imposition of a short period of shear flow with a shear rate = 10 s⁻¹ and a shear strain = 1000 su, where shear units = shear rate x elapsed time. The in situ SAXS

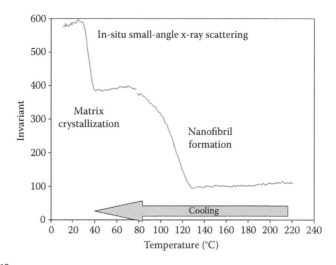

FIGURE 4.19
A plot of the invariant obtained from time-resolved SAXS measurements of a sample of PCL 3% w/w DBS as the sample was cooled from the melt to room temperature. (From Wangsoub, S. et al., unpublished work.)

measurements facilitate measurement of both the alignment and dimensions of the crystals. We observe a strong equatorial streak indicating that the crystals are highly extended and preferentially aligned parallel to the shear flow. Analysis of both the vertical and horizontal cross-sections gives an extended object greater than 500 Å in length with a radius of ~140 Å.[4] We have calculated the isotropic average of the pattern in Figure 4.2b and compared it with that calculated for the patterns recorded for an unsheared sample. These are found to be very similar showing that the shear flow has not altered the characteristics or fraction of the DBS crystals in the system other than the level of preferred alignment. In short, cooling the homogenous mixture has led to an isotropic distribution of highly extended nanoparticles. The application of shear flow leads to preferential alignment of the nanoparticles and this enables the width and length of the particles to be quantitatively evaluated in the melt using the in situ SAXS data. The in situ technique allows us to study the nanoparticles in the melt without the complication of a crystalline matrix. Moreover, the in situ techniques allow us to explore the effect of shear flow on the particles which form in the two phase region in more detail. Figure 4.20a shows the SAXS pattern for a 3% DBS/PCL sample subjected to a shear flow field of 10 s⁻¹ for 30,300 shear units.[4]

The pattern shows a high level of preferred alignment, but in comparison to the pattern taken at low shear strain, there additional peaks on the equatorial section. These can be seen more clearly on the equatorial cross-section plotted on a logarithmic scale in Figure 4.20b. We see these additional features as an indicator that the distribution of fibril radii has become narrower as the pattern approaches that of a monodisperse system of cylinders.

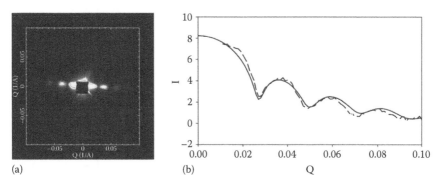

FIGURE 4.20
(a) SAXS pattern of the sample shown in Figure 4.18 but after prolonged shearing with a shear strain of 28,000 shear units and (b) the equatorial section of the pattern shown in 20a (broken line). The full line was calculated from a model of polydisperse cylinders. (From Wangsoub, S. and Mitchell, G.R., *Soft Mater.*, 5, 525, 2009. With permission.)

The scattering for a system of N cylinders with a uniform electron density $I(|\underline{Q}|, \alpha)$ is given by[27]

$$I\left(|\underline{Q}|, \alpha\right) = k\left[\frac{2J_0\left(|\underline{Q}|R\right)}{|\underline{Q}|R} \frac{\sin(|\underline{Q}|L\cos\alpha)}{|\underline{Q}|L\cos\alpha}\right]^2 \qquad (4.2)$$

where

J_0 is the cylindrical Bessel function of the first kind

R and L are the radius and the half-length of the cylinder

k is a scaling constant which takes account of the number of cylinders, their volume and the electron density difference between the crystal and the liquid

For a dilute system of particles without interparticle correlations, it is straightforward to extend this to a system with a distribution of particle sizes. We have used such a model and a Monte Carlo methodology to evaluate the SAXS data obtained at different shear strains and in particular to derive values for mean r and standard deviations σ of the distribution describing the radii of the extended particles.[4] The full line in Figure 4.20b corresponds to the best-fit model for a set of polydisperse cylinders. The fitting parameters are plotted in Figure 4.21.

There is a steady but modest increase in the mean radius of almost 4%. There is a marked narrowing of the distribution of the radii with increasing shear strain. The standard deviation levels off at a value which ~0.08 of the radius. This analysis of the SAXS data has revealed a most striking reduction in the dispersity of the radii of the particles.

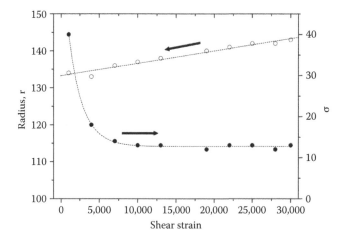

FIGURE 4.21

A plot of the mean radius and the standard deviation of the distribution of radii of cylinders used in the model to fit data similar to that shown in Figure 4.20 for different shear strains. (From Wangsoub, S. and Mitchell, G.R., *Soft Mater.*, 5, 525, 2009. With permission.)

The crystals that form on cooling into the biphasic region are highly anisotropic indicating a large difference in the growth rates for the fibril ends and the lateral surface. At any particular temperature there is equilibrium between the DBS dissolved in the PCL solvent and the DBS crystals. A typical size distribution for particles formed through growth is a log normal distribution which has been related to a growth rate proportional to the surface area of growth face and a log normal distribution of residence times of the solute molecules at the growth face.[28] We attribute the small growth in the radii of the DBS crystalline particles to a shear enhanced collision rate which offsets to a certain extent the already slow lateral growth. However, growth in itself is not a sufficient condition to narrow the size distribution. One possibility is that the shear flow coupled with the higher viscosity of a polymer matrix leads to breakage of the fibrils leading to redissolution of the DBS and subsequent recrystallization. Experiments with a lower molecular PCL matrix showed that a matrix with a molecular weight of 2,000 gave no preferred alignment while a molecular weight of 10,000 gave alignment but no narrowing of the size distribution within the same shear strain. Clearly the stress exerted by the polymer on the DBS crystals is a significant factor. The highly anisotropic morphology suggests that lateral coherence may not be high.

Figure 4.22 shows a SEM micrograph extracted DBS fibrils grown in an oligomer hydrocarbon and these suggest a twisted structure perhaps

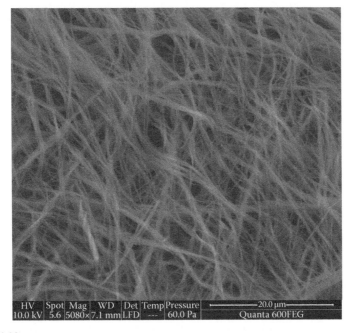

FIGURE 4.22
A SEM micrograph of DBS fibrils extract from an oligomer hydrocarbon matrix. (From Wangsoub, S. et al., submitted to *Phys. Chem.*)

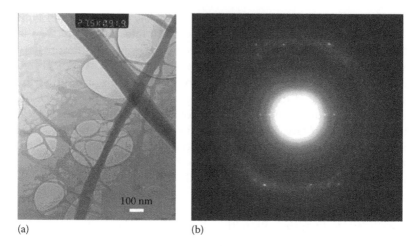

(a)   (b)

FIGURE 4.23

(a, b) TEM micrograph of fibrils shown in Figure 4.22 and an electron diffraction of the fibrils. (From Wangsoub, S. et al., submitted to *Phys. Chem.* With permission.)

arising from the chiral DBS molecule.[29] The twisted nature may place an upper limit on the lateral size of the fibril and this may make detachment more likely. Figure 4.23 shows a TEM micrograph of the fibrils shown in Figure 4.22 and an electron diffraction pattern from the fibrils. The diffraction patterns show sharp spots indicative of a highly aligned almost single crystal structure.[29] Clearly, these nanoparticles are highly ordered.

4.4 In Situ Scattering and Ex Situ Microscopy

A continuing thread in the examples provided earlier is the complementary nature of scattering techniques and microscopy. In this section, the value of in situ time-resolved scattering techniques is explored in connection with polymer processing. To achieve control of both the flow field and the temperature profile we used a specially designed shear cell which has been introduced in an earlier section and shown schematically in Figure 4.17 and mounted on the BM16 beam line at the ESRF in Figure 4.3. Figure 4.24 shows the geometry of x-ray scattering in the flow cell.

In an x-ray scattering experiment, the structure is probed by the scattering vector Q which is the difference between the incident and scattered beam vectors. Almost universally, SAXS experiments at synchrotron based beam lines are performed using a flat 2-D detector and thus the orientation of Q depends on the scattering angle 2θ. In fact, the scattering vector lies

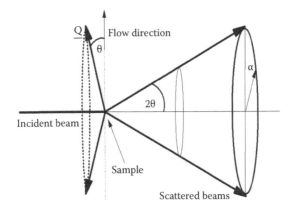

FIGURE 4.24
The geometry of x-ray scattering in the context of the shear cell.

on the surface of a cone with a semiangle 90-θ. For small-angle scattering, $2\theta_{max} \sim 2°$ and hence from a practical point of view, Q lies in the plane containing the flow axis. This means that the intensity variation around a ring at constant scattering angle maps directly to the variation in structure as a function of the angle, α. In contrast, the WAXS data may arise from a scattering angle 2θ in the range 20°–40° and hence there is a significant angle between Q and the flow direction resulting in a more complex mapping from detector to structure.

Most time-resolving experiments involve collecting data in snapshots ranging from 1 to 10 s, although shorter times measured in milliseconds are possible, dependent on the sample, source, and detector. For example, the PILATUS 2M detector provides a dynamic range of 20 bits with an image size of 1475 × 1679 pixels corresponding to an area of 254 mm × 289 mm. The shortest data accumulation time corresponds to a repetition frequency of 30 Hz. As a consequence, large amounts of data can be rapidly produced in which an individual image contains 6 Mb of data. Each SAXS pattern will contain a variety of structural parameters and extracting these parameters may involve considerable processing. To facilitate the automation of this analysis we have developed a software package XESA[30] which can operate interactively or through a script to enable the processing to be tailored to the requirements of the particular experiment. All of the analysis presented in this chapter has been processed using XESA.

Figure 4.25 provides a selection of data from a particular experiment concerned with the crystallization of iPP from a sheared melt.[31] The horizontal axis is plotted as the frame number, and the interval for each frame was 10 s. The diamond symbols show the value of the invariant (Equation 4.1) as the experiment proceeds. During this time, the temperature is dropping at a controlled rate of 10°C min[-1]. The square symbols show the derivative of the invariant with respect to time of frame number. The invariant is calculated from the whole pattern using Equation 4.1. In the early stages of crystallization, the value is proportional to

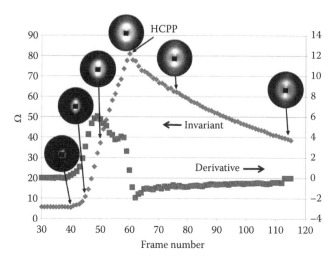

FIGURE 4.25

A plot of the invariant (diamonds) obtained from time-resolved SAXS measurements of a sample of iPP which had been sheared in the melt and cooled at 10°C/min. The frame number is proportional to time. The squares show the derivative of the invariant with respect to time. The images show the SAXS patterns corresponding to the indicated frame numbers. (From Moghaddam, Y. et al., submitted to *J. Macromol. Sci. B.*)

the volume fraction of crystals present. However, if the temperature lowers, the density difference between the crystalline and amorphous phases reduces and the polymer between the existing lamellae also crystallizes and these all serve to reduce the electron density difference. As a consequence, the invariant will drop even though the level of crystallinity may be constant or even increasing slightly as shown in Figure 4.25. If we take the first derivative of the invariant against time, this is a measure of the growth rate of the fraction of crystallinity and the maximum identifies the temperature at which the sample exhibits the highest growth rate. The curve reveals two maxima in the derivative curve suggesting two different crystal growth mechanisms. The first derivative plot is clearly related to a DSC curve, but of course here we have been able to study the crystal growth following the imposition of shear flow in the melt. In addition, considerable other information becomes available. Figure 4.25 shows SAXS patterns for specific frame numbers. The high level of preferred of lamellar orientation is indicated by the intense spots above and below the beam stop. We have used the azimuthual variation in the intensity to derive quantitative orientation parameters[32,33] and this is discussed in detail elsewhere.[31]

We introduced the system based on iPP together with 1% N′-Dicyclohexyl-2,6-Naphthalene dicarboxamide in Figure 4.4. The low molar mass additive self assembles into nanoparticles in a similar manner to that described in Section 3.4.[11] The data shown in Figure 4.4 was obtained for a sample sheared at 210°C and then cooled at 10°C min⁻¹. The SAXS pattern shows four maxima corresponding to orthogonal orientations of the lamellar crystals, parallel and perpendicular to the flow direction. It is clear from the greater intensity of the maxima which are normal to the flow direction that the set lamellar which grow parallel to the flow direction are formed in a greater proportion. Figure 4.26 shows the results from an equivalent experiment but where the temperature of shear in the melt was 170°C.

Examination of the SAXS pattern shows that the situation is now very different with the vast majority of the lamellar scattering is located above and below the beam stop and therefore arises from lamellar which grow normal to the flow direction. In situ scattering has allowed to follow this in real-time and understand the origin of the differences and this is described in greater detail elsewhere[11]. iPP can exhibit a number of crystal forms and WAXS recorded from the same samples shows that the high temperature

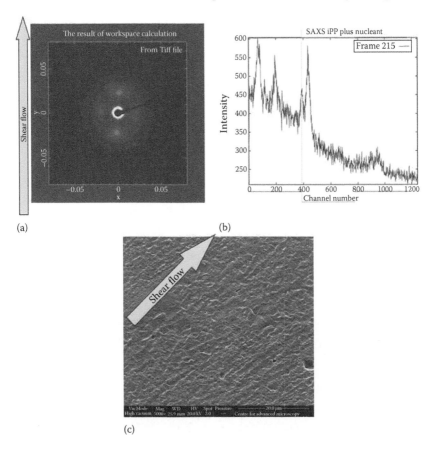

(a)　　　　　　　　　　　　　　　　(b)

(c)

FIGURE 4.26
Four (a) SAXS of a sample iPP + 1% N'-Dicyclohexyl-2,6-Naphthalene dicarboxamide sheared in the melt at 170°C and then cooled at 10°C/min; (b) WAXS recorded from the same sample; (c) SEM micrograph of the etched surface of the sample at the point of the sample corresponding to the scattering experiments. (From Mitchell, G.R. and Felisari, L. submitted to *Polymer.*)

phase exhibits the β form whereas, the lower temperature shear temperature leads to the α form as can be from the different peak positions in Figures 4.4b and 4.26b. Figure 4.4c shows the morphology of the etched surface and it is quite clear that the nanoparticles template the crystallization which leads to the lamellar orientation shown in both the SAXS pattern and the SEM micrograph. In contrast, Figure 4.26c shows an etched surface and the morphology typical of a row nucleated structure driven by the extension of chains in the shear flow. This variation of lamellar orientation, crystal phase, and morphology arises from a complex set of factors which includes temperature-dependent differential nucleation and growth and relaxation rates of extended chains.[11] The use of in situ x-ray scattering and ex-situ microscopy has proved critical in developing a comprehensive and quantitative model.

Figure 4.27 shows the results of an experiment to map out the dissolution of nanoparticles formed from DBS which were described in Section 3.4. To obtain this data, we have employed in situ SAXS techniques to evaluate the proportion

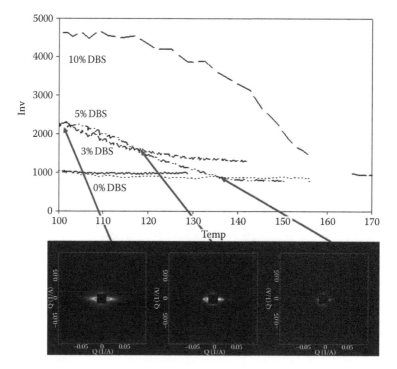

FIGURE 4.27

A plot of the invariant measured from SAXS patterns of PCL with differing proportions of DBS as shown in the Figure as a function of temperature on heating from 80°C during which time the sample was subjected to shear flow. The three SAXS patterns shown are taken from the set for 3% DBS at the temperatures indicated by the arrows. (From Wangsoub, S. et al., *Macromol. Chem. Phys.*, 206, 1826, 2005. With permission.)

of nanoparticles as a function of temperature.[21] Samples were sheared at 80°C and then heated at 10°C min⁻¹. Figure 4.27 plots the invariant evaluated from the sequence of SAXS data against the temperature. The SAXS patterns at three temperatures for the 3%DBS in PCL are shown. At a temperature of ~100°C there is a distinctive equatorial streak, the origin of which was discussed earlier. By 120°C, the intensity of the streak is reducing and by ~135°C the streak has disappeared and only the background level remains and dissolution is complete. Figure 4.27 shows that increasing the fraction of DBS leads to a higher dissolution temperature and the availability of such data were critical to generating the phase diagram shown in Figure 4.16.

In the SAXS experiments described earlier the scattering vector $\underline{Q}$ lies in the plane containing the flow direction and the vorticity vector. Shear flow is not axially symmetric and therefore it is of great value to probe the other planes especially the plane which contains the flow direction and the velocity gradient. For a parallel plate shear cell, this would require the incident beam to lie in the plane of the plates and this is clearly impractical. We have developed an alternative set of plates to be inserted in to the shear cell to provide a Couette cell system as shown in Figure 4.28a. The geometry is also shown in Figure 4.28b.

The scattering vector $\underline{Q}$ lies in the plane containing the flow direction and the velocity gradient. We have used this system to probe the behavior of a nanocomposite based on PCL and 12 nm graphene flakes.[34] Figure 4.28 shows a SAXS pattern taken at the end of a time-resolving sequence of SAXS images using the Diamond facility and in particular beam line I22, in which the samples have been sheared in the melt and then cooled to room temperature. Under these conditions, the PCL alone exhibits an isotropic distribution of lamellar orientations. We attribute this to the fact that any chains extended in the melt relax rapidly and hence the memory is lost at the point that polymer crystallization starts. However, for the PCL/grapheme sample, the SAXS pattern shows a high level of preferred orientation of the lamellar crystals shown by the intense spots above and below the beam stop. Moreover, the pattern exhibits a horizontal streak indicating the presence of extended objects parallel to the flow direction. By examining the sequence of patterns we established that the horizontal streak was present in the SAXS patterns taken in the melt after shear flow and therefore this arises from the graphene flakes. TEM shows that the graphene flakes have a distribution of flakes sizes centered around a few μm with a thickness ~12 nm. The SAXS pattern indicates that these flakes have been aligned in shear flow to lie in the plane containing the flow direction and the vorticity vector. This means that the flake is at a constant position in terms of the velocity gradient and therefore minimizes any torque on the flake. In other words, the flake is edge on to the incoming x-ray beam. We attribute the high alignment of the PCL lamellar crystals to the templating of the crystallization process by the graphene flakes.[34] We note that the same process was not observed in the case of clay platelets.

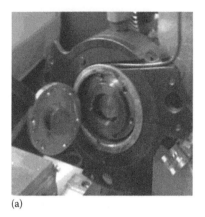

(a)

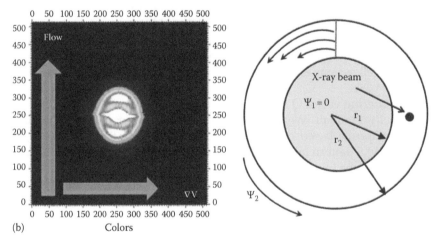

(b)   Colors

FIGURE 4.28

(a, b) A photograph of the modified shear cell to provide a Couette geometry, the SAXS pattern recorded at room temperature for a sample of PCL plus 1% w/w 12 nm grapheme flakes; a schematic of the Couette geometry. (From G.R., Mitchell, et al., to be submitted to *Macromol. Rapid Comm.*)

4.5 New Experimental Techniques

There is continual development of x-ray and neutron diffraction facilities which enable and support new science and new experiments. With Synchrotron Radiation Facilities, new sources and new beam lines produce x-ray beams which are more intense enabling a shorter time cycle, and finer beams allowing microfocus work and the mapping of structure within a sample. Developments in detector technologies mean that the greater intensity can be usefully deployed and the scattering data recorded in shorter snapshots. It is now a routine to record simultaneously combined SAXS and

WAXS data, albeit with the WAXS data only available over a relatively small angular range (Figure 4.3).

Pulsed neutron facilities offer particular advantages to the study of nano-composites as the utilization of a range of neutron wavelengths coupled with time-of-flight techniques means that a much wider range of Q can be accessed. Notable with respect to this, is NIMROD at the UK Pulsed Neutron Source ISIS. This is a diffractometer specifically designed for the study of the structure of liquids and disordered matter with a Q-range of between 0.1 and 100 Å^{-1}, enabling it to probe length scales ranging from the interatomic to the mesoscopic.[35] Moreover, the Q range can be accessed in a time-resolved manner to give unparalleled quality of data across the complete range required. Figure 4.29 shows data obtained at room temperature using NIMROD for a sample of deuterated poly (ε-caprolactone).[36,37]

The large peak at low Q arises from the lamellar crystals (~150 Å) in the polymer, while the "wide-angle" crystalline scattering peaks (~5 Å) occur at the midpoint on this logarithmic scale and at high Q we reach a length scale of 0.1 Å. The ability to obtain a single structure factor with this range of structural information opens immense possibilities in the study of structural reorganization accompany phase transitions such as crystallization. The inset to Figure 4.29 shows an example, again of deuterated poly (ε-caprolactone) of time-resolved following quenching from the melt to an intermediate temperature to evaluate the isothermal crystallization processes. In 2011 it proved possible to obtain data using Nimrod in ~40 s time slices which were of

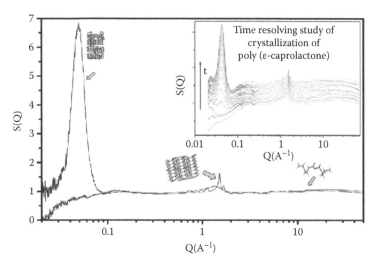

FIGURE 4.29

The structure factor $S(Q)$ recorded for a sample of perdeuterated PCL in the melt and at room temperature over a highly extended Q rang using the NIMROD diffractometer at the ISIS pulsed neutron facility. The inset shows scattering curves recorded in a time-resolving manner during the isothermal crystallization of the PCL. (From Mitchell, G.R. et al., submitted to *Macro Lett.*)

adequate quality to follow the crystallization process.[37] Clearly this single instrument is capable of measuring the complete pattern for a nanocomposite including the "small-angle" scattering from the nanoparticles and we will be reporting on such applications in the near future.

4.6 Summary

This work has shown that the addition of small quantities of nanoparticles to conventional thermoplastics can result in controlled morphologies but in other cases may simply result in a two phase mixture. Scattering and microscopy in partnership provide a powerful combination of quantitative tools to probe these complex systems. Scattering techniques can be adapted to provide in situ studies of nanocomposites subjected to controlled flow and temperature variations. Such procedures allow the evolution of structure during flow to be quantitatively evaluated and the subsequent process of crystallization to be followed. The in situ studies allow the structure to be critical evaluated at specific points during the processing cycle and the key stages identified. Ex situ microscopy provides a high resolution probe of the spatial variation of the structure. It is particularly advantageous to perform the microscopy studies on well characterized samples such as those available at the end of an in situ scattering experiment. The use of etching procedures considerably enhances the visualization of nanoparticles embedded in the polymer matrix.

Many nanoparticles are highly stable and these exist as nanoparticles prior to compounding with the polymer resin. They also remain as nanoparticles during the active use of the packaging as well as in the subsequent waste and recycling streams. We have shown that it is possible to construct nanoparticles within the polymer films during processing from organic compounds selected to present minimal or no potential health hazards. These can be manipulated during flow both with respect to their size and their orientation. Such particles can also be used to direct the subsequent crystallization of the polymer matrix with the potential for generating morphologies which provide a controlled permeability to oxygen and other gas molecules.

We have shown that the shape of the nanoparticle, whether it be flat, cylindrical, or spherical has a pronounced effect on the morphology and that by the correct selection of the shape and type of nanoparticle a range of tailored morphologies and structures can be achieved. In all cases, the characterization of the resultant nanostructured polymers presented a number of challenges. In terms of characterization, the coupled challenges of the nanoscale of the particles and the low fraction present in the polymer matrix were clearly uppermost. We have shown the substantial merits of a combined scattering (neutron and x-ray) and microscopy (SEM, TEM, AFM) approach and we commend this approach to other researchers.

Acknowledgments

Much of this work was performed in the framework of the EU NAPOLYNET and COST FA0940 programs, and I am very grateful to the participants in these programs for the provision of samples and stimulation discussion and collaborations and in particular Dr. Sossio Cimmino, Dr. Donatella Duraccio, Marilena Pezzuto, and Dr. Clara Silvestre at ICTP Pozzuoli. I thank my collaborators for permission to show figures which are unpublished or in press and these are indicated in the text. I am particularly grateful to Dr. Laura Felisari of the Centre for Advanced Microscopy at Reading for her help with the electron microscopy work. The chapter makes extensive use of data obtained at synchrotron and neutron facilities, and I thank Dr. Sigrid Bernstoff (Elettra), Dr. Francois Fauth (ESRF), Dr. Sergio Funari (Hasylab), Dr. Jen Hiller and Dr. Nick Terrill (Diamond), Dr. Daniel Bowron, Dr. Richard Heenan, Dr. Steve King, Dr. Sarah Rogers, and Dr. Ann Terry, (ISIS) for their help with the experiments.

References

1. C. Silvestre, D. Duraccio, and S. Cimmino, *Progress in Polymer Science*, 36 (2011) 1766.
2. H. Qian, E.S. Greenhalgh, M.S.P. Shaffer, and A. Bismarck, *Journal of Materials Chemistry*, 20 (2010) 4751.
3. C. Álvarez-Chávez, S. Edwards, R. Moure-Eraso, and K. Geiser, *Journal of Cleaner Production*, 23 (2012) 47.
4. S. Wangsoub and G.R. Mitchell, *Soft Matter*, 5 (2009) 525.
5. C. Riekel, M. Burghammer, R. Davies, R. Gebhardt, and D. Popov, *Lecture Notes in Physics*, 776 (2009) 91.
6. A. Nogales, R.H. Olley, and G.R. Mitchell, *Macromolecular Rapid Communications*, 24 (2003) 496.
7. G.R. Mitchell, *Macromolecular Rapid Communications*, (2012) to appear.
8. J.K. Hobbs, O.E. Farrance, and L. Kailas, *Polymer*, 50 (2009) 4281.
9. D.R. Rueda et al., *Review of Scientific Instruments*, 77 (2006) 033904.
10. A. Nogales, S.A. Thornley, and G.R. Mitchell, *Journal of Macromolecular Science-Physics*, B43 (2004) 1161.
11. G.R. Mitchell and L. Felisari, submitted to *Polymer*.
12. G.H. Michler, *Electron Microscopy of Polymers*, Springer, Berlin, Germany (2008) Chapter 13.
13. G.R. Mitchell, D. Duraccio, M. Pezzuto, and C. Silvestre, unpublished work.
14. E. Joussein, S. Petit, and J. Churchman, *Clay Minerals*, 40 (2005) 383.
15. R.H. Olley, *Science Progress*, 70 (1986) 17.
16. R.H. Olley and D.C. Bassett, *Polymer*, 23 (1982) 1707.
17. L.C. Sawyer and D.T. Grubb, *Polymer Microscopy*, Chapman & Hall, London, U.K. (1996).

18. R.H. Olley and G.R. Mitchell, unpublished data.
19. C.D. Green, A.S. Vaughan, G.R. Mitchell, and T. Liu, *IEEE Transactions on Dielectrics and Electrical Insulation*, 15 (2008) 134.
20. M. Pezzuto, L. Felisari, D. Duraccio, G.R. Mitchell, and C. Silvestre, unpublished work.
21. S. Wangsoub, R.H. Olley, and G.R. Mitchell, *Macromolecular Chemistry and Physics*, 206 (2005) 1826.
22. R. Shenhar, T.B. Norsten, and V.M. Rotello, *Advanced Materials*, 17 (2005) 657.
23. Y. Oaki, S. Hayashi, and H. Imai, *Chemical Communications*, (2007) 2841.
24. A.C. Balazs, T. Emrick, and T.P. Russell, *Science*, 314 (2006) 1107.
25. R.-J. Roe, *Methods of X-ray and Neutron Scattering in Polymer Science*, Oxford University Press, New York (2000).
26. S. Wangsoub, R.H. Olley, and G.R. Mitchell, unpublished work.
27. A. Guinier and G. Fournet, *Small-Angle Scattering of X-Rays*, Wiley, New York (1955) Chapter 2.
28. L.B. Kiss, J.S. Oderlund, G.A. Niklasson, and C.G. Granqvist, *Nanotechnology*, 10 (1999) 25.
29. S. Wangsoub, P.J.F. Harris, and G.R. Mitchell, submitted to *Physical Chemistry*.
30. G.R. Mitchell, *XESA Version 3.3 User Manual* (2011).
31. Y. Moghaddam, R.H. Olley, and G.R. Mitchell, submitted to *Journal of Macromolecular Science B* (2012).
32. R. Lovell and G.R. Mitchell, *Acta Crystallographica A*, 37 (1981) 135.
33. G.R. Mitchell, S. Saengsuwan, and S. Bualek-Limcharoen, *Progress in Colloid and Polymer Science*, 130 (2005) 149.
34. G.R. Mitchell, M. Domingos, and A. Tojeiria, to be submitted to *Macromolecular Rapid Communications*.
35. D.T. Bowron, A.K. Soper, K. Jones, S. Ansel, S. Birch J. Norris et al., *Review of Scientific Instruments*, 81 (2010) 033905.
36. G.R. Mitchell, *Neutron Diffraction of Polymers and Other Soft Matter in Essentials of Neutron Techniques for Soft Matter*, Eds. T. Imae, T. Kanaya, and M. Furusaka, Wiley, Tokyo, Burkina Faso (2010). ISBN-10: 0-470-40252-0.
37. G.R Mitchell, D. Lopez Garcia, F.J. Davis, and D. Bowron, submitted to *Macro Letters* (2012).

5

Plasma Technology for Polymer Food Packaging Materials

Ondřej Kylián, Andrei Choukourov,
Lenka Hanyková, and Hynek Biederman

CONTENTS

5.1 Introduction

Nonequilibrium plasmas are widely used in an already impressive range of technological applications, comprising cleaning, sterilization, or decontamination of surfaces; their patterning or functioning; deposition of protective or decorative coatings; and modification of surfaces for improvement of their bioresponsive properties or as sources of radiation. There are several reasons for this—plasma-based methods are relatively simple and cost effective, they overcome the need of wet chemistry with all the consequences related, for instance, to the safety and environmental aspects, and last, but not least, enable to deposit coatings on a wide range of substrate materials. The main intention of this chapter is to show that plasma technologies are of high interest also in the field of production of food packaging materials.

The contribution is organized as follows. In the first part of this chapter a brief overview of basic properties of plasma is given. In the second part, different possibilities for the production of barrier coatings on polymeric materials by nonequilibrium gas discharges are discussed with the emphasis on

SiO_x (Section 5.3.1) and hydrocarbon films (Section 5.3.2). In addition, an outlook for fabrication of nanocomposite functional films (Section 5.3.3) is mentioned. Subsequently, the possibility to modify properties of food packaging foils by plasma treatment is given (Chapter 4). Finally, a brief overview of plasma-based sterilization of polymeric foils and bottles is presented.

5.2 Basic Characterization of Nonequilibrium Plasma

Plasma, often considered as the fourth state of matter, in physics denotes a quasi-neutral ionized gas, i.e., a gas in which a fraction of particles is charged and the amount of positively and negatively charged particles is equal.[1–4] Due to the presence of charged species, plasma is a highly conductive gas and unlike the neutral gas responds quickly to electromagnetic fields.

The most common way to produce plasma for technological uses is based on the application of an external electric field (direct current (DC), radio frequency (RF), or microwave (MW)) to the gas either rarified or at atmospheric pressure. The DC electrical breakdown of the gas (e.g., Ne or Ar) under low pressure (~100 Pa) results in self-sustained discharge, and by using a convenient external resistor to limit current, one can reach the mode of DC glow discharge.[4] In this case, positive ions of the gas bombard the cathode and emit electrons that are accelerated in opposite direction until they reach sufficient energy for further ionization of the neutral gas. The highest potential drop is at the cathode that is seen as cathode dark space followed by negative glow and so-called positive column and very short anode region. The positive column is a part of DC discharge where the definition of plasma is fulfilled. The DC discharge can be used only in cases where cathode and anode are all the time conducting. If the surface of cathode is nonconducting, RF discharge must be used. This discharge is sustained because ion–electron pairs are generated in the discharge volume and this process is not surface (cathode) dependent. The aforedescribed plasmas are low ionized, i.e., 1 ion or electron per about 10^6 neutrals; however, they are still well electrically conducting. Therefore, if an external electrical field is applied, as mentioned earlier charged particles are accelerated in more general case (magnetic field may be present as well) by the Lorentz force and can gain sufficient energy for ionization, excitation, and dissociation of atoms or molecules of neutral gas. This leads to the production of molecules or radicals originally not present in the initial gas or gas mixture, which in turn initiates rather complex (plasmo)chemical reactions. For instance, if in the feed gas organic molecules are present, so-called plasma polymerization process takes place. In addition, radiative de-excitation of excited species leads to emission of light in a wide spectral range from infrared down to ultraviolet (UV) and vacuum-ultraviolet (VUV). As a result of this, energetic photons and different radicals or ions may reach a surface exposed to plasma, which can lead to

- Modification of its chemical structure (e.g., by incorporation of radicals into the surface)
- Modification of the morphology of treated materials (e.g., by their chemical etching caused by chemically active radicals)
- Growth of a film, whose chemical structure is strongly linked with the operational conditions used (working gas mixture, pressure, system geometry etc.)

However, it has to be stressed that plasma interacts solely with the topmost layer of treated materials and thus their bulk properties remain unaffected by the plasma.

Moreover, due to the considerable different masses of electrons and positive ions, the electrons gain much higher kinetic energy as compared to ions, whose kinetic energy remains low and comparable to the kinetic energy of neutral species that are unaffected by applied electrical fields. Because of kinetic energies (temperatures) of electrons on the one hand and positively charged ions and neutral species on the other hand, such created plasmas are not in the thermodynamic equilibrium and thus are called nonequilibrium plasmas or nonthermal plasmas. Differences in kinetic energies of species presented in plasmas have two important consequences for technological use of nonequilibrium plasmas:

- The temperature of nonequilibrium plasmas remains relatively low and, under appropriate conditions, close to the room temperature, which makes electrical discharges suitable for processing heat-degradable materials comprising also polymeric materials used for food and beverage packaging.
- Any electrically isolated object immersed into plasma becomes negatively charged with respect to plasma: in order to maintain the net current (i.e., the flux of negative electrons and positive ions) to be zero, the negative self-bias will appear on a surface, which limits the electron current. As a consequence of the presence of such virtual negative potential, positive ions are accelerated toward the surface in the zone close to it (so-called plasma sheath) and can gain relatively high energy (from several eV to hundreds of eV) that can be used to release material from the surface (so-called physical sputtering). This phenomenon is commonly used for the sputtering of nonconductive materials such as metal oxides or polymers.

Finally, it has to be noted that the plasma treatment or deposition may be performed both at reduced or at atmospheric pressures. Both approaches have certain advantages as well as drawbacks. Concerning the atmospheric plasma, no need of expensive vacuum equipment and possibility to adapt the treatment into existing production line is often highlighted. This is, however, counterbalanced by the loss of the process control, due to technical problems with the presence of unavoidable and poorly controllable traces of gaseous impurities in the plasma (e.g., water vapor). In addition, to maintain a low temperature of the process, relatively high gas flows are needed to achieve sufficient cooling of the system, which represents additional costs. Vice versa, low pressure plasma systems have a high degree of process control but their use is not possible without initial investment needed for the production of the vacuum system. The selection of atmospheric or low-pressure plasma thus depends on a particular application.

Reader interested in more details regarding the electrical discharges can find information, for instance, in excellent monographs of Von Engel,[1] Raizer,[2] Liberman and Lichtemberg,[3] or Biederman and Osada[4] or in review articles of Braithwaite,[5] Conrads and Schmidt,[6] or Kogelschatz et al.[7]

5.3 Deposition of Barrier Coatings

One of the parameters that determine the suitability of materials for food and beverage packaging is their high barrier character toward oxygen, water vapor, or carbon dioxide. These gases, if they permeate through the packages, may have an adverse impact on the quality of the packed foodstuff (e.g., odor formation, release of CO_2 from carbonized drinks) or their shelf life.

However, as can be seen in Figure 5.1, current technical polymers used for food packages (polypropylene (PP), polystyrene (PS), polyethylene (PE), polyethylene terephthalate (PET)) have relatively high oxygen and water vapor transition rates and thus their barrier properties have to be improved by deposition of an additional barrier layer. Historically, the first material that was used for coating polymeric materials to improve their barrier properties was aluminum, which was followed by other metals (e.g., Mg, Cu, Cr). However, the metal films deposited on polymeric materials have several principal drawbacks, which relate either to their limited transparency or to the fact that they are not recyclable and microwavable (e.g., [8]). As a consequence, metal coatings are substituted by films of other materials such as metal oxides, ceramics, or hydrocarbons. Although various materials were successfully applied including, for instance, aluminum oxide,[9–12] titanium or zinc oxides,[12] aluminum or titanium oxynitrides,[13–15] or indium tin oxide,[9] the most investigated and used materials are silicon oxide and diamond-like carbon films. The overview of the current state of the art of the production of thin barrier films from the two latter materials will be given in the subsequent text.

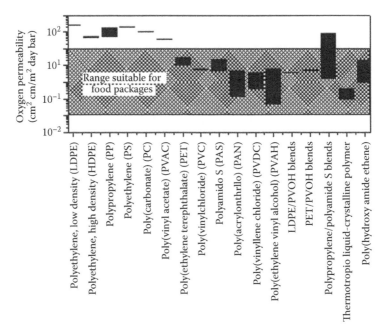

FIGURE 5.1
Summary of oxygen transition rates for different polymers. (Data adapted from the work of Leterrier, Y., *Prog. Mat. Sci.*, 48, 1, 2003.)

5.3.1 Deposition of SiO$_x$ Films

Probably the most studied type of barrier coatings deposited by plasma methods are coatings of silicon oxide that were originally industrialized in the late 1960s. This material, besides its excellent optical transparency, offers also very good barrier properties that come from the tight interstitial space of the Si-O lattice: As it was summarized in the work of Leterrier,[16] thin films of SiO$_x$ deposited on a wide range of polymeric materials lead to decrease of their oxygen transition rate (OTR) values typically by two orders of magnitude.

Although the silicon oxide coatings can be deposited by RF magnetron sputtering of SiO$_2$ target or by reactive magnetron sputtering of Si target, the prevailing method of deposition of such films is plasma-enhanced chemical vapor deposition (PEVCD) that offers much higher deposition rates. The PEVCD was performed using various reactor geometries (typical examples of capacitively coupled plasma reactor and microwave reactor are presented in Figure 5.2) and employing various organosilicon precursors (e.g., [17–29]), mostly hexamethyldisiloxane (HMDSO) or tetramethoxysilane (TMOS), whose chemical structures are depicted in Figure 5.3.

Although the aforementioned monomers contain oxygen, it has been found out by many researchers that oxygen addition into the working gas

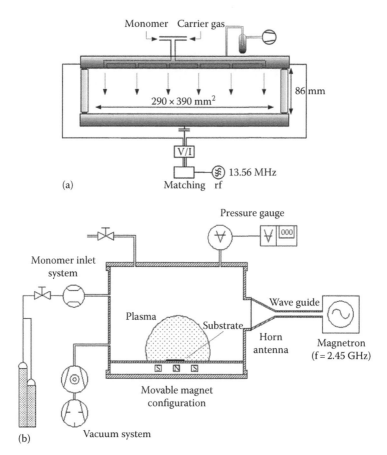

FIGURE 5.2

(a) Capacitively coupled plasma reactor (Reprinted from *Surf. Coat. Technol.*, 142–144, Hegemann, D., Brunner, H., and Oehr, C., 849. Copyright 2001, with permission from Elsevier) and (b) microwave plasma reactor (Reprinted from *Surf. Coat. Technol.*, 200, Walker, M., Meermann, F., Schneider, J., Bazzoun, K., Feichtinger, J., Schulz, A., Krüger, J., and Schumacher, U., 947. Copyright 2005, with permission from Elsevier.)

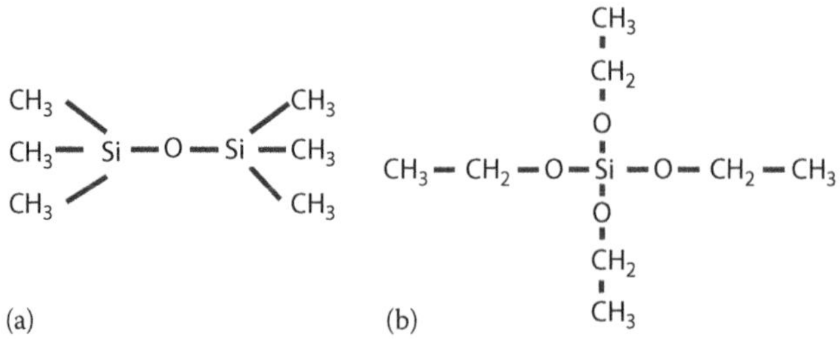

FIGURE 5.3

Chemical structures of (a) hexamethyldisiloxane and (b) tetramethoxysilane.

mixture is necessary for the growth of silicon oxide films that have good barrier properties. This is because of the fact that presence of oxygen leads to the reduction of "organic" part of the coating in favor of "inorganic" one as confirmed, for instance, by fourier transform infrared spectroscopy (FT-IR) measurements (e.g., [20,25]). This can be explained by the removal of hydrocarbons through an oxidation reaction from the deposition processes, for instance, in the case of HMDSO via a hypothetical scheme:[30]

$$Si_2O(CH_3)_6 + 16O \rightarrow 2SiO_2 + 3CO_2 + 3CO + 4H_2O + 5H_2$$

However, there exists an optimal ratio of O_2 and precursor, which varies with the system geometry and applied power. Based on the study of Erlat et al.,[18] at low oxygen admixtures, i.e., excess of the organosilicon monomer, unreacted hydrocarbon radicals will remain in the deposition chamber and contribute to the growth of SiO_x coatings as impurities and thus compromise their barrier properties. On the contrary, insufficient flow of monomer results in its premature depletion, which may lead to the inhomogeneous surface coverage and presence of more defect sites. In addition, oxygen influences also the morphology of SiO_x films. For instance, Teshima et al.[20] reported that addition of oxygen into TMOS led to much more compact coatings as compared to the situation, when only TMOS was used as can be seen in Figure 5.4.

Applied power and overall flow rate of the precursor have also a crucial importance. The power input per gas flow, W/F, determines the chemical composition of the SiO_x films prepared by PECVD (e.g., [31]). Concretely, at higher powers needed for the production of SiO_x films, the deposition is in so-called monomer-deficient regime, in which the precursor structure becomes relatively unimportant for the chemical structure of the resulting films due to the strong fragmentation of precursor molecules in the plasma. This was confirmed in the recent work of Coclite et al.,[32] who compared chemical structures of SiO_x films fabricated using different precursors (besides

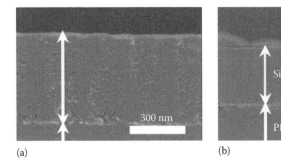

(a) (b)

FIGURE 5.4
Cross-sectional images of silica films deposited at (a) 0% of oxygen and (b) 50% of oxygen in TMOS. (Reprinted from *Thin Solid Films*, 420–421, Teshima, K., Inoue, Y., Sugimura, H., and Takai, O., 324. Copyright 2002, with permission from Elsevier.)

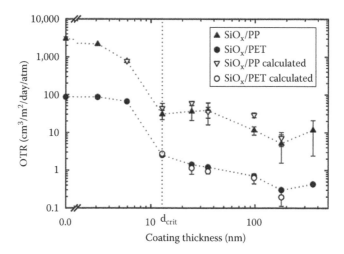

FIGURE 5.5
Oxygen transmission rate (OTR) as a function of the coating thickness d for SiO$_x$ on 30 μm PP foil and on 12 μm PET foil. (Reprinted from *Thin Solid Films*, 518, Körner, L., Sonnenfeld, A., and von Rohr, Ph.R., 4840. Copyright 2010, with permission from Elsevier.)

HMDSO and TMOS, divinyltetramethyldisiloxane and allyltrimethylsilane were used). These authors found out that all the films had similar elemental composition. However, the selection of the precursor had strong impact on the density and microstructure of the coatings as well as on their barrier properties. Explanation of this observation is still needed.

In addition, it was observed that the barrier improvement factor strongly depends on the thickness of the SiO$_x$ coating: As can be seen in Figure 5.5, the OTR rapidly decreases with increasing thickness of the deposited film, but as soon as it reaches a critical value, which is typically around 10 nm, no further improvement of the barrier properties is observed (e.g., [17,26]).

In order to explain this observation, detailed study of the growth mechanism of silicon oxide films on polymeric materials was conducted by Dennler et al.[33,34] These authors studied initial phase of the silicon oxide film growth by combination of Rutherford backscattering spectroscopy (RBS) to measure the film thickness, angle resolved x-ray photoelectron spectroscopy (XPS) for determination of surface chemical composition of growing films, and scanning electron microscopy combined with oxygen reactive ion etching to visualize and quantify the defects in the films of different thickness. Based on these investigations, authors ruled out the hypothesis that separated islands of silicon oxide appear on polymeric materials in the initial phases of the formation of SiO$_x$ film (Volmer–Weber mechanism of film formation): Even at thickness of SiO$_x$ films lower than 2 nm no island structure was observed and the films were continuous, which suggests layer-by-layer growth (Frank–van der Merwe growth mechanism). This finding was related to the activation of polymeric surfaces by active species and/or UV radiation emitted by plasma immediately after the plasma ignition, which leads to fast increase of the surface energy of polymers. The relatively high values of OTR at small thickness of the coatings were therefore explained by the presence of a vast number of small pinholes in the films in the initial stage of film formation, whose density dramatically decreases as the film grows.

Another interesting feature of SiO_x barrier coatings is dependence of their permeability on temperature. It was observed (e.g., [12]) that the activation energy of the oxygen permeation does not differ for polymeric materials with or without SiO_x film (it is worth to mention that similar finding was observed also for other barrier coatings based on metal oxides (e.g., [9]). This suggests that the permeation mechanism is not dependent on the deposited barrier layer and the rate-limiting permeation step is diffusion through the polymeric material. In other words, this means that oxygen does not chemically react with the barrier coating and traverses it through defects in it. This situation is, however, markedly different for water vapor, for which the permeation through the barrier coating is not only defect dominated, but it is also driven by chemical interaction of water molecules with material of barrier coating.

The main limitation of SiO_x coatings as gas permeation barriers is their relatively high brittleness, which often leads to the formation of cracks in the films that in turn hampers their functionality. To overcome this drawback an approach based on gradient profile of mechanical properties of SiO_x layer schematically depicted in Figure 5.6 was proposed by Schwarzer.[35]

Following this approach, the interfacial layer between polymeric substrate and silicon oxide coating has to have the Young's modulus close to one of the polymer to limit the interface stress. The Young's modulus in this scheme then increases with the thickness of the coating to the value of the

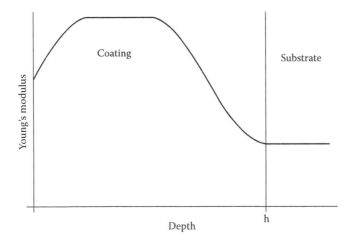

FIGURE 5.6
"Optimal" coating structure as proposed by Schwarzer. (Reprinted from *Surf. Coat. Technol.*, 133–134, Schwarzer, N., 397. Copyright 2000, with permission from Elsevier.)

modulus of SiO_x film. In this region, the coating acts as a gas barrier. Finally, the Young's modulus decreases again toward the free surface of the coating. This limits the onset of superficial defects and crack.

Such gradient structure may be fabricated, e.g., by variation of oxygen flow in the working gas mixture. This approach was recently adopted, for instance, by Patelli et al.[36] for production of multilayer structures combining "organic" and inorganic layers.

Nowadays, there is an increasing interest to prepare SiO_x coatings using also atmospheric plasmas (e.g., [37–45]). This is motivated by several technological benefits: Atmospheric pressure deposition reduces considerably the costs of the process; since it avoids usage of vacuum systems, it is suitable for the in-line production and, in the case of dielectric barrier discharge (DBD) systems, allows coating over large areas. However, there are two main obstacles connected with the PECVD performed using DBD systems. First, although there exists spatially homogeneous diffuse mode of DBD, the most common form of DBD is the filamentary mode, which is not suitable for deposition of homogeneous coatings. Second, the operation at higher pressures often leads to dust formation, which is not compatible with desired film properties. In spite of this, various systems were already introduced that allow overcoming aforementioned complications, e.g., by combination of electronic stabilization and pulsed operation of plasma presented by Starostine et al.[39] This system was proven to be capable of producing SiO_x coatings, whose properties may be controlled by oxygen fraction in the $Ar/N_2/O_2/HMDSO$ working gas mixture or by the duty cycle of the power pulsing.[39] Later on, the same group demonstrated that this system may be operated using directly air with only 1% admixture of Ar instead of nitrogen/oxygen/argon mixture, which dramatically increases the cost efficiency of the film deposition process.[41] Finally, Premkumar et al.[42] reported that 225 nm thick films deposited on polyethylene-2,6-napthalate foils using this system and tetraethylorthosilicate (TEOS) as precursor have OTR values below 5×10^{-3} cm^3 m^{-2} day^{-1}, which makes them able to compete with the films prepared at low pressure systems.

Finally, there are nowadays attempts to improve not only barrier properties of deposited SiO_x coatings. One interesting example that will be mentioned here is the production of anti-fog barrier foils. Fog occurs on surfaces, when water vapor condensates and creates microdroplets on a surface that scatters light and thus reduces light transmission. It is well known that this effect may be reduced, when the surface is highly hydrophilic (very low water contact angle). In this case, the condensed water vapors form a homogenous film instead of microdroplets. An interesting approach to produce anti-fog barrier coatings was presented by Maechler et al.,[46] who used a multilayer structure deposited on polycarbonate (PC) film. This structure consisted of a $SiO_xC_yN_z$:H film deposited onto PC using mixture of $HMDSO/N_2$. This film acted as a gradient interlayer between fully polymeric material and subsequent SiO_x inorganic gas barrier layer deposited in the mixture of $HMDSO/N_2O/N_2$. Next layer was again $SiO_xC_yN_z$:H one deposited in $HMDSO/N_2$

atmosphere. This film contained amino groups that were used for subsequent grafting of the anti-fog film, which was spin-casted from poly(ethylene-maleic anhydride) followed by spin-casting of poly(vinyl alcohol). It has been reported that such multilayer structure maintained optical transparency of original PC film (a decrease of 5%–6% was observed), but assured its anti-fog character that was due to high surface density of –OH groups.

5.3.2 Deposition of Diamond-Like Carbon Films

Diamond-like carbon (DLC) films that consist of dense amorphous carbon or hydrocarbons (denoted often as a-C:H) are extensively studied due to their unique properties such as, for instance, chemical inertness, corrosion resistance, high electrical resistivity, as well as excellent tribological performance (e.g., review articles of Grill[47,48] or Robertson[49]). Due to these properties, DLC films are already used in a wide range of applications, where they act as wear-resistant protective coatings for metals, optical components, or electronic components. In addition, it has been shown recently that DLC coatings are also biocompatible[50,51] and pose excellent barrier properties, which made DLC a highly interesting material also in the food and beverage packaging field. In fact, DLC coatings represent an interesting alternative to silicon oxide films that have much longer tradition but have serious limitations in terms of their mechanical properties. In contrast to SiO_2 coatings, DLC films combine high hardness comparable to ceramic materials with high elasticity of polymer-like materials.[52] Furthermore, DLC films are more compatible with the recyclation process as compared to metal oxides or silicon oxide.

DLC films are produced typically by PEVCD. Acetylene (C_2H_2) or methane (CH_4) are mostly used as monomers to be polymerized, and plasma is ignited at reduced pressures either by microwaves[53–55] or by applied RF power (e.g., [56–60]), eventually with biasing the substrate holder.[61] Examples of different setups that were successfully used for DLC film deposition on PET foils or inner surfaces of PET bottles are presented in Figure 5.7 that were adopted from the works of Ogino and Nagatsu[53] and Shirakura et al.[59]

Barrier enhancement factors for PET foils coated by DLC films reported in literature differ significantly depending on the deposition conditions as can be seen in Table 5.1. This wide range of values of gas permeability that spans over two orders of magnitude suggests that the actual deposition process has strong impact on the resulting barrier properties.

The first parameter that was reported to affect quality of deposited DLC in terms of their barrier properties is the film density. For instance, Vasquez-Boruski et al.[57] showed that an increase of the film density, which was regulated by applying different negative bias onto the sample holder, resulted in a decrease of the barrier property of the films. This behavior was ascribed to the rise of the intrinsic stress inside denser coatings, which led to the formation of microcracks spread over the coating as can be seen in Figure 5.8. The formation of the cracks was greatly reduced when silicon was incorporated

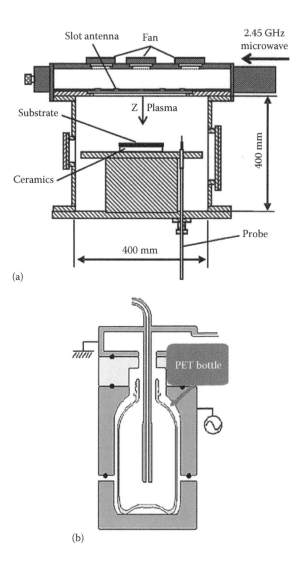

FIGURE 5.7
Schematics of (a) MW plasma reactor and (b) capacitively coupled reactor. (Reprinted from *Thin Solid Films*, 515, Ogino, A. and Nagatsu, M., 3597. Copyright 2007; *Thin Solid Films*, 494, Shirakura, A., Nakaya, M., Koga, Y., Kodama, H., Hasebe, T., and Suzuki, T., 84. Copyright 2006, with permission from Elsevier.)

into the films. This led to the higher hydrogenation of a-C:H coatings that made them more polymer-like and almost free of microcracks and improved their barrier properties: addition of approximately 8% of silicon increased barrier enhancement factor of DLC films for both oxygen and water vapor by more than 1 order of magnitude.[58,63]

The second parameter that was identified as playing a key role for the reduction of OTR of DLC coatings is crystallinity of the films. For instance,

TABLE 5.1

Barrier Enhancement for Oxygen of PET Foils Coated
with DLC Films

Film Thickness (nm)	Gas Mixture	Barrier Enhancement for Oxygen[a]	Ref.
207	He/CH_4	150	[53]
10–40	C_2H_2	26.3	[59]
362	C_2H_2	15	[57]
290	C_2H_2	160	[61]
330	C_2H_2/Ar	160	[61]
700	C_2H_2/CH_4	70	[61]
N/A	C_2H_2/Ar	1.6–2	[62]
50	C_2H_2	20	[55]

[a] Defined as a ratio of OTR measured on uncoated film to OTR
value measure on film coated by DLC films.

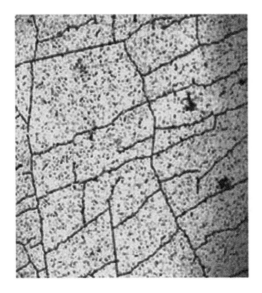

FIGURE 5.8
Optical microscopy images with a magnification × 10 of 130 nm thick a-C:H layer on PET
substrate. (Reprinted from *Carbon*, 43, Abbas, G.A., Roy, S.S., Papakonstantinou, P., and
McLaughlin, J.A., 303. Copyright 2005, with permission from Elsevier.)

Yoshida et al.[61,64] found out an inverse relationship between OTR and the ratio
of graphite content in the film to fraction of amorphous carbon as measured
by means of Raman spectroscopy. This finding can also explain the observed
variation of OTR with the substitution of C_2H_2 by CH_4 in the deposition process, since the use of CH_4 reduces the size of crystallite fraction of the films.[61]

Another aspect of great importance for good performance of DLC barrier coatings is their adhesion to polymeric substrates. As shown by Cruz et al.[60] an example of recycled PET, short pretreatment by oxygen plasma improved significantly adhesion of DLC coatings on such materials. This phenomenon was attributed both to the increase of PET surface roughness caused by its etching by oxygen radicals and to the incorporation of oxygen containing groups to the polymeric surface.

Finally, it is worth to mention that DLC films were recently found to have not only advantageous barrier properties but also an antibacterial character. For instance, Zhou et al.[54] reported 66% reduction in the number of *Escherichia coli* on DLC coatings as compared to stainless steel. Similar results were presented also by Marciano et al.,[65] who demonstrated the killing of 37.6%, 54.2%, 17%, and 38.7% of *E. coli*, *Pseudomonas aeruginosa*, *Salmonella*, and *Staphylococcus aureus*, respectively, after 3 h of incubation on DLC film.

5.3.3 Deposition of Nanocomposite Functional Coatings

Recent development in the field of food and beverage packaging materials tends from the fabrication of passive barrier coatings to production of active coatings that can interact with the packaged items or even to the development of so-called smart packages that can monitor the conditions of packaged foodstuff. Also in these fields, plasma-based technologies may be highly advantageous. This is because of the fact that the rapid development in the field of plasma technologies allows reliable and cost effective production of various nanocomposite materials that can answer to the demands of the market. One of the most prominent examples are Ag/fluorocarbon or Ag/hydrocarbon plasma polymer nanocomposites that were shown to exhibit very good antibacterial character (e.g., [66–68]), which is connected with the release of Ag^+ ions. For instance, Körner et al.[67] used plasma polymerization combined with Ag sputtering for production of nanocomposites of Ag with a:C:H:O plasma polymers. These authors have shown that by variation of the applied power or working gas mixture (CO_2/C_2H_4) it was possible to adjust Ag content in the coatings. This consequently had an effect on the release of Ag^+ ions, which is rather advantageous for tailoring properties of the coatings to requirements of specific applications. Furthermore, the antibacterial effect was found to be even enhanced by the addition of small amounts of Au.[66] This phenomenon may be explained by the enhanced formation of Ag^+ ions because in the galvanic pair, silver–gold, Ag is more active than Au.

Promising direction in the field of production of functional nanocomposite materials is connected with the use of nanoclusters or nanoparticles produced by gas aggregation nanocluster sources (GAS). The first GAS was introduced by Sattler in 1980.[69] In these nanocluster sources, a metal is evaporated into a rather high pressure of inert gases and clusters are created by a homogeneous nucleation, which happens in an aggregation chamber. Created nanoclusters are subsequently blown away from the aggregation chamber through an orifice into the low pressure deposition chamber. Novel possibilities of nanocluster production were brought by the replacement of an evaporator with a planar magnetron.[70] In this case, considerable amount of clusters are positively or negatively charged. This facilitates electromagnetic manipulation with clusters and allows their acceleration in the direction of the substrate or mass filtration. Such sources were successfully applied for production of various metallic nanoclusters (e.g., Cu, Co, Ni, Pd, Ag, Ti (e.g., [71–75]). Some examples showing Pt and Ti nanoclusters are presented in Figure 5.9. More details about GAS designs can be found in reviews.[76,77]

The principal advantage of the GAS systems over other approaches designed for nanocluster production based on a "wet" chemistry is the possibility to combine them with other vacuum-based coating technologies, which allows effective production of diverse nanocomposite films with well-controlled properties. A typical example of the setup that has been recently used for the deposition of Ag/C:H nanocomposites is schematically presented in Figure 5.10. Another advantage of GAS is that the amount and the size of nanoclusters incorporated into the matrix can be controlled independently, which is not possible in the case of co-evaporation or co-sputtering.

Another interesting feature of metal/plasma polymer nanocomposite coatings prepared by means of nanocluster beams combined with plasma polymerization is reduced oxidation of the metal inclusions. An example

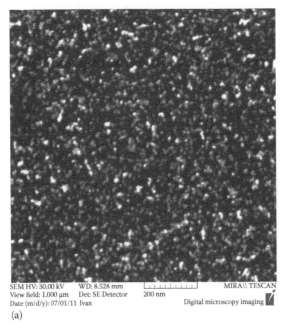

SEM HV: 30.00 kV   WD: 8.528 mm   ⌊ ┴ ┴ ┴ ┴ ┴ ┴ ┴ ┴ ┘   MIRA\\ TESCAN
View field: 1.000 μm   Det: SE Detector   200 nm
Date (m/d/y): 07/01/11 Ivan   Digital microscopy imaging

(a)

FIGURE 5.9
SEM images of (a) Pt nanoclusters.

(continued)

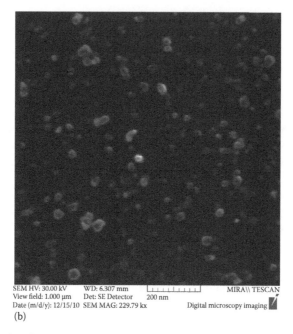

SEM HV: 30.00 kV   WD: 6.307 mm   MIRA\\ TESCAN
View field: 1.000 μm   Det: SE Detector   200 nm
Date (m/d/y): 12/15/10 SEM MAG: 229.79 kx   Digital microscopy imaging
(b)

FIGURE 5.9 (continued)
SEM images of (b) Ti nanoclusters deposited by a gas aggregation cluster source.

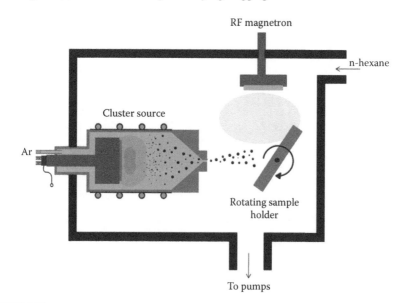

FIGURE 5.10
Schematics of experimental setup used for deposition of nanocomposite films. (Adapted from *Thin Solid Films*, Polonskyi, O., Solař, P., Kylián, O., Drábik, M., Artemenko, A., Kousal, J., Hanuš, J., Pešička, J., Matolínová, I., Kolíbalová, E., Slavínská, D., and Biederman, H. *Thin Solid Films* 520 (2012) 4155. Copyright 2011, with permission from Elsevier.)

of this was given in the work of Polonskyi et al.[78] for the deposition of Ag/plasma polymer nanocomposites. It is assumed that the reduced oxidation is due to the fact that the nanoclusters are already formed in the cluster source, which limits their oxidation due to the residual water vapors as compared to the case when metal is deposited on the substrate in atomic form.

5.4 Plasma Modification of Polymeric Materials

As stated earlier, plasma can be used not only for the deposition of thin films but also for modification of surfaces of polymeric materials. In this section, we will give one example of successful applications of plasma treatment relevant to the processing of food packaging materials: printability improvement.

Packaging materials have, besides their protective function, also importance for marketing and communication (e.g., they should provide relevant information to consumers related to the foodstuff composition, expiry date, etc.). As a consequence of this, there is a requirement for good printability and ink adhesion on polymeric materials used as food packages. These two parameters are closely related to the surface wettability—higher wettability (i.e., lower liquid contact angle) enhances the ink spreading over the surface and promotes its adhesion to it. However, most of the polymers commonly used for the fabrication of packages have relatively high contact angles and therefore their surfaces have to be modified prior to their printing. Although the modification of surfaces may be performed utilizing diverse methods (e.g., by the chemical way), application of nonequilibrium plasma brings certain advantages. First, the plasma interacts with the topmost layer of treated materials only and thus does not induce undesirable changes of their original bulk properties. Second, plasma treatment may be performed in common gases (Ar, He, oxygen, nitrogen, air etc.), which makes the process chain cheap and more environmentally friendly as compared to the chemical methods. Third, as will be shown later, application of plasma is very time effective.

The modification of the wettability of polymeric materials may be undertaken both at reduced pressures and at atmospheric pressures. Both of these approaches were proven to be effective for wettability enhancement of polymeric materials (e.g., [79–82]). For instance Deshmukh and Bhat [81] reported an increase of ink adhesion on PE films from almost 0% up to 80% after 5 s of plasma treatment in low-pressure plasma sustained in air. This improvement of printability was ascribed both to the increase of the surface wettability and to the increase of the effective contact area for ink spreading. Whereas the first effect is due to the incorporation of polar groups to the surface of polymeric material, the latter is connected with etching of the surface of polymeric foil that leads to its roughening.

However, it has to be noted as well that the changes in surface wettability are not temporally stable: Wettability decreases with storage time of treated samples in open air either because of accumulation of air born impurities or by rearrangement of the chemical structure of the surface with time. As a consequence of this, the polymeric materials have to be printed soon after their treatment.

5.5 Sterilization of Polymeric Packaging Materials by Plasma

Another aspect relevant to the processing of food packaging materials is their sterilization: surfaces coming into direct contact with food or beverages have to be free of any contamination of biological origin. This is necessary both for food preservation and for the safety of consumers. The food packages are traditionally sterilized by chemical agents (e.g., hydrogen peroxide) or by autoclaving (hot steam sterilization). However, these techniques have principal drawbacks connected either with the use of toxic substances that have to be removed from the surfaces after the sterilization step is finished or by impossibility to sterilize heat-degradable polymeric materials. The use of other common sterilization techniques based on ionizing radiation (gamma ray sterilization or sterilization by energetic electron beams) is also limited, which is partly due to the concerns about consumer responses to use of such technologies.[83] Therefore, alternative methods allowing cost and time effective sterilization of surfaces of heat-degradable polymeric materials are of high interest. From this point of view, one of the most promising possibilities is application of nonthermal plasmas, which has been already proven to be capable not only to sterilize bacterial spores (e.g., [84–89]), but also to remove diverse biomolecules potentially risky for consumers, such as proteins (e.g., [87–92]) or bacterial endotoxins (e.g., [87–89,93,94]).

The advantage of plasma-based sterilization lies in the unique nature of nonequilibrium plasma that generates high fluxes of charged particles, radicals, and UV photons on the one hand and allows to maintain the process temperature relatively low on the other hand.

The pathway of sterilization or removal of biological contamination from surfaces is strongly linked with the treatment conditions (operational pressure, applied power, system geometry, driving frequency etc.). However, basic mechanisms leading to the sterilization or decontamination of surfaces were already described in the literature. For the inactivation of bacteria or bacterial spores, the key role is often played by UV radiation emitted by the plasma. UV radiation is capable of penetrating the cell walls and inducing irreversible damages to their genetic material. This effect may be accelerated by the erosion of the spores or removal of material shielding the cells from the direct view of UV radiation[87] (for an example, see Figure 5.11).

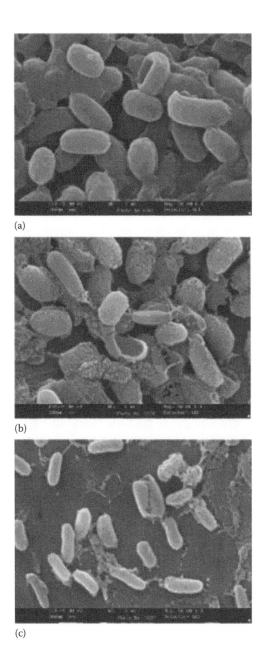

(a)

(b)

(c)

FIGURE 5.11

SEM pictures of treated spores in different gas mixtures: (a) pure N_2, (b) O_2/N_2 20:80, and (c) pure O_2 (treatment time 10 min, power 500 W, 20% DC, 5 ms time on, 13.3 Pa, 10 sccm). (From Rossi, F., Kylián, O., and Hasiwa, M.: *Plasma Process. Polym.* 2006. 3. 431. Copyright Wiley-VCH Verlag GmbH & Co. KGaA. Reprinted with permission.)

Two principal processes for the erosion or removal of biological contamination were described: chemical etching, which is a process driven by the chemical interactions of polymers with chemically active species such as oxygen atoms, nitrogen atoms, or OH radicals, or by chemical sputtering, which is a joint effect of impact of energetic ions and chemically active radicals.[89,95–97] Besides these processes that seem to be dominant in the low-pressure plasmas, additional mechanisms were proposed for the treatment at atmospheric pressures. The most plausible one is connected with hydrogen peroxide, which is produced in the plasma sustained in gases containing water vapor. Hydrogen peroxide relatively easily penetrates through the cell walls and is subsequently converted into highly reactive and toxic OH radicals by Fenton reaction inside the cells (e.g., [98]).

Although the most attention in the field of plasma-based sterilization was devoted to the treatment of medical tools, instruments, and accessories, the results of the numerous studies can be easily applied also to the treatment of food packages.

For instance, Heise et al.[99] designed a cascaded dielectric barrier discharge (CDBD) that combines high intensity of UV light radiation with treatment by DBD plasma. This setup is based on the introduction of a second dielectric barrier into the gap of common DBD. This volume in between two dielectric barriers is filled by appropriate gas, which serves for the emission of intense UV radiation generated by the plasma sustained by applied power. The second part of this setup behaves as a classical DBD and is used for the production of plasma radicals.

The CDBDs were tested by Muranyi et al.[100] for a wide range of vegetative gram-positive and gram-negative bacteria, bacterial endospores, as well as conidiospores of *Aspergillus niger* deposited on PET foils. Significant reduction of the number of viable cells or spores was observed after 3 s treatment for all tested strains. The values of logarithmic reduction (defined as a logarithm of ratio of initial number of bacteria or spores to number of survived ones) ranged from 3.8 observed for *A. niger* up to more than 6.9 for *S. aureus*.

Another interesting approach for sterilization of inner surfaces of PET bottles was introduced by Deilmann et al.[22] This arrangement is based on a pulsed ignition of plasma by applied microwave power. It has been observed that number of spores of *Bacillus atropheaus* decreased by 5 orders of magnitude by treatment lasting 5 s.

Acknowledgments

This work was supported by the grant LD11032 from the program COST CZ and partly by the research plan MSM021620834, both financed by the Ministry of Education of the Czech Republic.

References

1. A.H. Von Engel, *Ionised Gases*, Oxford University Press, Oxford, U.K. (1964).
2. Y.P. Raizer, *Gas Discharge Physics*, Springer, Berlin, Germany (1991).
3. M.A. Lieberman and A.J. Lichtenberg, *Principles of Plasma Discharges for Materials Processing*, Wiley Interscience, New York (1994).
4. H. Biederman and Y. Osada, *Plasma Polymerization Processes*, Elsevier, Amsterdam, the Netherlands (1992).
5. N.St.J. Braithwaite, *Plasma Sources Sci. Technol.*, 9 (2000) 527.
6. H. Conrads and M. Schmidt, *Plasma Sources Sci. Technol.*, 9 (2000) 441.
7. U. Kogelschatz, B. Eliasson, and W. Egli, *J. Phys. IV*, 7 (1997) 47.
8. M. Benmalek and H.M. Dunlop, *Surf. Coat. Technol.*, 76–77 (1995) 821.
9. U. Moosheimer and Ch. Bichler, *Surf. Coat. Technol.*, 116–119 (1999) 812.
10. C.H. Bichler, T. Kerbstadt, H.-C. Langowski, and U. Moosheimer, *Surf. Coat. Technol.*, 112 (1999) 373.
11. B.M. Henrya, A.G. Erlat, A. McGuigan, C.R.M. Grovenor, G.A.D. Briggs, Y. Tsukahara, T. Miyamoto, N. Noguchi, and T. Niijima, *Thin Solid Films*, 382 (2001) 194.
12. J. Fahlteich, M. Fahland, W. Schönberger, and N. Schiller, *Thin Solid Films*, 517 (2009) 3075.
13. A.G. Erlat, B.M. Henry, J.J. Ingram, D.B. Mountain, A. Mc Guigan, R.P. Howson, C.R.M. Grovenor, G.A.D. Briggs, and Y. Tsukahara, *Thin Solid Films*, 388 (2001) 78.
14. M.C. Lin, L.S. Chang, and H.C. Lin, *Appl. Surf. Sci.*, 254 (2008) 3509.
15. A.G. Erlat, B.M. Henry, C.R.M. Grovenor, A.G.D. Briggs, R.J. Chater, and Y. Tsukahara, *J. Phys. Chem. B*, 108 (2004) 883.
16. Y. Leterrier, *Prog Mat. Sci.*, 48 (2003) 1.
17. A.S. da Silva Sobrinho, M. Latrèche, G. Czeremuszkin, J.E. Klemberg-Sapieha, and M.R. Wertheimer, *J. Vac. Sci. Technol.*, A 16 (1998) 3190.
18. A.G. Erlat, R.J. Spontak, R.P. Clarke, T.C. Robinson, P.D. Haaland, Y. Tropsha, N.G. Harvey, and E.A. Vogler, *J. Phys. Chem. B*, 103 (1999) 6047.
19. D. Hegemann, H. Brunner, and C. Oehr, *Surf. Coat. Technol.*, 142–144 (2001) 849.
20. K. Teshima, Y. Inoue, H. Sugimura, and O. Takai, *Thin Solid Films*, 420–421 (2002) 324.
21. M. Walker, F. Meermann, J. Schneider, K. Bazzoun, J. Feichtinger, A. Schulz, J. Krüger, and U. Schumacher, *Surf. Coat. Technol.*, 200 (2005) 947.
22. M. Deilmann, H. Halfmann, S. Steves, N. Bibinov, and P. Awakowicz, *Plasma Process. Polym.*, 6 (2009) S695.
23. A. Bieder, V. Gondoin, Y. Leterrier, G. Tornare, Ph.R. von Rohr, and J.-A.E. Månson, *Thin Solid Films*, 515 (2007) 5430.
24. J. Schneider, M.I. Akbar, J. Dutroncy, D. Kiesler, M. Leins, A. Schulz, M. Walker, U. Schumacher, and U. Stroth, *Plasma Process. Polym.*, 6 (2009) S700.
25. J. Jiang, M. Benter, R. Taboryski, and K. Bechgaard, *J. Appl. Polymer Sci.*, 115 (2010) 2767.
26. L. Körner, A. Sonnenfeld, and Ph.R. von Rohr, *Thin Solid Films*, 518 (2010) 4840.
27. S.R. Kim, M.H. Choudhury, W.H. Kim, and G.H. Kim, *Thin Solid Films*, 518 (2010) 1929.

28. A. Francescangeli, F. Palumbo, and R. d'Agostino, *Plasma Process. Polym.*, 5 (2008) 708.
29. M. Deilmann, M. Grabowski, S. Theiß, N. Bibinov, and P. Awakowicz, *J. Phys. D Appl. Phys.*, 41 (2008) 135207.
30. L. Zajíčková, V. Buršíková, V. Peřinam, A. Macková, D. Subedi, J. Janča, and S. Smirnov, *Surf. Coat. Technol.*, 142–144 (2001) 449.
31. D. Hegemann, H. Brunner and C. Oehr, *Surf. Coat. Technol.*, 174–175 (2003) 253.
32. A.M. Coclite, A. Milella, R. dAgostino, and F. Palumbo, *Surf. Coat. Technol.*, 204 (2010) 4012.
33. G. Dennler, A. Houdayer, M. Latrèche, Y. Ségui, and M.R. Wertheimer, *Thin Solid Films*, 382 (2001) 1.
34. G. Dennler, A. Houdayer, Y. Ségui, and M.R. Wertheimer, *J. Vac. Sci. Technol.*, A 19 (2001) 2320.
35. N. Schwarzer, *Surf. Coat. Technol.*, 133–134 (2000) 397.
36. A. Patelli, S. Vezzù, L. Zottarel, E. Menin, C. Sada, A. Martucci, and S. Costacurta, *Plasma Process. Polym.*, 6 (2009) S665.
37. R. Morenta, N. DeGeyter, S. Van Vlierberghe, P. Dubruel, C. Leys, L. Gengembre, E. Schacht, and E. Payen, *Prog. Org. Coat.*, 64 (2009) 304.
38. P. Scopece, A. Viaro, R. Sulcis, I. Kulyk, A. Patelli, and M. Guglielmi, *Plasma Process. Polym.*, 6 (2009) S705.
39. S. Starostine, E. Aldea, H. de Vries, M. Creatore, and M.C.M. van de Sanden, *Plasma Process. Polym.*, 4 (2007) S440.
40. P.A. Premkumar, S.A. Starostin, H. de Vries, R.M.J. Paffen, M. Creatore, T.J. Eijkemans, P.M. Koenraad, and M.C.M. van de Sanden, *Plasma Process. Polym.*, 6 (2009) 693.
41. S.A. Starostin, P.A. Premkumar, M. Creatore, H. de Vries, R.M.J. Paffen, and M.C.M. van de Sanden, *Appl. Phys. Lett.*, 96 (2010) 061502.
42. P.A. Premkumar, S.A. Starostin, M. Creatore, H. de Vries, R.M.J. Paffen, P.M. Koenraad, and M.C.M. van de Sanden, *Plasma Process. Polym.*, 7 (2010) 635.
43. V. Raballand, J. Benedikt, S. Hoffmann, M. Zimmermann, and A. von Keudell, *J. Appl. Phys.*, 105 (2009) 083304.
44. D. Trunec, L. Zajíčková, V. Buršíková, F. Studnička, P. St'ahel, V. Prysiazhnyi, V. Peřina, J. Houdková, Z. Navrátil, and D. Franta, *J. Phys. D Appl. Phys.*, 43 (2010) 225403.
45. S. Martin, F. Massines, N. Gherardi, and C. Jimenez, *Surf. Coat. Technol.*, 177–178 (2004) 693.
46. L. Maechler, C. Sarra-Bournet, P. Chevallier, N. Gherardi, and G. Laroche, *Plasma Chem. Plasma Process.*, 31 (2011) 175.
47. A. Grill, *Wear*, 168 (1993) 143.
48. A. Grill, *Diamond Relat. Mater.*, 8 (1999) 428.
49. J. Robertson, *Mater. Sci. Eng. R Reports*, 37 (2002) 129.
50. F.Z. Cuia, D.J. Li, *Surf. Coat. Technol.*, 131 (2000) 481.
51. M. Allen, B. Myer, and N. Rushton, *J. Biomed. Mat. Res.*, 58 (2001) 319.
52. H. Dimigen and C.-P. Kiages, *Surf. Coat. Technol.*, 49 (1991) 543.
53. A. Ogino and M. Nagatsu, *Thin Solid Films*, 515 (2007) 3597.
54. H. Zhou, L. Xu, A. Ogino, and M. Nagatsu, *Diam. Relat. Mater.*, 17 (2008) 1416.
55. N. Boutroy, Y. Pernel, J.M. Rius, F. Auger, H.J. von Bardeleben, J.L. Cantin, F. Abel, A. Zeinert, C. Casiraghi, A.C. Ferrari, and J. Robertson, *Diam. Relat. Mater.*, 15 (2006) 921.

56. S. Vasquez, C.A. Achete, C.P. Borges, D.F. Franceschini, F.L. Freire, and E. Zanghellini, *Diam. Relat. Mater.*, 6 (1997) 551.
57. S. Vasquez-Borucki, W. Jacob, and C.A. Achete, *Diam. Relat. Mater.*, 9 (2000) 1971.
58. G.A. Abbas, S.S. Roy, P. Papakonstantinou, and J.A. McLaughlin, *Carbon*, 43 (2005) 303.
59. A. Shirakura, M. Nakaya, Y. Koga, H. Kodama, T. Hasebe, and T. Suzuki, *Thin Solid Films*, 494 (2006) 84.
60. S.A. Cruz, M. Zanin, P.A.P. Nascente, and M.A. Bica de Moraes, *J. Appl. Polym. Sci.*, 115 (2010) 2728–2733.
61. M. Yoshida, T. Tanaka, S. Watanabe, M. Shinohara, J.-W. Lee, and T. Takagi, *Surf. Coat. Technol.*, 174–175 (2003) 1033.
62. J. Li, C. Gong, X. Tian, S. Yang, R.K.Y. Fu, and P.K. Chu, *Appl. Surf. Sci.*, 255 (2009) 3983.
63. G.A. Abbas, J.A. McLaughlin, and E. Harin-Jones, *Diam. Relat. Mater.*, 13 (2004) 1342.
64. M. Yoshida, T. Tanaka, M. Shinohara, S. Watanabe, J.W. Lee, and T. Takagi, *J. Vac. Sci. Technol.*, A 20 (2002) 1802.
65. F.R. Marciano, L.F. Bonetti, N.S. Da-Silva, E.J. Corat, and V.J. Trava-Airoldi, *Appl. Surf. Sci.*, 255 (2009) 8377.
66. V. Zaporojtchenko, R. Podschun, U. Schürmann, A. Kulkarni, and F. Faupel, *Nanotechnology*, 17 (2006) 4904.
67. E. Koerner, M.H. Aguirre, G. Fortunato, A. Ritter, J. Ruhe, and D. Hegemann, *Plasma Process. Polym.*, 7 (2010) 619.
68. E. Sardella, P. Favia, R Gristina, M. Nardulli, and R. d'Agostino, *Plasma Process. Polym.*, 3 (2006) 456.
69. K. Sattler, J. Mühlbach, and E. Recknagel, *Phys. Rev. Lett.*, 45 (1980) 821.
70. H. Haberland, M. Mall, M. Mosseler, Y. Qiang, T. Rainers, and Y. Turner, *J. Vac. Sci. Technol.*, A 12 (1994) 2925.
71. S.R. Bhattacharyya, D. Datta, T.K. Chini, D. Ghose, I. Shyjumon, and R. Hippler, *Nucl. Instrum. Methods Phys. Res. B*, 267 (2009) 1432.
72. A. Marek, J. Valter, S. Kadlec, and J. Vyskočil, *Surf. Coat. Technol.* 205 (2011) S573.
73. M. Maicas, M. Sanz, H. Cui, C. Aroca, and P. Sanchez, *J. Magn. Magn. Mater.*, 322 (2010) 3485.
74. A.I. Ayesh, S. Thaker, N. Qamhieh, and H. Ghamlouche, *J. Nanopart. Res.*, 13 (2011) 1125.
75. M. Drábik, A. Choukourov, A. Artemenko, J. Kousal, O. Polonskyi, P. Solař, O. Kylián, J. Matoušek, J. Pešička, I. Matolínová, D. Slavínská, and H. Biederman, *Plasma Process. Polym.*, 8 (2011) 640.
76. K. Wegner, P. Piseri, H. Vahedi Tafreshi, and P. Milani, *J. Phys. D Appl. Phys.*, 39 (2006) R439.
77. C. Binns, *Surf. Sci. Rep.*, 44 (2001) 1.
78. O. Polonskyi, P. Solař, O. Kylián, M. Drábik, A. Artemenko, J. Kousal, J. Hanuš, J. Pešička, I. Matolínová, E. Kolíbalová, D. Slavínská, and H. Biederman, *Thin Solid Films* 520 (2011) 4155.
79. J.B. Lynch, P.D. Spence, D.E. Baker, and T.A. Postlethwaite, *J. Appl. Polym. Sci.*, 71 (1999) 319.
80. R.R. Deshmukh and N.V. Bhat, *Mat. Res. Innovat.*, 7 (2003) 283.
81. R.R. Deshmukh and A.R. Shetty, *J. Appl. Polym. Sci.*, 104, (2007) 449.

82. Q. Chen, Y. Zhang, E. Han, and Y. Ge, *Plasma Sources Sci. Technol.*, 14 (2005) 670.
83. J. Wan, J. Coventry, P. Swiergon, P. Sanguansri, and C. Versteeg, *Trends Food Sci. Technol.*, 20 (2009) 414.
84. M. Moisan, J. Barbeau, S. Moreau, J. Pelletier, M. Tabrizian, and L'H. Yahia, *Int. J. Pharm.*, 226 (2001) 1.
85. S. Lerouge, A.C. Fozza, M.R. Wertheimer, R. Marchand, and L'H. Yahia, *Plasma Polym.*, 5 (2001) 31.
86. T.C. Montie, K. Kelly-Wintenberg, and J.R. Roth, *IEEE Trans. Plasma Sci.*, 28 (2000) 41.
87. F. Rossi, O. Kylián, and M. Hasiwa, *Plasma Process. Polym.*, 3 (2006) 431.
88. F. Rossi, O. Kylián, H. Rauscher, M. Hasiwa, and D. Gilliland, *New J. Phys.*, 11 (2009) 115017.
89. A. von Keudell, P. Awakowicz, J. Benedikt, V. Raballand, A. Yanguas-Gil, J. Opretzka, C. Flotgen, R. Reuter, L. Byelykh, H. Halfmann, K. Stapelmann, B. Denis, J. Wunderlich, P. Muranyi, F. Rossi, O. Kylián, M. Hasiwa, A. Ruiz, H. Rauscher, L. Sirghi, E. Comoy, C. Dehen, L. Challier, and J.P. Deslys, *Plasma Process. Polym.*, 7 (2010) 327.
90. X.T. Deng, J.J. Shi, and M.G. Kong, *J. Appl. Phys.*, 101 (2007) 074701.
91. O. Kylián, H. Rauscher, D. Gilliland, F. Brétagnol, and F. Rossi, *J. Phys. D Appl. Phys.*, 41 (2008) 095201.
92. H.C. Baxter, G.A. Campbell, A.G. Whittaker, A.C. Jones, A. Aitken, A.H. Simpson, M. Casey, L. Bountiff, L. Gibbard, and R.L. Baxter, *J. Gen. Virol.*, 86 (2005) 2393.
93. O. Kylián, M. Hasiwa, and F. Rossi, *IEEE Trans. Plasma Sci.*, 34 (2006) 2606.
94. M. Hasiwa, O. Kylián, T. Hartung, and F. Rossi, *Innate Immun.*, 14 (2008) 89.
95. V. Raballand, J. Benedikt, J. Wunderlich, and A. von Keudell, *J. Phys. D Appl. Phys.*, 41 (2008) 115207.
96. O. Kylián, J. Benedikt, L. Sirghi, R. Reuter, H. Rauscher, A. von Keudell, and F. Rossi, *Plasma Process. Polym.*, 6 (2009) 255.
97. H. Rauscher, O. Kylián, J. Benedikt, A. von Keudell, and F. Rossi, *ChemPhysChem*, 11 (2010) 1382.
98. D. Dobrynin, G. Fridman, G. Friedman, and A. Fridman, *New J. Phys.*, 11 (2009) 115020.
99. M. Heise, W. Neff, O. Franken, P. Muranyi, and J. Wunderlich, *Plasmas Polym.*, 9 (2004) 23.
100. P. Muranyi, J. Wunderlich, and M. Heise, *J. Appl. Microbiol.*, 103 (2007) 1535.

6

Polypropylene and Polyethylene-Based Nanocomposites for Food Packaging Applications

Donatella Duraccio, Clara Silvestre, Marilena Pezzuto,
Sossio Cimmino, and Antonella Marra

CONTENTS

6.1 Introduction

6.1.1 Roles of Food Packaging

The principal roles of food packaging are to protect food products from outside influences and damage, to contain the food, and to provide consumers with ingredient and nutritional information.[1,2] Traceability, convenience, and tamper indication are secondary functions of increasing importance. The goal of food packaging is to contain food in a cost-effective way that satisfies industry requirements and consumer desires, maintains food safety, and minimizes environmental impact.

Food packaging can retard product deterioration, retain the beneficial effects of processing, extend shelf life, and maintain or increase the quality and safety of food. In doing so, packaging provides protection from three major classes of external influences: chemical, biological, and physical.

Chemical protection minimizes compositional changes triggered by environmental influences such as exposure to gases (typically oxygen), moisture (gain or loss), or light (visible, infrared, or ultraviolet). Many different packaging materials can provide a chemical barrier. Glass and metals provide a nearly absolute barrier to chemical and other environmental agents, but few packages are purely glass or metal since closure devices are added to facilitate both filling and emptying. Closure devices may contain materials that allow minimal levels of permeability. For example, plastic caps have some permeability to gases and vapors, as do the gasket materials used in caps to facilitate closure and in metal can lids to allow sealing after filling. Plastic packaging offers a large range of barrier properties but is generally more permeable than glass or metal.

Biological protection provides a barrier to microorganisms (pathogens and spoiling agents), insects, rodents, and other animals, thereby preventing disease and spoilage. In addition, biological barriers maintain conditions to control senescence (ripening and aging). Such barriers function via a multiplicity of mechanisms, including preventing access to the product, preventing odor transmission, and maintaining the internal environment of the package.

Physical protection shields food from mechanical damage and includes cushioning against the shock and vibration encountered during distribution. Typically developed from paperboard and corrugated materials, physical barriers resist impacts, abrasions, and crushing damage, so they are widely used as shipping containers and as packaging for delicate foods such as eggs and fresh fruits. Appropriate physical packaging also protects consumers from various hazards. For example, child-resistant closures hinder access to potentially dangerous products. In addition, the substitution of plastic packaging for products ranging from shampoo to soda bottles has reduced the danger from broken glass containers.

6.1.2 Materials Used in Food Packaging

Package design and construction play a significant role in determining the shelf life of a food product. The right selection of packaging materials and technologies maintains product quality and freshness during distribution and storage. Today's food packages often combine several materials to exploit each material's functional or aesthetic properties. As research to improve food packaging continues, advances in the field may affect the environmental impact of packaging. Over the world several agencies regulate packaging material for food contact. In the United States, the Food and

Drug Administration (FDA) regulates packaging materials under section 409 of the federal Food, Drug, and Cosmetic Act. The primary method of regulation is through the food contact notification process that requires that manufacturers notify FDA 120 d prior to marketing a food contact substance (FCS) for a new use. An FCS is "any substance intended for use as a component of materials used in manufacturing, packing, packaging, transporting or holding of food if the use is not intended to have a technical effect in such food." All FCSs that may reasonably migrate to food under conditions of intended use are identified and regulated as food additives unless classified as generally recognized as safe substances. Materials that have traditionally been used in food packaging include glass, metals (aluminum, foils and laminates, tinplate, and tin-free steel), paper and paperboards, and plastics. Moreover, a wider variety of plastics have been introduced in both rigid and flexible forms. There are several advantages to using plastics for food packaging. Fluid and moldable plastics can be made into sheets, shapes, and structures, offering considerable design flexibility. Because they are chemically resistant, plastics are inexpensive and lightweight with a wide range of physical and optical properties. In fact, many plastics are heat sealable, easy to print, and can be integrated into production processes where the package is formed, filled, and sealed in the same production line. The major disadvantage of plastics is their variable permeability to light, gases, vapors, and low molecular weight molecules. There have been some health and environment concerns regarding residual monomer and additive components in plastics, including stabilizers and plasticizers. Some of these concerns are based on studies using very high intake levels; others have no scientific basis. According to the most regulations, any substance that can reasonably be expected to migrate into food is classified as an indirect food additive subject to FDA regulations. A threshold of regulation defined as a specific level of dietary exposure that typically induces toxic effects and therefore poses negligible safety concerns may be used to exempt substances used in food contact materials from regulation as food additives. FDA revisits the threshold level if new scientific information raises concerns. Furthermore, FDA advises consumers to use plastics for intended purposes in accordance with the manufacturer's directions to avoid unintentional safety concerns.

Despite these safety concerns, the use of plastics in food packaging has continued to increase due to the low cost of materials and functional advantages (such as thermosealability, microwavability, optical properties, and unlimited sizes and shapes) over traditional materials such as glass and tinplate.[3]

Multiple types of plastics are being used as materials for packaging food, including polyolefin, polyester, polyvinyl chloride, polyvinylidene chloride, polystyrene, polyamide, and ethylene vinyl alcohol. Although more than 30 types of plastics have been used as packaging materials,[4] polyolefins and polyesters are the most common.

6.2 Current Applications of Polyolefins in Food Packaging Sector

Polyethylene (PE) and polypropylene (PP) are the two most widely used plastics in food packaging. PE and PP both possess a successful combination of properties, including flexibility, strength, lightness, stability, moisture and chemical resistance, and easy processability, and are well suited for recycling and reuse.

There are two basic categories of PE: high density and low density. High-density polyethylene (HDPE) is stiff, strong, tough, resistant to chemicals and moisture, permeable to gas, easy to process, and easy to form. It is used to make bottles for milk, juice, and water; cereal box liners; margarine tubs; and grocery, trash, and retail bags. Low-density polyethylene (LDPE) is flexible, strong, tough, easy to seal, and resistant to moisture. Because LDPE is relatively transparent, it is predominately used in film applications and in applications where heat sealing is necessary. Bread and frozen food bags, flexible lids, and squeezable food bottles are examples of LDPE. PE bags are sometimes reused (both for grocery and nongrocery retail). Of the two categories of PE, HDPE containers, especially milk bottles, are the most recycled among plastic packages.[5]

Harder, denser, and more transparent than PE, PP has good resistance to chemicals and is effective at barring water vapor. Its high melting point (160°C) makes it suitable for applications where thermal resistance is required, such as hot-filled and microwavable packaging. Popular uses include yogurt containers and margarine tubs. When used in combination with an oxygen barrier such as ethylene vinyl alcohol (EVOH) or polyvinylidene chloride (PVdC),[6] PP provides the strength and moisture barrier for ketchup and salad dressing bottles. The competitive costs of PP plastics combined with their versatile properties have made these plastics the preferred type of packaging for a wide range of foodstuffs in all the common forms of food packaging: pots, containers, tubs, bottles, pouches, and wrapping films. It is the most widely used plastics material for rigid-type food packaging, with the exception of beverage bottles, where PET (polyethylene terephthalate) is the leader, and milk bottles, where the plastic type is usually HDPE.[7] Films are produced as either cast or biaxially oriented polyprolylene (OPP) films. The gas barrier properties (oxygen and carbon dioxide) of OPP films are improved with coatings and multilayer structures. Sealability is obtained either with coatings or lamination with PE or propylene–ethylene copolymers. OPP films have good barrier properties against loss of food aromas, but the coatings and the multilayer structures with barrier resins provide improvements where necessary. Gas barrier properties and barriers to ultraviolet (UV) light are also obtained with metallized (aluminum) surface treatments or by lamination with aluminum foils.

Coated and laminated films are used in the form of bags or pouches, as sealed wrapping, as overwraps with the food product on plastic trays, in cartonboard containers, or as lids on containers. Films are pigmented or pearlized to provide opacity or colors. For some uses in which the food product gives off moisture, such as chilled fresh salads, the film surfaces are treated with anti-mist agents to prevent condensing moisture obscuring the food products. OPP films also are used as bottle labels, because unlike paper, they eliminate mold growth, do not easily rip, have good abrasion resistance, and do not come off bottles when chilled in iced water.

In recent years, PP has replaced other plastics in a number of applications. A typical replacement is for regenerated cellulose films (cellophane) for wrapping confectionery. Both the "crinkle" and dead-twist properties of cellophane can now be reproduced with PP films. PP, with its good resistance to oils and fats, is also now the principal plastic type used for margarine tubs. The exterior surfaces of most food packaging films can be printed to provide product identification and information for the consumer. Surface treatments, such as corona discharge or acrylic coatings, are necessary to ensure good print adhesion.[8] For some packaging types, the PP plastic surfaces also require anti-static agents to prevent attraction of dirt and maintain good product appearance. PP plastics are available in thicker forms for pots and containers with significantly improved clarity, although still not equaling that of crystal polystyrene. Other applications include jars that can be sterilized or pasteurized for hot filling. Further extended uses of food-contact PP plastics include kitchenware such as reusable domestic and commercial food-storage containers and disposable beverage glasses.

PP and the copolymer plastics also find extensive use for caps and closures for bottles, pots, and containers.[8] They can be used for microwave heating/cooking of foods such as ready meals. PP plastics are used for re-usable trays and crates for retail bulk delivery of foodstuffs, including bread products, fruits, and vegetables.

6.3 Limits of Polyolefins for Food Packaging Application: The Importance of Nanotechnology

Polyolefins have replaced conventional materials (metals, ceramics, and paper) in packaging applications due to their functionality, lightweight, ease of processing, and low cost. However, despite their enormous versatility, a limiting property of these polymeric materials in food packaging is their inherent permeability to gases and vapors, including oxygen, carbon dioxide, and organic vapors. The most frequently used strategies to enhance barrier properties are the use of polyolefin blends, coating articles with high barrier materials, and the use of multilayered films containing a high barrier film. An effective high barrier material is aluminum foil. Thin coatings of aluminum can be applied to films and containers by several vapor deposition technologies. Multilayers are formed by embedding a thin layer of a high barrier material within layers of structural polymers. Coatings and multilayers are effective but their application is limited by the level of adherence between the materials involved. Polyolefins can also be added with suitable fillers to form composites of enhanced barrier properties. Composites typically consist of a polymer matrix (polyolefin) or continuous phase and a discontinuous phase or filler.[9] Unfortunately, while multilayer films and polymer blending have yielded packaging materials with acceptable gas barrier properties that lack the limitations inherent to many monolayer films composed of ultrahigh barrier polymers, they possess higher production and material costs, require the use of additional additives and adhesives that complicate their regulation by federal agencies, and entail added difficulty when it comes to recycling. As a result, there is still a significant push in the polymer industry to generate monolayer films of polyolefins with improved mechanical and gas barrier properties.

Polymer nanocomposites (PNCs) are the latest materials aimed at solving the aforementioned problems. Polymer nanotechnology is actually developed mainly to improve barrier performance pertaining to gases such as oxygen and carbon dioxide. Once sure from a safety point of view and produced at a competitive ratio of cost/performances, the new polyolefin nanocomposites will be very attractive for extensive applications. The use of polymer nanotechnology can in fact extend and implement all the principal functions of the package (containment, protection and preservation, marketing, and communication).[5,10–21] PNCs are created by dispersing an inert, nanoscale filler throughout a polymeric matrix. Filler materials can include clay and silicate nanoplatelets,[22–26] silica (SiO_2) nanoparticles,[27–30] carbon nanotubes,[31–38] graphene,[39–41] starch nanocrystals,[42,43] cellulose-based nanofibers or nanowhiskers,[44–52] chitin or chitosan nanoparticles,[53–56] and other inorganics.[57–60] Though enhancing polymer barrier properties is the most obvious application of PNCs in the food industry, PNCs are also stronger,[31,58–60] more flame resistant,[57,58,61] and possess better thermal properties (e.g., melting points, degradation, and glass transition temperatures)[62–64] than control polymers, which contain no nanoscale filler; alterations in surface wettability and hydrophobicity have also been reported.[65] Detailed information on polyolefin-based nanocomposites can be found in Refs. [27,66–87]. Some of these physical property enhancements can be particularly impressive. Packaging with improved barrier properties extends the shelf life of food by preventing humidity or substances such as oxygen, ethylene, aroma, or unusual flavors interacting with the food. Avoiding contact between these substances and the food decreases the risk of adverse reactions that could reduce the organoleptic and/or safety and quality of the product.

In addition to protecting the food from dirt or dust, oxygen, light, pathogenic microorganisms, moisture, and a variety of other destructive or harmful substances, the packaging must also be safe under its intended conditions of use, inert, cheap to produce, lightweight, easy to dispose of or reuse, able to withstand extreme conditions during processing or filling, impervious to a host of environmental storage and transport conditions, and resistant to physical abuse. In the end, PNCs should offer the food packaging industry better downgauging opportunities, in addition to cost savings and waste reductions, due to the smaller amounts of polymer that need to be used to attain packaging materials with identical or even better mechanical attributes and contribution to decrease CO_2 emissions.[11,13–16,88–91]

This chapter provides an overview of the latest innovations in food packaging based on polyolefins. The most important results obtained for polyolefin nanocomposites will be presented according to their primary functions/applications in food packaging systems.

6.4 Polyolefin Nanocomposite Applications

The most promising nanoscale fillers for nanocomposites are nanoplatelets composed of nanoclays or other silicate materials. The popularity of nanoclays in food contact applications derives from their low cost, effectiveness, high stability, and (alleged) benignity. The prototypical clay utilized is montmorillonite (MMT) [$(Na,Ca)_{0.33}(Al,Mg)_2(Si_4O_{10})(OH)_2 \cdot nH_2O$], a soft 2:1 layered phyllosilicate clay comprised of highly anisotropic platelets separated by thin layers of water (figure shown elsewhere in this book). The platelets have an average thickness of 1 nm and average lateral dimensions ranging from a few tens of nm to several mm. Each platelet contains a layer of aluminum or magnesium hydroxide octahedra sandwiched between two layers of silicon oxide tetrahedra. The faces of each platelet have a net negative charge, which causes the interstitial water layer (known as the gallery) to attract cations (Ca^{2+}, Mg^{2+}, Na^+, etc.) and allows for the construction of multilayer polymer assemblies under appropriate conditions. Individual MMT clay platelets possess surface areas in excess of 750 m^2/g and aspect ratios on the order of 100–500.[91]

The permeability of polymeric materials to gases is determined by the adsorption rate of gas molecules into the matrix at the atmosphere/polymer boundary and the diffusion rate of adsorbed gas molecules through the matrix.[92–94] The adsorption rate is generally dependent on the rate of formation of free volume holes in the polymer created by random (Brownian) or thermal motions of the polymer chains, and diffusion is caused by jumps of molecular gas molecules to neighboring (empty) holes. Thus, the permeability of polymer films is dependent on free volume hole sizes, degree of polymer motion, and specific polymer–polymer and polymer–gas interactions, all of which can be affected by intrinsic polymer chemistry as well as external properties such as temperature and pressure. Of course, the overall rate of gas diffusion is also directly dependent on the film thickness. The dispersal of nano-sized fillers into the polymer matrix affects the barrier properties of a homogeneous polymer film in two specific ways. The first way is by creating a tortuous path for gas.[94] Because the filler materials are essentially impermeable inorganic crystals, gas molecules must diffuse around them rather than taking a (mean) straight line path that lies perpendicular to the film surface. The result is a longer mean path for gas diffusion through the film in the presence of fillers. Essentially, the tortuous path allows the manufacturer to attain larger effective film thicknesses while using smaller amounts of polymer. The effect of dispersed nanomaterials on the mean path length for gas diffusion has been modeled theoretically.

The simplest model, first proposed by Nielsen, assumes that fillers are evenly dispersed throughout the matrix and take the shape of rectangular platelets of uniform size and supposes that the tortuosity of the path is the only factor influencing the gas diffusion rate.[95] The Nielsen model is valid only for small loading percentages (<10%), as higher loadings result in particle agglomeration, which in turn effectively reduces the mean particle aspect ratio[94] and may affect other properties of the system such as the amount of polymer available to intercalate into the nanoclay galleries and the proportion of "interphase" regions in which the nanoclay surface and any organic modifiers interact directly with the polymeric host material.[96] Improvements on the Nielsen model include adjustments for random positioning of the filler throughout the matrix,[97–99] as well as filler shape (e.g., hexagonal[100] or disk[101,102]), size uniformity,[100,103] angular orientation with respect to the lateral dimension of the film,[104,105] degree of agglomeration or stacking,[103,106] and high nanoclay filler contents.[96] Temperature effects have also been studied.[96] In general, all of these models predict that large volume fractions or large particle aspect ratios are required to reduce the gas permeability by an appreciable degree. These theoretical considerations have been reviewed in detail[94] and some of the more widely utilized models have recently been tested experimentally over a full range of nanoclay filler content and modified accordingly.[96]

While tortuosity is usually the primary mechanism by which nanofillers impact the barrier properties of polymer nanocomposites (PNCs), this is not always the case. The second way that nanoparticulate fillers influence the barrier properties is by causing changes to the polymer matrix itself at the interfacial regions. If the polymer–nanoparticle interactions are favorable, polymer strands located in close proximity to each nanoparticle can be partially immobilized. The result is that gas molecules traveling through these interfacial zones have attenuated hopping rates between free volume holes or altered density and/or size of holes.[94] In addition, the presence of surfactants or other additives used to efficiently incorporate the filler into the matrix can also affect the diffusivity or solubility of permeants. The effects of the interfacial regions have been found to be particularly important in polymer matrices that possess very high native gas permeabilities, such as polyolefins.[107] Attempts have been made to model the effect of the interfacial regions[94,108,109] on the diffusivity properties of migrant gases through polymer films, but the relevant parameters are not always easily measurable. In any case, each PNC system is different and properties can only be predicted generally. It is also worth mentioning that gas transport properties can also be modified in the absence of exogenous nanofillers; semicrystalline polymers such as PET and PE have gas permeabilities that are directly related to their degree of crystallinity due to the fact that nanocrystalline regions within the polymer matrix increase tortuosity of gas diffusion and effect changes to the gas transport regions of the interfacial regions.[110,111] In other words, crystalline regions of semicrystalline polymers act as nanoscale fillers.

6.4.1 Barrier Applications

The structural characteristics contribute to MMT's excellent utility as a filler material for polyolefins, typically giving rise to impressive increases in polymer strength and barrier properties with only a few wt.% added to the polymer matrix. However, because they have such large surface energies, clay nanoplatelets tend to stick together, particularly when dispersed in polyolefins that are nonpolar polymers. Agglomeration of clay platelets leads to tactoid structures (microcomposites) with reduced aspect ratios and, according to the Nielsen model, reduced barrier efficiencies. Four morphological arrangements can be achieved: non-intercalated nanocomposites, intercalated nanocomposites, exfoliated nanocomposites, and flocculated nanocomposites. Intercalated nanocomposites are normally interlayered by a few molecular chains of the polymer. In some cases, silicate layers are flocculated due to the hydroxylated edge–edge interaction of the silicate layers. The exfoliated nanocomposites consist of individual nanometer-thick layers suspended in a polymer matrix and are a result of extensive penetration of the polymer in the silicate layers with the spacing between layers expanded up to 10 nm or more.[2,59]

The manner in which the polyolefin–clay nanocomposite is fabricated can play a large role in how the clay platelets are distributed throughout the matrix and, therefore, in the barrier properties of the resulting materials. In Figure 6.1, Transmission Electron Microscopy (TEM) images of intercalated (6.1b) and exfoliated (6.1c) PP/clay nanocomposites, obtained by changing the way in which the materials are prepared, are reported.[112]

Melt processing and in situ polymerization processing techniques may be more or less suitable for clay/polyolefin systems,[25,113,114] and an excellent review[115] discusses these relationships extensively. Unfortunately, efficient delamination of platelets to form fully exfoliated morphologies is hindered by the fact that clay particles are hydrophilic and polyolefins are hydrophobic.[25,59]

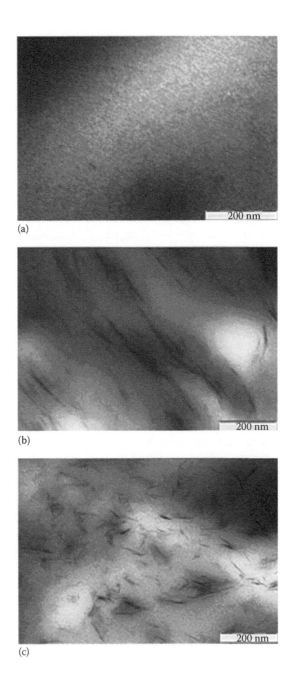

FIGURE 6.1

TEM micrograph of (a) PP control (without nanoparticles), (b) PP + 5% MMT (C15A) (nonexfoli-
ated), and (c) PP + 5% MMT (C15A) (exfoliated). The nanocomposites have been obtained in
different processing conditions. (Reproduced from *Eur. Polym. J.*, 43, Pereira de Abreu, D.A.,
Paseiro Losada, P., Angulo, I., and Cruz, J.M., 2229. Copyright 2007, with permission from
Elsevier.)

Good dispersibility of nanoclay platelets in hydrophobic matrices is typically achieved by functionalizing the polar clay surface with organic ammonium ions bearing long aliphatic chains.[62,107,116–118] Not surprisingly, the barrier properties depend on the type of organic compatibilizer utilized due to varying effects on nanoparticle morphology. For example, a systematic demonstration of the influence of the modifier on degree of MMT exfoliation and, hence, gas barrier properties is provided by Osman et al., who utilized quaternary ammonium modifiers bearing either one, two, three, or four long alkyl (octadecyl) chains.[119] When MMT/PE nanocomposites were fabricated using modifiers having more numerous long alkyl chains, the clay platelets generally had larger d-spacing values (inter-platelet separation) due to steric interactions and thus better gas barriers. MMT clays with better cation exchange capacities also exhibited better degrees of exfoliation within the polymer matrix due to similar steric considerations. When all of the datasets are combined, a clear inverse correlation between d-spacing and oxygen transmission rate is apparent. The Osman study is a great example of the power of chemistry and nanotechnology: A clear understanding of the factors involved, combined with the unique chemical properties of nanoscale particles, can lead to impressive control over the physical properties of macroscopic materials.

Anyway, complete miscibility between polyolefin and clay (two heterogeneous phases) could not be achieved only by exchanging the inorganic cations of the aluminosilicates with alkylammonium ions to obtain organically modified clay (OM). For the preparation of PE–OM nanocomposites, two approaches have been applied: in situ intercalative polymerization and melt compounding.[116,119–125] Although the first offers more homogeneous composites with a better dispersion of the silicate layers, the second is usually preferred because the in situ polymerization leads to polymers of low molecular weight. The dispersion attained by the in situ polymerization method is also thermodynamically unstable. Theoretical studies showed that an interplay of entropic and energetic factors governs the exfoliation process and that the miscibility between the two phases is a function of the Flory–Huggins interaction parameter ($\chi = 0$ is required).[126–128] For alkyl OM–PE composites, it has been predicted that only an intercalated structure is possible even if $\chi < 0$. The exfoliation of the OM and the stability of the dispersion were also correlated to the solubility parameters of the two phases.[129]

To promote the dispersion of clays in polyolefins, the addition of end functionalized polymer chains in small quantities has been proposed.[130,131] The functional group is expected to anchor to the clay surface and the polymer chain provides a favorable enthalpy and entropy of mixing with the polymer matrix. Simulation studies also showed that increased polymer–filler attractive interactions (functionalized polymer matrix) may create bridges between adjacent silicate layers, leading to poor intercalation.[132] The use of maleic anhydride–grafted PP oligomers (PP-g-MA), as an additive (compatibilizer) to prepare PP–OM nanocomposites has been described by

Kawasumi et al. and studied in detail by Reichert et al.[66,67] and studied in detail by Reichert et al.[133] The silicate exfoliation and properties of the composites were correlated to the MA graft density and weight fraction of the "compatibilizer." The different synthetic routes for the preparation of PP–MMT nanocomposites have been reviewed by Manias et al.[134] In these composites, the elastic modulus and the yield stress were appreciably enhanced, although PP-g-MA oligomers were used in high concentrations (20 wt.%). Ammonium-terminated PP has been synthesized and used to surface modify MMT, leading to an exfoliated structure but the properties of the composites were not described.[134] Similarly, maleic anhydride–grafted PE (PE-g-MA) has been used to prepare PE–OM nanocomposites but much lesser studies were devoted to PE than to PP.[123–125] PE-g-MA was added in concentrations up to 30 wt.% and the same conclusions were drawn as for PP, that is, the filler dispersion improved with increasing MA graft density and with increasing weight fraction of PE-g-MA.

Hotta and Paul[125] studied the influence of PE-g-MA addition on the mechanical and permeation properties of linear low-density PE composites with octadecyl- and dioctadecyl-OMs. The exfoliation, tensile modulus, and yield strength were enhanced with increasing ratio of "compatibilizer" to OM. The dioctadecyl-OM composites showed better dispersion and mechanical properties than those of the octadecyl-OM. The gas permeability was also decreased by ca. 40% in the dioctadecyl-OM nanocomposites at 0.069 inorganic weight fraction and 13 wt.% PE-g-MA.

6.4.2 Antimicrobial Applications

Metal nanoparticles, metal oxide nanomaterials, and carbon nanotubes are the most used nanoparticles to develop antimicrobial active polymer nanocomposites. These particles function on direct contact, but they can also migrate slowly and react preferentially with organics present in the food. Silver, gold, and zinc nanoparticles are the most studied metal nanoparticles with antimicrobial function, with silver nanoparticles already found in several commercial applications. Silver, that has high temperature stability and low volatility, at the nanoscale is known to be an effective antifungal and antimicrobial and is claimed to be effective against 150 different bacteria.[136–142] In addition, Ag nanoparticles are toxic to fungi (e.g., *Candida albicans*,[142,143] *Aspergillus niger*,[142] *Trichophyton mentagrophytes*,[143] and yeast isolated from *Bovine mastitis*[144]), algae (e.g., *Chlamydomonas reinhardtii*[145]), and phytoplankton (e.g., *Thalassiosira weissflogii*[146]) and are inhibitory to at least two viruses (HIV[147] and monkeypox[148]). More recently, the antimicrobial properties of nano-ZnO and MgO have been discovered. Compared to nanosilver, the nanoparticles of ZnO and MgO are expected to provide a more affordable and safe food packaging solution in the future. Titanium dioxide (TiO_2) particles are promising too.[149–151] Unlike Ag, the antimicrobial activity of TiO_2 nanoparticles is photocatalyzed and thus TiO_2-based antimicrobials are only active in the presence of UV light. Carbon nanotubes (CNTs) could be used not only for improving the properties of polymer matrix but also for the antibacterial properties. Direct contact with aggregates of CNTs was demonstrated to be fatal for *E. coli*, possibly because the long and thin CNTs puncture the microbial cells, causing irreversible damages.[152] The application of CNT at the moment is stopped as several studies suggest that CNTs are cytotoxic to human cells, at least when in contact to skin.[153]

Numerous Ag/polyolefin nanocomposites have been reported in the literature. Sanchez-Valdes et al., for example, coated a five layer (PE/tie/PA-6/tie/PE; PA-6 = polyamide six and tie = maleic anhydride–grafted PE) plastic film with an Ag/PE nanocomposite layer and found antimicrobial activity against the fungus *A. niger*, a common food contaminant.[154] Moreover, they found that the activity was dependant on the coating method: methods that gave rise to a rougher surface (and hence more surface area for silver-ion release) had higher activity than those that resulted in a smoother surface. Munstedt and coworkers published several studies on Ag/PP nanocomposites (Ag particle size 800 nm), which possessed antimicrobial activity against *E. coli* and *S. aureus* as well as the fungus *C. albicans*, polychaete worms (*S. spirorbis*), sea squirts (*C. intestinalis*), and algae (*U. intestinalis*).[155–157] The antimicrobial activity was also found to be dependent on factors that affect silver-ion release rate, such as degree of polymer crystallinity,[155] filler type (i.e., silver particles, zeolites, etc.),[156] hydrophobicity of the matrix,[157] and particle size (i.e., nanocomposite vs. microcomposite).[158] Ag/polyolefin nanomaterials have been tested with real food systems to determine the effect of Ag nanoparticles' antimicrobial properties on food shelf life (see paragraph 5).[166,170–174]

Nanomaterials containing nano-ZnO-based light catalyst, claimed to sterilize in indoor lighting, have been recently introduced. It is reported that ZnO exhibits antibacterial activity that increases with decreasing particle size.[159] This activity does not require the presence of UV light (unlikeTiO_2), but it is stimulated by visible light.[160] The exact mechanism of action is still unknown. ZnO nanoparticles have been incorporated in a number of different polymers including PP[161–163] and LDPE,[164] as UV light absorber. The nanoparticles are also able to improve the thermal stability of polymers.

6.5 Current Commercial Status of Polyolefin Nanomaterials, Safety, and Outlook

Because polymer nanomaterials for packaging are relatively inexpensive to manufacture, there are already numerous companies that have made them commercially available. Several trademarked product lines exist, such as Aegis™ (Honeywell Polymers), Durethan® (LANXESS Deutschland GmbH), Imperm® (ColorMatrix Corp.), nanoTuff™ (Nylon Corporation of America), and NanoSeal™ (NanoPack, Inc.). A more complete summary of active companies in the field is provided elsewhere.[165] Nanocor™ offers a wide variety of surface-modified MMT clay and masterbatchs suitable for polyolefins. In particular, the clay products Nanomer® can be blended into polyolefin with good dispersion and property improvements via melt-compounding process. The loading level is commonly in the range of 4%–6% for mechanical improvement and 1%–4% for flame retardation. NanoMax® is a series of nanomer-polyolefin resin masterbatch products. They are the first nanoclay products to feature a convenient pellet form. These masterbatch products are produced through melt compounding based on patented technologies (US 6,462,122 and 6,632,868). They offer excellent processability and can be used in a wide variety of equipment, including extruders, mixers, and even injection molders.

Anyway, given the number of studies that cite food packaging as a likely endpoint for polyolefin nanomaterial research, the number of researchers who have investigated these materials in shelf life or safety experiments using real food components is surprisingly small. The most tested materials with real food are Ag-based nanocomposites. More relevant to behavior with real food systems is a 2010 study[166] that showed that nanopackaging, obtained by blending LDPE and nanopowder (nano-Ag, kaolin, anatase TiO_2, and rutile TiO_2), with higher barrier and mechanical properties was successfully synthesized and then were applied to the preservation of strawberry fruit during 4°C storage.

The study showed that this type of nanopackaging (i) reduced fruit decay rates (Figure 6.2), (ii) inhibited the increase of total soluble solid and Malondialdehyde (MDA) production (a secondary end product of

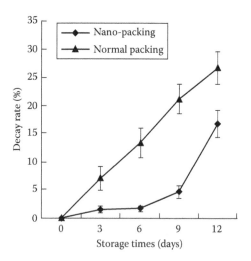

FIGURE 6.2
Effects of nanopacking and normal packing on decay rate of strawberry fruit during 4°C storage. (From Yang, F.M., Li, H.M., Li, F., Xin, Z.H., Zhao, L.Y., Zheng, Y.H., and Hu, Q.H.: *J. Food Sci.* 2010. 75. C236. Copyright Wiley-VCH Verlag GmbH & Co. KGaA. Reproduced with permission.)

polyunsaturated fatty acid oxidation[167]), (iii) maintained more ascorbic acid and (iv) inhibited the activity of two enzymes responsible for browning reaction: Polyphenoloxidase (PPO) and Pyrogallol peroxidase (POD): Polyphenoloxidase (PPO) and Pyrogallol peroxidase (POD).[168,169]

Furthermore, they have the advantages of simple processing and feasibility to be industrialized in contrast with other storages, some of which are time consuming, high cost, and off-flavor altering. These nanopackaging may provide an attractive alternative to keep the quality of strawberries during extended storage. Nevertheless, it should be noted that this research was exploratory in nature. More research will be needed to elucidate the definite mechanism by which nano-packaging functions during storage to facilitate the application of nanotechnology in a broader range in the future.

In another work, Valipoor Motlagh et al.[170] studied the effects of LDPE-Ag packages on the several factors of dried barberry. All in all, the following results were achieved: (i) LDPE-Ag packages showed antimicrobial effects on barberry compared with pure LDPE packages except in the low concentration of silver particles of about 0.02%; (ii) by the increase of the concentration of silver particles in LDPE-Ag packages, a decrease in microbial growth in the barberry was obtained; and (iii) LDPE-Ag packages helped to keep the aroma, taste, and total acceptance factors of the barberry in comparison with pure LDPE packages except in the low concentration of silver particles of about 0.02% (Figure 6.3). So, LDPE-Ag packages with more than 1% concentration of silver particles well preserved the quality of barberry and increased its shelf life in comparison with pure LDPE packages. LDPE-Ag packages will probably solve the problem of barberry packaging. It should be noted that further studies are still required to understand the barrier properties of LDPE-Ag packages and the possible health problems caused by silver particle penetration into barberry.

Other studies include (i) Chinese jujube fruit stored in food storage bags composed of Ag/nanoparticulate TiO_2/polyethylene. The nano-packing material had a quite beneficial effect on physicochemical and sensory quality compared with normal packing material. After 12-day storage, fruit softening, weight loss, browning and climatic evolution of nano-packing were significantly inhibited; (ii) Orange juice stored at 4°C in LDPE films incorporated with TiO_2 and 10 nm nanosilver mixture. The juice exhibited statistically significant reduction in *Lactobacillus plantarum* growth over a time period of 112 days[172] (Figure 6.4).

In 2007, Avella et al. found that total microbial and mold counts on apple slices decreased significantly over a period of 10 days when packaged in $CaCO_3$/iPP films, as opposed to apples stored in neat isotactic Polyproplylene (iPP), which experienced an increase of total mesophilic microflora over the same time period. The study also showed that the apples stored in the packaging ripened better due to ethylene gas retention and exhibited less oxidation than those stored in conventional PP packages.[173] Another food packaging study showed that oriented PP films coated with TiO_2 nanoparticles inhibited *E. coli* growth on fresh cut lettuce in vitro and in actual tests. PP films coated

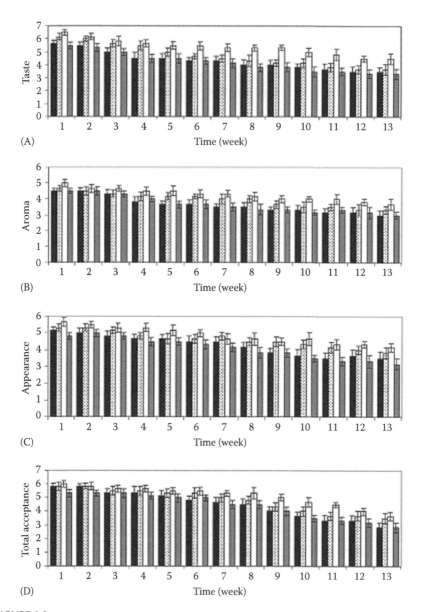

FIGURE 6.3

Average results for samples packaged in LDPE-Ag films with 0.02% (■), 1% (▨), and 2% (□) concentration of silver particles and pure LDPE packages (▨) up to 13 weeks: (A) taste; (B) aroma; (C) appearance; (D) total acceptance of barberry. (From Valipoor Motlagh, N., Hamed Mosavian, M.T., Mortazavi, S.A., and Tamizi, A.: *J. Food Sci.* 2012. 71. E2. Copyright Wiley-VCH Verlag GmbH & Co. KGaA. Reproduced with permission.)

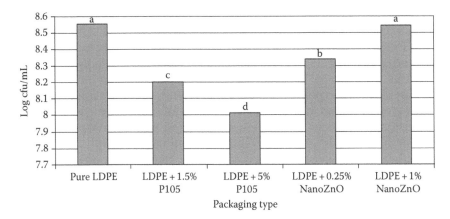

FIGURE 6.4
Effect of packaging type on inactivation of *Lactobacillus plantarum*. P105 is a powder composed of 95% TiO$_2$ and 5% metal nanosilver). (Reproduced from *Food Control*, 22, Emamifar, A., Kadivar, M., Shahedi, M., and Soleimanian-Zad, S., 408. Copyright 2011, with permission from Elsevier.)

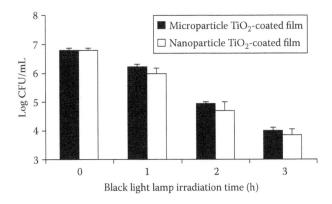

FIGURE 6.5
Inactivation of *Escherichia coli* with microparticles and nanoparticles of TiO$_2$-coated OPP packaging film under UV light. (Reproduced from *Int. J. Food Microbiol.*, 123, Chawengkijwanich, C. and Hayata, Y., 288, Copyright 2008, with permission from Elsevier.)

with TiO$_2$ exhibited the bacterial activity when exposed to UV light.[174] In this work it is also reported that the cell numbers of *E. coli* in the *in vitro* test were similar by using micro or nanoparticles TiO$_2$ (Figure 6.5) indicating that the bactericidal effect is independent of the particle size.

It is important to keep in mind that in determining the health impact of a new food contact substance, toxicity information needs to be contextualized by a determination of how readily this substance can become released from packaging materials into various foods substances. Concerning MMT, a 2010 study[175] showed that Uvitex OB, a commonly used additive to polyolefins approved for food contact use in Europe (2002/72/EEC), has very low release into oil-based and aqueous-based food simulants from a MMT/wheat-gluten polymer nanocomposite film, compared to up to 60% loss in LLDPE films subjected to the same test. This MMT/wheat-gluten film also permitted aluminum and silicon migration into food simulants that was allegedly well within the limits set forth by European regulations, although the authors of this study were careful to point out that a more difficult quantification of nanoparticle migration should ideally be performed. While the aforementioned study demonstrates that polyolefin nanomaterials (and polymer nanomaterials in general) may slow down the migration of potentially harmful additives into foods, the body of safety research is at this time fragmentary and incomplete.

A theoretical treatment has predicted that MMT particles with surface modification embedded in various polymer matrices are unlikely to migrate into foods from a polymer nanocomposite food contact material in any detectable quantities.[176] Nevertheless, more comprehensive experimental studies need to be done especially since some food and beverage companies are already utilizing these materials in their products. More importantly, the availability of clay nanoparticle toxicology data is still lacking, as are developments in strategies to detect and categorize clay and other nanoparticles in complex food matrices.

One study determined that exfoliated silicate nanoclays exhibited low cytotoxicity and genotoxicity, even when part of a diet fed to rats (measured acute oral toxicity, median lethal dose, LD50 > 5700 mg/kg body weight under the conditions probed)[177]; however, the authors of this study only tested a single clay type and morphology, so it is unclear whether it can be applied in a general sense. Furthermore, it has recently been shown[175] that PNC films exhibit enhanced migration of nanoclay components into food simulants when the films undergo high-pressure treatment, a food preservation/sterilization method that is becoming increasingly popular; a follow-up study[178]demonstrated that MMT clays undergo undetermined structural or chemical changes under pressures as low as 300 MPa and concluded that such changes should be taken into consideration when bringing MMT–polymer composites into contact with food.

For spherical particles, very little work has been done to assess the ability to migrate through rigid polymer environments and cross over the packaging/food interface. Šimon et al.[176] used a physicochemical approach to theorize that embedded Ag nanoparticles may diffuse from food packaging into foods at detectable levels only when the particle radius is very small (~1 nm), when the packaging is comprised of a polymer with relatively low dynamic viscosity (e.g., polyolefins such as LDPE, HDPE, and PP) and when there are no significant interactions between the particles and the polymer. Though broad-scope experiments assessing this prediction have not yet been published, at least one study[172] has shown evidence of migration into a food substance, in this case orange juice, from LDPE packaging materials incorporating Ag or ZnO antimicrobial nanoparticles.

A more pressing concern, however, may be food contact materials in which the Ag nanoparticles are located on the material surface, such that they come into direct contact with the food matrix. Even so, systematic attempts to study relationships between particle characteristics, polymer type, food pH/polarity, and, especially, environmental conditions relevant to food production, storage, and packaging use (e.g., temperature, pressure, humidity, light exposure, and storage time) are decidedly lacking, making it difficult to broadly assess this important aspect of the safety of Ag-based food contact materials at this time. As a result of these considerations, while polyolefins and polymer (in general) nanomaterials may represent the next revolution in food packaging technology, there are still steps that need to be taken in order to ensure that consumers are protected from any potential hazards these materials pose.

References

1. R. Coles, *Introduction*. In: R. Coles, D. McDowell, and M.J. Kirwan, eds. *Food Packaging Technology*, Blackwell Publishing, CRC Press, London, U.K. (2003), p. 1.
2. C. Silvestre, D. Duraccio, and S. Cimmino, *Prog Polym Sci*, 36 (2011) 1766.
3. A. Lopez-Rubio, E. Almenar, P. Hernandez-Munoz, J.M. Lagaron, R. Catala, and R. Gavara, *Food Rev Int*, 20 (2004) 357.
4. O.W. Lau and S.K. Wong, *J Chrom A*, 882 (2000) 255.
5. K. Marsh and B. Bugusu, *J Food Sci*, 72 (2007) R39.
6. G.A. Giles ed., *Handbook of Beverage Packaging*, Sheffield Academic Press, Sheffield, U.K. (1999), p. 115.
7. W.E. Brown, *Plastics in Food Packaging*, Marcel Decker, Inc., New York (1992), p. 111.
8. A.L. Brady and K.S. Marsh eds., *Wiley Encyclopaedia of Packaging Technology*, 2nd edn., John Wiley & Sons, New York (1997), pp. 215, 415–422, and 765–768.
9. F.L. Matthews and R.D. Rawlings, *Composite Materials: Engineering and Science*, Chapman & Hall, London, U.K. (1994), p. 137.
10. T.V. Duncan, *J Colloid Interface Sci*, 363 (2011) 1.
11. J.M. Lagaron, E. Gimenez, M.D. Sánchez-García, M.J. Ocio, A. Fendler, Novel nanocomposites to enhance quality and safety of packaged foods. In: *Food Contact Polymers*, Rapra Technologies, Shawbury, U.K., paper 19 (2007), p. 1.
12. Q. Chaudhry, M. Scotter, J. Blackburn, B. Ross, A. Boxall, L. Castle, R. Aitken, and R. Watkins, *Food Addit Contam*, 25 (2008) 241.
13. M.C. De Azedero Henriette, *Food Res Int*, 42 (2009) 1240.
14. C.I. Moraru, C.P. Panchapakesan, Q. Huang, P. Takhistov, S. Liu, and J.L. Kokini, *Food Technol*, 57 (2003) 24.
15. P. Sanguansri and M.A. Augustin, *Trends Food Sci Technol*, 17 (2006) 547.
16. G.L. Robertson, *Food Packaging: Principles and Practice*, 2nd edn., CRC Press, Taylor & Francis Group, Boca Raton, FL (2006).
17. A.L. Brody and K.S. Marsh, *Encyclopedia of Packaging Technology*, 2nd edn., John Wiley & Sons Inc., New York (1997).

18. W. Soroka, *Fundamentals of Packaging Technology*, 4th edn., DEStech Publications, Naperville, IL (2009).
19. A.L. Brody, B. Bugusu, J.H. Han, C.K. Sand, and T.H. Mchugh, *J Food Sci*, 73 (2008) R107.
20. A.K. Mohanty, M. Misra, and H.S. Nalwa, eds. *Packaging Nanotechnology*, American Scientific Publisher, Valencia, CA (2009).
21. A.L. Brody, *Food Technol*, 60 (2003) 92.
22. J.J. Luo and I.M. Daniel, *Compos Sci Technol*, 63 (2003) 1607.
23. E. Thostenson, C. Li, and T. Chou, *J Compos Sci Technol*, 65 (2005) 491.
24. D. Schmidt, D. Shah, and E.P. Giannelis, *Curr Opin Solid State Mater Sci*, 6 (2002) 205.
25. M. Alexandre and P. Dubois, *Mater Sci Eng Rep*, 28 (2000) 1.
26. F. Hussain, D. Derrick, A. Haque, and A.M. Shamsuzzoha, *J Adv Mater*, 37 (2005) 16.
27. C.L. Wu, M.Q. Zhang, M.Z. Rong, and K. Friedrich, *Compos Sci Technol*, 62 (2002) 1327.
28. V. Vladimirov, C. Betchev, A. Vassiliou, G. Papageorgiou, and D. Bikiaris, *Compos Sci Technol*, 66 (2006) 2935.
29. S. Tang, P. Zou, H. Xiong, and H. Tang, *Carbohydr Polym*, 72 (2008) 521.
30. X. Jia, Y. Li, Q. Cheng, S. Zhang, and B. Zhang, *Eur Polym J*, 43 (2007) 1123.
31. X. Zhou, E. Shin, K.W. Wang, and C.E. Bakis, *Compos Sci Technol*, 64 (2004) 2425.
32. W. Chen, X. Tao, P. Xue, and X. Cheng, *Appl Surf Sci*, 252 (2005) 1404.
33. Y. Bin, M. Mine, A. Koganemaru, X. Jiang, and M. Matsuo, *Polymer*, 47 (2006) 1308.
34. H. Zeng, C. Gao, Y. Wang, P.C.P. Watts, H. Kong, X. Cui, and D. Yan, *Polymer*, 47 (2006) 113.
35. S. Morlat-Therias, E. Fanton, J.L. Gardette, S. Peeterbroeck, M. Alexandre, and P. Dubois, *Polym Degrad Stab*, 92 (2007) 1873.
36. J.Y. Kim, S.I. Han, and S.H. Kim, *Polym Eng Sci*, 47 (2007) 1715.
37. K. Prashantha, J. Soulestin, M.F. Lacrampe, P. Krawczak, G. Dupin, and M. Claes, *Compos Sci Technol*, 69 (2009) 1756.
38. J.Y. Kim, S.I. Han, D.K. Kim, and S.H. Kim, *Compos. Part A: Appl Sci Manuf*, 40 (2009) 45.
39. T. Ramanathan, A.A. Abdala, S. Stankovich, D.A. Dikin, M. Herrera-Alonso, R.D. Piner, D.H. Adamson, H.C. Schniepp, X. Chen, R.S. Ruoff, S.T. Nguyen, I.A. Aksay, R.K. Prud'homme, and L.C. Brinson, *Nature Nanotechnol*, 3 (2008) 327.
40. K. Wakabayashi, C. Pierre, D.A. Dikin, R.S. Ruoff, T. Ramanathan, L.C. Brinson, and J.M. Torkelson, *Macromolecules*, 41 (2008) 1905.
41. C. Borriello, A. De Maria, N. Jovic, A. Montone, M. Schwarz, and M.V. Antisari, *Mater Manuf Process*, 24 (2009) 1053.
42. Y. Chen, X. Cao, P.R. Chang, and M.A. Huneault, *Carbohydr Polym*, 73 (2008) 8.
43. E. Kristo and C.G. Biliaderis, *Carbohydr Polym*, 68 (2007) 146.
44. H.M.C. Azeredo, L.H.C. Mattoso, D. Wood, T.G. Williams, R.J. Avena-Bustillos, and T.H. McHugh, *J Food Sci*, 74 (2009) N31.
45. H.M.C. Azeredo, L.H.C. Mattoso, R.J. Avena-Bustillos, G.C. Filho, M.L. Munford, D. Wood, and T.H. McHugh, *J Food Sci*, 75 (2010) N1.
46. C. Bilbao-Sainz, R.J. Avena-Bustillos, D.F. Wood, T.G. Williams, and T.H. McHugh, *J Agric Food Chem*, 58 (2010) 3753.
47. A. Dufresne, J.Y. Cavaille, and W. Helbert, *Macromolecules*, 29 (1996) 7624.

48. W. Helbert, J.Y. Cavaille, and A. Dufresne, *Polym Compos*, 17 (1996) 604.
49. P. Podsiadlo, S.-Y. Choi, B. Shim, J. Lee, M. Cuddihy, and N.A. Kotov, *Biomacromolecules*, 6 (2005) 2914.
50. M.A.S.A. Samir, F. Alloin, J.Y. Sanchez, and A. Dufresne, *Polymer*, 45 (2004) 4149.
51. K. Oksman, A.P. Mathew, D. Bondeson, and I. Kvien, *Compos Sci Technol*, 66 (2006) 2776.
52. X. Cao, Y. Chen, P.R. Chang, M. Stumborg, and M.A. Huneault, *J Appl Polym Sci*, 109 (2008) 3804.
53. Y. Lu, L. Weng, and L. Zhang, *Biomacromolecules*, 5 (2004) 1046.
54. J. Sriupayo, P. Supaphol, J. Blackwell, and R. Rujiravanit, *Polymer*, 46 (2005) 5637.
55. M.R. de Moura, F.A. Aouada, R.J. Avena-Bustillos, T.H. McHugh, J.M. Krochta, and L.H.C. Mattoso, *J Food Eng*, 92 (2009) 448.
56. M.R. de Moura, M.V. Lorevice, L.H.C. Mattoso, and V. Zucolotto, *J Food Sci*, 76 (2011) N25.
57. D. Yang, Y. Hu, L. Song, S. Nie, S. He, and Y. Cai, *Polym Degrad Stab*, 93 (2008) 2014.
58. Y.C. Li, J. Schulz, and J.C. Grunlan, *ACS Appl Mater Int*, 1 (2009) 2338.
59. S.S. Ray and M. Okamoto, *Prog Polym Sci*, 28 (2003) 1539.
60. H.M. Park, X. Li, C.Z. Jin, C.Y. Park, W.J. Cho, and C.-S. Ha, *Macromol Mater Eng*, 287 (2002) 553.
61. S.S. Ray and M. Bousmina, *Prog Mater Sci*, 50 (2005) 962.
62. S. Kumar, J.P. Jog, and U. Natarajan, *J Appl Polym Sci*, 89 (2003) 1186.
63. Y. Yoo, S.S. Kim, J.C. Won, K.Y. Choi, and J.H. Lee, *Polym Bull*, 52 (2004) 373.
64. F. Bertini, M. Canetti, G. Audisio, G. Costa, and L. Falqui, *Polym Degrad Stab*, 91 (2006) 600.
65. Q. Zhou, K.P. Pramoda, J.-M. Lee, K. Wang, and L.S. Loo, *J Colloid Interface Sci*, 355 (2011) 222.
66. M. Kawasumi, M. Hasegawan, A. Usuki, and A. Okada, *Macromolecules*, 30 (1997) 6333.
67. N. Hasegawa, M. Kawasumi, M. Kato, A. Usuki, and A. Okada, *J Appl Polym Sci*, 67 (1998) 87.
68. M. Kato, A. Usuki, and A. Okada, *J Appl Polym Sci*, 66 (1997) 1781.
69. N. Hasegawa, H. Okamato, M. Kato, and A. Usuki, *J Appl Polym Sci*, 78 (2000) 1918.
70. P.K. Mallick and Y.X. Zhou, *J Mater Sci*, 38 (2003) 3183.
71. Y. Tang, *Polym Degrad Stabil*, 82 (2003) 127.
72. M. Mehrabzadeh and M.R. Kamal, *Polym Eng Sci*, 44 (2004) 1152.
73. M. Zhang and U. Sundararaj, *Macromol Mater Eng*, 291 (2006) 697.
74. M. Okamoto, P.H. Nam, M. Maiti, T. Kotaka, T. Nakayama, M. Takada, M. Ohshima, A. Usuki, N. Hasegawa, and H. Okamoto, *Nano Lett*, 1 (2001) 503.
75. P.H. Nam, M. Okamoto, P. Maiti, T. Kotaka, T. Nakayama, M. Takada, M. Ohshima, N. Hasegawa, and A. Usuki, *Polym Eng Sci*, 42 (2002) 1907.
76. R. Andrews and M.C. Weisenberger, *Curr Opin Solid State Mater Sci*, 8 (2004) 31.
77. W.E. Dondero and R.E. Gorga, *J Polym Sci Part B Polym Phys*, 44 (2006) 864.
78. A.R. Bhattacharyya, T.V. Sreekumar, T. Liu, S. Kumar, L.M. Ericson, R.H. Hauge, and R.E. Smalley, *Polymer*, 44 (2003) 2373.
79. W.Z. Tang, M.H. Santare, and S.G. Advani, *Carbon*, 41 (2003) 2779.
80. E. Assouline, A. Lustiger, A.H. Barber, C.A. Cooper, E. Klein, E. Wachtel, and H.D. Wagner, *J Polym Sci Part B: Polym Phys*, 41 (2003) 520.

81. J. Sandler, G. Broza, M. Nolte, K. Schulte, Y.M. Lam, and M.S.P. Shaffer, *J Macromol Sci Phys*, B42 (2003) 479.

82. J.C. Kearns and R.L. Shambaugh, *J Appl Polym Sci*, 86 (2002) 2079.

83. E.M. Moore, D.L. Ortiz, V.T. Marla, R.L. Shambaugh, and B.P. Grady, *J Appl Polym Sci*, 93 (2004) 2926.

84. H.S. Xia, H. Xia, Q. Wang, K. Li, and G.H. Hu, *J Appl Polym Sci*, 93 (2004) 378.

85. B. Johnson, M.H. Santare, and S. Advani, Manufacturing and performance of carbon nanotube/high density polyethylene composites. In: *International Conference on Flow Processes in Composite Materials*, July, Bristol, U.K. (2004).

86. Y.P. Zheng, Y. Zheng, and R.C. Ning, *Mater Lett*, 57 (2003) 2940.

87. D. Ma, T.A. Hugener, R.W. Siegel, A. Christerson, E. Martensson, C. Onneby, and L. Schadler, *Nanotechnology*, 16 (2005) 724.

88. R. Coles, D. McDowell, and M.J. Kirwan, eds., *Food Packaging Technology*. Blackwell Publishing Ltd, Oxford, U.K. (2003), p. 65.

89. A. Brody, Packaging by the numbers. *Food Technol*, 62 (2008) 89.

90. J.M. Lagaron, D. Cava, L. Cabedo, R. Gavara, and E. Gimenez, *Food Addit Contam*, 22 (2005) 994.

91. A. Arora and G.W. Padua, *J Food Sci*, 75 (2010) R43.

92. P. Mercea, Models for diffusion in polymers. In: O.G. Piringer and A.L. Baner, eds. *Plastic Packaging*, 2nd edn., Wiley-VCH GmbH & Co. KGaA, Weinheim, Germany (2008).

93. B. Finnigan, Barrier polymers. In: K.L. Yam, ed. *The Wiley Encyclopedia of Packaging Technology*. John Wiley & Sons, Inc., New York (2009), p. 103.

94. G. Choudalakis and A.D. Gotsis, *Eur Polym J*, 45 (2009) 967.

95. L.E. Nielsen, *J Macromol Sci Part A: Pure Appl Chem*, 1 (1967) 929.

96. E. Dunkerley and D. Schmidt, *Macromolecules*, 43 (2010) 10536.

97. E.L. Cussler, S.E. Hughes, W.J. Ward, and R. Aris, *J Membr Sci*, 38 (1988) 161.

98. W.T. Brydges, S.T. Gulati, and G. Baum, *J Mater Sci*, 10 (1975) 2044.

99. N.K. Lape, E.E. Nuxoll, and E.L. Cussler, *J Membr Sci*, 236 (2004) 29.

100. G.D. Moggridge, N.K. Lape, C. Yang, and E.L. Cussler, *Prog Org Coat*, 46 (2003) 231.

101. G.H. Fredrickson and J. Bicerano, *J Chem Phys*, 110 (1999) 2181.

102. A.A. Gusev and H.R. Lusti, *Adv Mater*, 13 (2001) 1641.

103. E. Picard, A. Vermogen, J.-F. Gerard, E. Espuche, *J Membr Sci*, 292 (2007) 133.

104. R.K. Bharadwaj, *Macromolecules*, 34 (2001) 9189.

105. R.D. Maksimov, S. Gaidukov, J. Zicans, and J. Jansons, *Mech Compos Mater*, 44 (2008) 505.

106. S. Nazarenko, P. Meneghetti, P. Julmon, B.G. Olson, and S. Qutubuddin, *J Polym Sci Part B Polym Phys*, 45 (2007) 1733.

107. E. Picard, H. Gauthier, J.F. Gerard, and E. Espuche, *J Colloid Interface Sci*, 307 (2007) 364.

108. A. Sorrentino, M. Tortora, and V. Vittoria, *J Polym Sci Part B: Polym Phys*, 44 (2006) 265.

109. R. Qiao and L.C. Brinson, *Compos Sci Technol*, 69 (2009) 491.

110. Z. Mogri and D.R. Paul, *Polymer*, 42 (2001) 2531.

111. A. Hiltner, R.Y.F. Liu, Y.S. Hu, and E. Baer, *J Polym Sci Part B: Polym Phys*, 43 (2005) 1047.

112. D.A. Pereira de Abreu, P. Paseiro Losada, I. Angulo, and J.M. Cruz, *Eur Polym J*, 43 (2007) 2229.

113. S. Ray, S.Y. Quek, A. Easteal, and X.D. Chen, *Int J Food Eng*, 2 (2006) Article 5.
114. A. Sorrentino, G. Gorrasi, and V. Vittoria, *Trends Food Sci Technol*, 18 (2007) 84.
115. J. Jordan, K.I. Jacob, R. Tannenbaum, M.A. Sharaf, and I. Jasiuk, *Mater Sci Eng A*, 393 (2005) 1.
116. M.A. Osman, J.E.P. Rupp, and U.W. Suter, *Polymer*, 46 (2005) 1653.
117. M.A. Osman, M. Ploetze, and U.W. Suter, *J Mater Chem*, 13 (2003) 2359.
118. X. Li and C.S. Ha, *J Appl Polym Sci*, 87 (2003) 1901.
119. M.A. Osman, J.E.P. Rupp, and U.W. Suter, *J Mater Chem*, 15 (2005) 1298.
120. J.S. Bergman, H. Chen, E.P. Giannelis, M.G. Thomas, and G.W. Coates, *Chem Commun*, 21 (1999) 2179.
121. J. Heinemann, P. Reichert, R. Thomann, and R. Mülhaupt, *Macromol Rapid Commun*, 20 (1999) 423.
122. M. Alexandre, P. Dubois, T. Sunb, J.M. Garcesb, and R. Jérôme, *Polymer*, 43 (2002) 2123.
123. M. Kato, H. Okamoto, N. Hasegawa, A. Tsukigase, and A. Usuki, *Polym Eng Sci*, 43 (2003) 1312.
124. G. Liang, J. Xu, S. Bao, and W. Xu, *J Appl Polym Sci*, 91 (2004) 3974.
125. S. Hotta and D.R. Paul, *Polymer*, 45 (2004) 7639.
126. R.A. Vaia and E.P. Giannelis, *Macromolecules*, 30 (1997) 7990.
127. A.C. Balazs, C. Singh, E. Zhulina, and Y. Lyatskaya, *Acc Chem Res*, 32 (1999) 651.
128. V.V. Ginzburg, C. Singh, and A.C.Balazs, *Macromolecules*, 33 (2000) 1089.
129. D.L. Ho and C.J. Glinka, *Chem Mater*, 15 (2003) 1309.
130. A.C. Balazs, C. Singh, and E. Zhulina, *Macromolecules*, 31 (1998) 8370.
131. V.V. Ginzburg and A.C.Balazs, *Adv Mater*, 12 (2000) 1805.
132. A. Sinsawat, K.L. Anderson, R.A. Vaia, and B.L. Farmer, *J Polym Sci, Part B Polym Phys*, 41 (2003) 3272.
133. P. Reichert, H. Nitz, S. Klinke, R. Brandsch, R. Thoman, and R. Mülhaupt, *Macromol Mater Eng*, 275 (2000) 8.
134. E. Manias, A. Touny, L. Wu, K. Strawhecker, B. Lu, and T.C. Chung, *Chem Mater*, 13 (2001) 3516.
135. Z.M. Wang, H. Nakajima, E. Manias, and T.C. Chung, *Macromolecules*, 36 (2003) 8919.
136. R. Kumar and H. Münstedt, *Biomaterials*, 26 (2005) 2081.
137. S.Y. Liau, D.C. Read, W.J. Pugh, J.R. Furr, and A.D. Russell, *Lett Appl Microbiol*, 25 (1997) 279.
138. A. Panâcek, L. Kvitek, R. Prucek, M. Kolař, R. Vecêřova, N. Pizurova, V.K. Sharma, T. Nevěêna, and R. Zbořil, *J Phys Chem B*, 110 (2006) 16248.
139. L. Kvitek, A. Panâcek, J. Soukupova, M. Kolař, R. Vecêřova, R. Prucek, M. Holecova, and R. Zbořil, *J Phys Chem C*, 112 (2008) 5825.
140. D.M. Eby, N.M. Schaeublin, K.E. Farrington, S.M. Hussain, and G.R. Johnson, *ACS Nano*, 3 (2009) 984.
141. A.M. Fayaz, K. Balaji, M. Girilal, R. Yadav, P.T. Kalaichelvan, and R. Venketesan, *Nanomed Nanotechnol Biol Med*, 6 (2010) 103.
142. S. Egger, R.P. Lehmann, M.J. Height, M.J. Loessner, and M. Schuppler, *Appl Environ Microbiol*, 75 (2009) 2973.
143. K.J. Kim, W.S. Sung, S.K. Moon, J.S. Choi, J.G. Kim, and D.G. Lee, *J Microbiol Biotechnol*, 18 (2008) 1482.

144. J.S. Kim, E. Kuk, K.N. Yu, J.H. Kim, S.J. Park, H.J. Lee, S.H. Kim, Y.K. Park, Y.H. Park, C.Y. Hwang, Y.K. Kim, Y.S. Lee, D.H. Jeong, and M.H. Cho, *Nanomed Nanotechnol Biol Med*, 3 (2007) 95.

145. E. Navarro, F. Piccapietra, B. Wagner, F. Marconi, R. Kaegi, N. Odzak, L. Sigg, and R. Behra, *Environ Sci Technol*, 42 (2008) 8959.

146. E. Navarro, A. Baun, R. Behra, N.B. Hartmann, J. Filser, A.J. Miao, A. Quigg, P.H. Santschi, and L. Sigg, *Ecotoxicology*, 17 (2008) 372.

147. J.L. Elechiguerra, J.L. Burt, J.R. Morones, A. Camacho-Bragado, X. Gao, H.H. Lara, and M.J. Yacaman, *J Nanobiotechnol*, 3 (2005) 1.

148. J.V. Rogers, C.V. Parkinson, Y.W. Choi, J.L. Speshock, and S.M. Hussain, *Nanoscale Res Lett*, 3 (2008) 129.

149. W. Kangwansupamonkon, V. Lauruengtana, S. Surassmo, and U. Ruktanonchai, *Nanomed Nanotechnol Biol Med*, 5 (2009) 240.

150. D.B. Hamal, J.A. Haggstrom, G.L. Marchin, M.A. Ikenberry, K. Hohn, and K.J. Klabunde, *Langmuir*, 26 (2010) 2805.

151. H. Kong, J. Song, and J. Jang, *Environ Sci Technol*, 44 (2010) 5672.

152. S. Kang, M. Pinault, L.D. Pfefferle, and M. Elimelech, *Langmuir*, 23 (2007) 8670.

153. D.B. Warheit, B.R. Laurence, K.L. Reed, D.H. Roach, G.A.M Reynolds, and T.R. Webb, *Toxicol Sci*, 77 (2004) 117.

154. S. Sanchez-Valdes, H. Ortega-Ortiz, L.F. Ramos-de Valle, F.J. Medellin Rodriguez, and R. Guedea-Miranda, *J Appl Polym Sci*, 111 (2009) 953.

155. R. Kumar and H. Munstedt, *Polym Int*, 54 (2005) 1180.

156. R. Kumar, S. Howdle, and H. Munstedt, *J Biomed Mater Res Part B*, 75B (2005) 311.

157. C. Radheshkumar and H. Munstedt, *React Funct Polym*, 66 (2006) 780.

158. C. Damm, H. Munstedt, and A. Rosch, *Mater Chem Phys*, 108 (2008) 61.

159. O. Yamamoto, *J Inorg Mater*, 3 (2001) 643.

160. N. Jones, B. Ray, K.T. Ranjit, and A.C. Manna, *Microbiol Lett*, 279 (2008) 71.

161. S. Chandramouleeswaran, S.T. Mhaske, A.A. Kathe, P.V. Varadarajan, V. Prasad, and N. Vigneshwaran, *Nanotechnology*, 18 (2007) 8.

162. N. Lepot, M.K. Van Bael, H. Van den Rul, J. D'Haen, R. Peeters, D. Franco, and J. Mullens, *J Appl Polym Sci*, 120 (2011) 1616.

163. H. Zhao and R.K.Y. Li, *Polymer*, 47 (2006) 3207.

164. G. Droval, I. Aranberri, A. Bilbao, L. German, M. Verelst, and J. Dexpert-Ghys, *e-Polymer*, (2008) 128.

165. M. Sanchez-Garcia, J.M. Lagaron, Nanocomposite packaging materials. In: K.L. Yam, ed. *The Wiley Encyclopedia of Packaging Technology*, 3rd edn., John Wiley & Sons, Inc., New York (2009), p. 807.

166. F.M. Yang, H.M. Li, F. Li, Z.H. Xin, L.Y. Zhao, Y.H. Zheng, Q.H. Hu, *J Food Sci*, 75 (2010) C236.

167. D.M. Hodges, J.M. Delong, C.F. Forney, and R.F. Prange, *Planta*, 207 (1999) 604.

168. C.K. Ding, K. Chachin, Y. Ueda, and Y. Imahori, *J Agric Food Chem*, 46 (1998) 4144.

169. M. Chisari, R.N. Barbagallo, and G. Spagna, *J Agric Food Chem*, 55 (2007) 3469.

170. N. Valipoor Motlagh, M.T. Hamed Mosavian, S.A. Mortazavi, and A. Tamizi, *J Food Sci*, 71 (2012) E2.

171. H. Li, F. Li, L. Wang, J. Sheng, Z. Xin, L. Zhao, H. Xiao, Y. Zheng, and Q. Hu, *Food Chem*, 114 (2009) 547.

172. A. Emamifar, M. Kadivar, M. Shahedi, and S. Soleimanian-Zad, *Food Control*, 22 (2011) 408.
173. Avella, G. Bruno, M.E. Errico, G. Gentile, N. Piciocchi, A. Sorrentino, and M.G. Volpe, *Packag Technol Sci*, 20 (2007) 325.
174. C. Chawengkijwanich and Y. Hayata, *Int J Food Microbiol*, 123 (2008) 288.
175. M. Mauricio-Iglesias, S. Peyron, V. Guillard, and N. Gontard, *J Appl Polym Sci*, 116 (2010) 2526.
176. P. Šimon, Q. Chaudhry, and D. Bakoš, *J Food Nutr Res*, 47 (2008) 105.
177. P.R. Li, J.C. Wei, Y.F. Chiu, H.L. Su, F.C. Peng, and J.J. Lin, *ACS Appl Mater Int*, 2 (2010) 1608.
178. M. Mauricio-Iglesias, N. Gontard, and E. Gastaldi, *Appl Clay Sci*, 51 (2011) 174.

7

Some Aspects Concerning the Nanomaterials from Renewable Resources Use in Food Packaging

Cornelia Vasile, Georgeta Cazacu, Raluca Petronela Dumitriu, Raluca Nicoleta Darie, and Irina Elena Răschip

CONTENTS

7.1 Introduction

Increased and indiscriminate use of plastic packaging films, which are petroleum based, has led to ecological problems due to their total nonbiodegradability. Continuous use of plastics in any form or shape has to be restricted and may even be gradually abandoned to protect and conserve environment. Such awareness, of late by one and all, has led to a paradigm shift to look for packaging films and processes that are biodegradable and, therefore, compatible with the environment. Such an approach also leads to natural resource conservation with an underpinning on a pollution-free environment. Thus, the concept of biodegradability enjoys both user-friendly and eco-friendly attributes, and the raw materials are essentially derived from either replenishable agricultural feed stocks (cellulose, starch, and proteins) or marine food processing industry wastes (chitin/chitosan). Their total biodegradation to environmentally friendly benign products (CO_2, H_2O/quality compost) is the turning point that needs to be capitalized upon. Polymer cross-linking and graft copolymerization of natural polymers with synthetic monomers are other alternative approaches of value to using biodegradable packaging films. Although complete replacement of synthetic plastics may be impossible to achieve and perhaps even unnecessary, at least for a few specific applications, attention and efforts are required in the days to come. Though expensive, biopackaging meets tomorrow's need for packaging, especially for a few value-added products. It offers an attractive route to waste management, as well.[1] Most materials currently used for food packaging are nondegradable, generating environmental problems. Several biopolymers have been exploited to develop materials for eco-friendly food packaging. However, the use of biopolymers has been limited because of their usually poor mechanical and barrier properties, which may be improved by adding reinforcing compounds (fillers), forming composites. Most reinforced materials present poor matrix–filler interactions, which tend to improve with decreasing filler dimensions. The use of fillers with at least one nanoscale dimension (nanoparticles) produces nanocomposites. Nanoparticles have proportionally larger surface area than their microscale counterparts, which favors the filler–matrix interactions and the performance of the resulting material. Besides nanoreinforcements, nanoparticles can have other functions when added to a polymer, such as antimicrobial activity, enzyme immobilization, biosensors, etc.[2]

Today, nanocomposite research is widespread and is conducted by companies and universities across the globe. Plastic suppliers who have already commercialized nanocomposite materials include Basel USA, Bayer, Dow Chemical, Eastman Chemical, Mitsubishi Gas Chemical, Nanocore, Triton Systems, Honeywell, and RTP Co. ("Cutting Edge," 2001). Most of these efforts are currently focused on either polyolefins or nylons, but, in theory, the clay nanoparticles could be used in any resin family.

Antimicrobial packaging has attracted much attention from the food industry because of the increase in consumer demand for minimally processed and preservative-free products. Reflecting this demand, preservative agents (preferably natural preservatives) must be applied at the lowest effective level possible.[3] However, the application of antibacterial substances directly onto a food has some limitations because the active substances can be neutralized, evaporated, or they may inadequately diffuse into the bulk of the food.[4,5] The growth of microorganisms on the cut surfaces is a main cause of food spoilage for fresh-cut produce. The incorporation of antimicrobial agents into packaging can create an environment inside the package that may delay or prevent the growth of microorganisms on the product's surface and, hence, lead to an extension of its shelf life.

The consumers' concerns on the addition of chemical additives to food have driven the food industry and food research toward the search for natural antimicrobial compounds.[6]

In food packaging, a major emphasis is on the development of high barrier properties against the migration of oxygen, carbon dioxide, flavor compounds, and water vapor. Decreasing water vapor permeability is a critical issue in the development of biopolymers as sustainable packaging materials. The nanoscale plate morphology of clays and other fillers promotes the development of gas barrier properties. Challenges remain in increasing the compatibility between clays and polymers and reaching complete dispersion of nanoplates. Nanocomposites may advance the utilization of biopolymers in food packaging.[7]

Chitosan, a natural polysaccharide derived from chitin, keeps attracting much interest because of its biodegradability, nontoxicity, hydrophilicity, biocompatibility, antibacterial property, efficiency, and low cost.[8] Chitosan refers to a family of polymers derived from chitin that have been de-N-acetylated to make them soluble in aqueous acidic solutions.[9] Free-standing edible films from chitosan have been prepared and their water vapor and ethylene permeabilities as well as tensile strength determined.[10] It has also been reported that chitosan exhibits antifungal properties.[11] Vacuum-packaged chitosan-coated grilled pork showed negligible microbial growth and was found to be organoleptically acceptable throughout the storage period. Chitosan coating along with vacuum packaging provided a type of active packaging to maintain quality and extend the shelf life of grilled pork.[12]

Chitosan is the second most abundant polysaccharide on earth and its antimicrobial properties are already known. According to Brody et al.[13] the antimicrobial effect of chitosan occurs when organisms are in direct contact with the active sites of chitosan. When antimicrobial agents are incorporated into the film, they diffuse out of the film, thus improving its antimicrobial efficacy. Zivanovic et al.[14] applied chitosan-oregano essential oil (EO) in comparison with chitosan films on inoculated bologna meat samples stored for 5 days at 10°C. Pure chitosan films reduced *Listeria monocytogenes* by 2 logs, whereas the films with 1% and 2% oregano EO decreased the numbers of *L. monocytogenes* by 3.6–4 logs and *Escherichia coli* by 3 logs. Pranoto et al.[15] incorporated garlic oil, potassium sorbate, and nisin in chitosan films. The activity of the antimicrobial films was tested against the food pathogenic bacteria, *E. coli*, *Staphylococcus aureus*, *Salmonella typhimurium*, *L. monocytogenes*, and *Bacillus cereus*. They found that the pure chitosan film had no inhibitory effect. Incorporation of 100 L of garlic oil/g, 100 mg potassium sorbate/g, or nisin at 51,000 IU/g of chitosan had antimicrobial activity against *S. aureus*, *L. monocytogenes*, and *B. cereus*.

Chitosan/methyl cellulose and chitosan/methyl cellulose films incorporating vanillin as a natural antimicrobial agent provided an inhibitory effect against *E. coli* bacteria and *Saccharomyces cerevisiae* yeast. Vanillin-containing film was more efficient than neat chitosan/methyl cellulose in reducing the microorganism levels (higher log reduction) to a greater extent, but over a longer time. In a low pH fruit, vanillin was more effective at inhibiting microorganisms, but a significant reduction of vitamin C/L-ascorbic acid content in pineapple wrapped in vanillin film was observed. Vanillin film increased the intensity of yellow color of pineapple.[16]

Matrices based on covalently cross-linked *chitosan hydrogels* have been selected because they have many applications from food additives to pharmaceuticals[17–19] and biomedical purposes. Covalent cross-linking leads to the formation of a permanent network allowing the free diffusion of water and/or bioactive compounds without dissolution and permits drug release by diffusion and enhancing the mechanical properties of the hydrogel. Chitosan hydrogels (C) also are biocompatible, nontoxic, and biodegradable.

Xanthan gum is classified E 415 in the European List of Permitted Food Additives. According to JECFA (Joint WHO/FAO Expert Committee on Food Additives), it has the status of ADI-non-specified (acceptable daily intake), i.e., no quantitative limitation is stated, and, as such, xanthan gum is recognized as a nontoxic additive for human consumption.

Lignins are complex phenolic polymers occurring in higher plant tissues and are the second most abundant terrestrial polymer after cellulose. Due to their very complex structure, lignins are amorphous polymers with rather limited industrial use. They are usually seen as waste products of pulp and paper industry and often used as fuel for the energy balance of the pulping process. The major problem identified with natural fibers during incorporation in hydrophobic polymers is their poor compatibility. To alleviate this problem, various polymer interface modifications have been proposed, which result in improvement of performance of the resulting composite.[20,21]

For the last decades, a great deal of research was devoted to the development of lignin-containing polymeric materials. A way of lignin use consists in the incorporation of small amounts of lignin in order to take advantage of lignin structure and to stabilize the material against photo and thermooxidation.[22,23] In the last decades, more and more information that lignins could play an important role in the biological systems acting as antioxidant, antibacterial, and antiviral components has appeared. Indeed, lignins are hindered phenolic polymers that can exert antioxidant properties.[24] Presence of phenolic fragments with functional groups containing oxygen (–OH, –CO, –COOH, –SO$_3$H) imparts antioxidant and antimicrobial activities, which depend on the extraction method of lignin.[25,26]

Nelson et al.[27] reported that Alcell lignin, an organosolv lignin, can inhibit the growth of *Pseudomonas aeruginosa*, *E. coli*, and *S. aureus* colonies. Nada[28] studied antimicrobial activity of lignin from cotton stalk against *B. subtilis*, *B. mycoids*, *E. coli*, and *A. niger*, and did not observe significant antimicrobial effect against gram negative bacteria but an important activity was manifested against gram positive bacteria.

Lignin is an effective free radical scavenger that stabilizes the reactions induced by oxygen and its radical species,[29] and it can also inhibit the activity of generation of superoxide anion radicals by acting as a natural antioxidant.[30]

Boeriu et al.[31] followed the ability of lignins to react with the free ABTS[+*] [ABTS—2,2'-azino-bis(3-ethylbenzo-thiazoline-6-sulphonate)] cation radicals generated by an enzymatic system consisting of peroxidase and hydrogen peroxide. All lignin reacted with ABTS[+*], but their efficiency was lower (20%–30%) than that measured for the commercial antioxidant BHT (butylated hydroxytoluene). Lignin samples from annual plants (abaca, sisal, and jute) and softwoods show the highest scavenging activity against the ABTS radicals. Ugartondo et al.[32] studied the antioxidant effect of industrial lignins (lignosulfonate [LS], curan lignin, steam explosion lignin, and bagasse lignin) against lipid peroxidation induced by H_2O_2 using human red blood cells (RBCs), showing that the main factor decreasing the radical scavenging activity is the high molecular weight of lignin.

Madad et al.[33] showed that the commercial sodium lignosulfonates (SLSs) and their fractions exhibit antioxidant activity, which is evaluated by TEAC (Trolox equivalent antioxidant capacity) and ORAC (oxygen radical absorbance capacity) assays. Authors observed that SLS fractions present various antioxidant activities in comparison with the most known antioxidant molecules, such as vitamin C, vitamin E, and rutin. These variations could be attributed to the structural difference observed between SLS fractions; the presence of high amount of syringyl units in some SLS fractions enhanced their antioxidant activity, as confirmed by Zhou.[34]

Vanillin (4-hydroxy-3-methoxybenzaldehyde) is one of the most popularly used flavoring components extracted from the seedpods of *Vanilla planifolia* and is widely used in foods, beverages, cosmetics, drugs, ice creams, biscuits, chocolates, confectionary, desserts, etc.[35–37] It has also been reported to exhibit multifunctional effects such as antimutagenic, antiangiogenetic, anti-colitis, anti-sickling, and antianalgesic effects. Natural and synthetically produced (nature identical) vanillin has been widely used by the food industry as a *GRAS flavoring agent* of a variety of food products, and its organoleptic feature is well accepted by consumers.[36,37] Vanillin, present in the essential oil fraction of the vanilla bean, is structurally similar to eugenol (2-methoxy-4-(2-propenyl) phenol) from cloves and is known to be *antimycotic and bacteriostatic.*

Concentrations of vanillin used in food and beverage products range widely from 0.3 to 33 mM.[38] Vanillin is efficacious for the treatment of chronic hypoacidic gastritis and chronic non-acid gastritis. By the inhibition occurring in the central nervous system, vanillin also influences the craving to consume food. In this context, it is supposed that vanillin is able to increase the concentration of the neurotransmitter serotonin in the brain.[17] Increased brain serotonin concentration, however, leads demonstratively to a reduced craving to consume food.

Vanillin is structurally similar to eugenol from cloves and is known to be *antimycotic*[35] and *bacteriostatic*.[39] Recent reports have shown that vanillin can be effective in *inhibiting bacteria*, yeasts, and molds,[39,40–42] the *antimicrobial activity of vanillin* depending on the time of exposure, concentration, and the target organism.

Vanillin has been used to inhibit *E. coli* O157:H7 and *L. monocytogenes* in "Granny Smith" apple juice.[43] Rupasinghe et al.[44] reported that total aerobic counts of fresh-cut apple slices decreased from 4.3 log cfu/g fresh weight (untreated) to 1.6 log/cfu by using NatureSeal (an antibrowning agent) plus 12 mM vanillin after 19 d at 4°C. Cerrutti et al.[45] treated strawberry puree with a mild heat treatment combined with 3000 mg/L vanillin and 500 mg/L ascorbic acid. They found that there was the inhibition of native and inoculated flora growth for at least 60 days storage at room temperature. Penney et al.[46] found that 2000 mg/L vanillin suppressed fungal and total microbial growth in yogurt significantly over the 3 week period.

Vanillin inhibits pathogenic and spoilage microorganisms in vitro and aerobic microbial growth in fresh-cut apples.[47] Several studies have demonstrated the antimicrobial activity of vanillin against yeasts and molds,[41,42,45,48–51] although the reports on antibacterial properties of vanillin are limited.[39,40]

Based on the studies conducted using *E. coli*, *Lactobacillus plantarum*, and *Listeria innocua*, the inhibitory activity of vanillin resides primarily in its ability to detrimentally affect the integrity of the cytoplasmic membrane, with the resultant loss of ion gradient and pH homeostasis and inhibition of respiratory activity. Vanillin is effective in inhibiting yeast and molds in vitro[42,50,51] and in fruit puree or juice.[45,48,49] Vanillin (12 mM) inhibited the growth of four food spoilage yeasts, *Saccharomyces cerevisiae*, *Zygosaccharomyces rouxii*, *Debaryomyces hansenni*, and *Zygosaccharomyces bailii*, in culture media and apple puree for 40 days storage at 27°C.[41] Incorporation of vanillin (3–7 mM) into fruit-based agars inhibited the growth of four *Aspergillus* species for 2 months.[52] When combined with 2 mM potassium sorbate, 3 mM vanillin could inhibit the growth of three *Penicillium* species, *Penicillium digitatum*, *Penicillium glabrum*, and *Penicillium italicum*, grown in potato dextrose agar (pH 3.5, *aw* 0.98) for 1 month.[42]

Fitzgerald et al.[49] reported that two yeast strains, *S. cerevisiae* and *Candida parapsilosis*, inoculated at a level of ~10^4 cfu/mL in apple juice and peach-flavored soft drink were inhibited by vanillin at 20 and 10 mM concentrations, respectively, over an 8 week storage period at 25°C.

The dose-dependent effect of vanillin on selected pathogenic, indicator, and spoilage microorganisms, such as *E. coli*, *Enterobacter. aerogenes*, *P. aeruginosa*, *S. newport*, *C. albicans*, *L. casei*, *P. expansum*, and *S. cerevisiae*, was investigated. All microorganisms tested were inhibited by vanillin at concentrations between 6 and 18 mM, with the exception of *P. expansum* (MIC > 18 mM). It was demonstrated that incorporation of vanillin (12 mM) in the post-cut dipping solution of fresh-cut apple slices could inhibit the microbial growth during the 19 day post-cut storage by 37% and 66% in studied apple slices.

Vanillin is an *effective antioxidant* in complex foods containing polyunsaturated fatty acids and its incorporation into dried foods (e.g., cereals) has shown a greater keeping quality than similar products left untreated.[53]

Taking in view the specific properties of these materials from renewable resources, this chapter deals with some aspects of the potential applications of chitosan, xanthan, and lignin and in obtaining antimicrobial and antioxidant formulations.

7.2 Experimental

7.2.1 Materials

Because the flavor compounds in food are mostly volatiles, loading method was applied to produce a powdered flavor in order to prevent the flavor loss to prolong the product's shelf life and make it convenient to use.[54–56]

Table 7.1 lists the materials used in this study and their characteristics.

7.2.2 Preparation Procedures

7.2.2.1 Hydrogel Preparation and Purification

CS–Cloisite® 15A composite hydrogel synthesis: CS–Cloisite® 15A composites cross-linked with glutaraldehyde (GA) in the same ratio (CS–Cloisite® 15A solution/GA 1:0.3 v/v) have been prepared following two steps: (1) synthesis of CS–Cloisite® 15A composites with different concentrations of the Cloisite® 15A (5 wt%, 7 wt%, 9 wt%) and (2) CS–Cloisite® 15A nanocomposites cross-linking with GA. Shortly, CS was dissolved in 1 wt% aqueous acetic acid solution at room temperature and left overnight with continuous mechanical stirring to obtain a 1% (w/v) solution. Separate, Cloisite® 15A clay was dispersed in 1 wt% aqueous acetic acid solution resulting in a clay solution (also at room temperature and left overnight with continuous mechanical stirring). Then the two prepared solutions were mixed for 4 h. The 5% (w/v) aqueous GA solution (in 1:0.3 v/v ratio) was added to CS–Cloisite® 15A solution under vigorous stirring at room temperature. After 1 h, the viscous solution was poured into Petri dishes and dried at room temperature overnight to form the hydrogel. Cross-linking took place at room temperature for 4 h in a dark space to protect system from oxidative/photodegradation of GA.[63] The hydrogels obtained were extensively washed with twice-distilled water to remove the excess of cross-linking agent (GA is easily water soluble) and then freeze-dried by means of a Labconco FreeZone device and stored until further use. The removal of unreacted cross-linking agent and other impurities was tested by pH measurements and biocompatibility tests both in vitro and in vivo as previously reported.[64] It has been established that the

TABLE 7.1

Materials Used in This Study and Their Characteristics

Sample	Code	Characteristics
Chitosan	CS	Medium molecular weight chitosan (MCS), M_n = 400,000 Da, $\eta \sim$ 200 MPas, Brookfield solution of 1 wt% in 1% acid acetic Powder, with a deacetylation degree of 90% Structure:
Cloisite® 15A	Cloisite® 15A	Cloisite® 15A nanoclay was purchased from Southern Clay Products, Inc. (Gonzales, TX). It is a natural Mt modified with quaternary ammonium salt containing organic modifier, dimethyl dehydrogenated tallow [2M$_2$HT], where HT is hydrogenated tallow with approximate composition 65% C18, 30% C16, 5% C14 [57] Specific gravity is 1.66 g/cm^3 and bulk density 172.84 kg/m^3 [58] Particle size ranges from 2 to 13 μm
Xanthan gum (Keltrol CG)—E 415	X	Molecular weight is been reported to be $M_w \sim 3.5 \times 10^6$ g/mol White or lightly yellow solid powder Structure:

Aroma: Vanillin V

White or lightly yellow solid powder
Molar mass = 152.15 g/mol, density = 1.056 g/cm^3, melting point = 80–81°C, boiling point = 285°C
Structure:

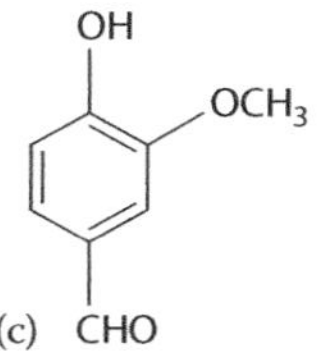

(c) CHO

Isotactic
polypropylene iPP

Malen-P F 401 type (Petrochemia Plock S.A. Poland)
Isotactic index ~95%
Melting point 170°C; MFI 2.4–3.2 g/10 min
Decomposition interval of 205°C–430°C, volatile compounds ~0.3 wt%
Vicat softening temperature ~148°C
Structure:

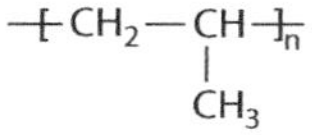

(continued)

TABLE 7.1 (continued)

Materials Used in This Study and Their Characteristics

Sample	Code	Characteristics
Annual fiber crops lignin (Granit, Lausanne, Switzerland)	GL	Elemental composition: C = 50.15%, H = 5.97%, O = 43.88% Functional groups: OCH_3 = 14.8%, OH_T = 11.7%, OH_{ph} = 2.43 mmol/g lignin Structure of main structural units: (d) (a) (b) (c) Monomeric units in lignin: (a) trans-p-coumaryl alcohol, (b) trans-coniferyl alcohol (guaiacyl unit), (c) trans-sinapyl alcohol (syringyl unit)
Annual plant alkaline lignin	AL	Commercial lignin Protobind 1000, Granit SA, Lausanne, Switzerland Klason lignin content of 75% and 1.6% ash; Mw = 55 400 Da; A_2 = 8.54 10^{-4} mL mol/g^2 [59] Limited solubility in the alkaline solutions (0.1 M NaOH) Preponderance of the condensation of the phenolic structures [60]

Aspen steam explosion lignin	AEL	ATO, Holland: elemental composition: C = 60.36%, H = 6.16%, O = 33.54%, functional groups OCH_3 = 21.4%, OHT = 7.5%, OHph = 2.4 mmol/g. Mw = 25,500 Da; A_2 = 1.5 10^{-3} mL mol/g² limited solubility in the alkaline solutions (0.1 M NaOH) [59] Preponderance of the condensation of the phenolic structures [60]
Ammonium lignosulfonate	LSA	Hardwood lignosulfonate (oak/beech) SC Celohart SA Zarnesti, Romania: OCH_3 functional groups = 9.5%, elemental composition: C = 48.78%, H = 5.10%, N = 6.63%, S = 9.81%. Mw = 1800 Da; 2nd virial coefficient = -4.1×10^{-3} mL mol/g². The negative value of the 2nd virial coefficient indicates that the polymer particles have a slight preference toward particle aggregation than to salvation; particle size ranges from 1 to 100 to more than 100 nm, indicating also the formation of the lignin aggregates due to its lipohydrophilic character, the presence of the hydrophobic aromatic skeleton and hydrophilic sulfonic groups [61]
Irganox 1076	Irg	Octadecyl-3-(3,5-di-tert.butyl-4-hydroxyphenyl)-propionate (Ciba Specialty Chemicals) sterically hindered phenolic antioxidant; molecular mass 531 g/mol; melting interval of 50°C–55°C
Compatibilizing agents:		
Bismaleimide functionalized polypropylene	PP-BMI	Synthesis in "P. Poni" Institute of Macromolecular Chemistry, Iasi, Romania, grafting degree = 6.25×10^{-5}—31×10^{-5} moles BMI/100 g sample PP-BMI (titration method) [62]
Ethylene-propylene rubber grafted with maleic anhydride	EP-MA	Trade name Exxelor VA 1803, supplied by Exxon Chemical. The EP-MA compound 0.7 wt% of MA, 43 wt% ethylene, 57 wt% propylene. Melt flow rate (230°C/10 kg): 22. Density: 0.86 g/cm³. Glass transition temperature: 57°C

chitosan hydrogels had no inhibitory effect on cell growth and the hemolytic action is <1%, which means good blood compatibility.

Xanthan—Lignin hydrogel synthesis: The hydrogels containing xanthan gum and annual fiber crops lignin were prepared in various mixing ratios of the two polymers in the presence of NaOH and epichlorohydrin (EPC). The hydrogels have been washed repeatedly with twice-distilled water until neither lignin nor EPC was detected in the washing waters by UV spectroscopy.

7.2.2.2 Polyolefin Film Processing

Blends containing polypropylene (PP), lignin with/without Irganox 1076, PP-BMI, or EP-MA have been obtained by means of a Haake Rheocord 9000 mixer in the following conditions: mixing temperature of 175°C, mixing time of 15 min, and rotational speed of 60 rpm. The amount of lignin was 5 wt% in respect of the initial amount of PP in the blend, while that of Irganox 1076 was of 0.5 wt%. The components were dried before mixing in a vacuum oven for 24 h at 80°C. The films (0.5 mm thickness) obtained by pressing are relatively homogeneous and compact, colored from light-yellow to brown, and transparent, with a smooth surface.

7.2.3 Investigation Methods

7.2.3.1 Aroma Loading and In Vitro Release Studies

The aroma loading of the hydrogel matrices was carried out by mixing vanillin with dried matrices in powdered form, and then a certain quantity of the appropriate solvent (maximum amount of liquid uptake during swelling) was added and left to swell at room temperature at least for one hour, while the aroma penetrates and/or attaches into matrices. The active ingredient concentration in solution was 18 mg/mL. At the end, the aroma-loaded samples were freeze-dried using a Labconco FreeZone device.

In vitro release studies have been conducted by a standard dissolution setup.[65] The dissolution medium was twice-distilled water. During dissolution testing, the media was maintained at $37 \pm 0.5°C$. Aliquots of the medium of 1 mL were withdrawn periodically at predetermined time intervals and analyzed at λ_{max} value of 228 using a HP 8450A UV–visible spectrophotometer. In order to maintain the solution concentration, the sample is carefully and totally reintroduced in the circuit after analyzing.

The concentrations of the active ingredient were calculated based on calibration curves prior determined for vanillin aroma at specific maximum absorption wavelengths.

A simple, semi-empirical equation using Korsmeyer and Peppas model was used to kinetically analyze the data regarding the aroma release from studied matrices system, which is applied at the initial stages (approximately 60% fractional release).[66]

$$\frac{M_t}{M_\infty} = k_r t^{n_r} \tag{7.1}$$

where

M_t/M_α represents the fraction of the aroma released at time t

M_t and M_∞ are the absolute cumulative amount of aroma released at time t and at infinite time (in this case, maximum release amount in the experimental conditions used at the plateau of the release curves), respectively

k_r is a constant incorporating characteristics of the macromolecular matrix

Aroma n_r is the diffusion exponent, which is indicative of the release mechanism.

In the previous equation, a value of $n_r = 0.5$ indicates a Fickian diffusion mechanism of the aroma from the matrix, while a value $0.5 < n_r < 1$ indicates an anomalous or non-Fickian behavior. When $n_r = 1$, a case II transport mechanism is involved, while $n_r > 1$ indicates a special case II transport mechanism.[67,68]

7.2.3.2 Mechanical Properties

Stress-strain measurements (Young modulus, strength at break and elongation at break characteristics) have been determined by means of an Instron Single Column Systems' tensile testing machine (model 3345) according to SR EN ISO 527:1996; cross-head speed of 10 mm/min.

7.2.3.3 Oxidation Induction Period

Thermo-oxidative stability of the samples was examined by isothermal differential scanning calorimetry (DSC). The oxidation induction period (OIP) was determined at 240°C in air by means of a Pyris Diamod DSC calorimeter (Perkin Elmer USA).

The protection factor (PF) was calculated with the following relation:[69]

$$PF = \frac{t_i(PP_{stab})}{t_i(PP)} \tag{7.2}$$

where $t_i(PP_{stab})$ and $t_i(PP)$ are the oxidation induction period of PP under study sample and control PP, respectively.

For calculation of the antioxidant effectiveness (AEX), the following equation was used:

$$AEX = \frac{PF-1}{X}$$

$$(7.3)$$

where X is the concentration of antioxidant in wt%.

7.2.3.4 Antimicrobial Activity

The inhibition of *E. coli* (SR ISO SR ISO 16649), *L. monocytogenes* (SR EN ISO 11290), *and S. enteritidis* (SR EN ISO 6579) colonies' growth at 37°C for 24 h and 48 h has been followed according to mentioned standard methods at the Veterinary Laboratory, Animal Health Department, Iasi, Romania.

7.3 Results

7.3.1 Vanillin Release from Chitosan Nanocomposite Hydrogels

The most commonly used clay mineral in the preparation of polymer nano-composites is montmorillonite (Mt), which is the major constituent of bentonite. It is well known that the filler anisotropy, i.e., large length to diameter ratio (aspect ratio), promotes the reinforcement. Due to unique structure of Mt, the mineral particle thickness can be only one nanometer although the length and width can be hundreds of nanometers, with a majority of particles after purification in the 200–400 nm range. Bionanocomposites are a promising class of hybrid materials derived from natural and synthetic biodegradable polymers and organic/inorganic fillers.[70]

Chitosan (CS)–montmorillonite (Mt) nanocomposites have a great potential in the biomedical field. Mt addition to quaternized CS enhanced the drug encapsulation and slowed down the bioactive component release.[71]

Biomedical nanocomposites based on Cloisite® 15A and poly(urethane urea) exhibited an increased modulus with increasing Cloisite® 15A content while maintaining the polymer strength and ductility.[72]

Polymer nanocomposites demonstrated good barrier properties due to the tortuous diffusion pathways that small molecules must travel in order to clear the material.[73] This property can be used toward the development of sustained drug release applications. The release kinetics was suggested to be dependent on the aspect ratio and degree of dispersion of the nanoparticle.[74]

Attractions between the negatively charged silicate surfaces and the bioactive principles resulted in slow release rate, while repulsive interactions increased the rate of drug elution.[75] The efficiency of Cloisite® 15A

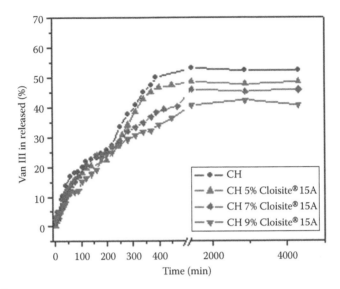

FIGURE 7.1
Vanillin release profiles from CH/Cloisite® 15A hydrogels.

incorporation in CS hydrogels for retarded release of some drugs has been previously reported.[76]

The release profiles of vanillin from glutaraldehyde cross-linked chitosan hydrogels containing different amounts of Cloisite® 15A at 37°C in twice-distilled water are presented in Figure 7.1.

Almost similar release profiles were recorded for the investigated samples during the first 200 min, subsequently being clearly differentiated in accordance with the clay amount introduced. A surprising feature of the release profiles of vanillin was the step-way manner of release, obvious for CS and CS 5 wt% Cloisite® 15A samples. At higher concentration of Cloisite® 15A, the vanillin release occurs slowly in a single step.

The sample containing 9 wt% Cloisite® 15A reaches only a release percent of 40 wt% compared with neat CH hydrogel from which the release is 52 wt%. The obtained results show that clay nanoparticles present in the chitosan nanocomposite network hinder vanillin release, as these samples are suitable for a sustained release.

The kinetic parameters for vanillin release were calculated using Korsmeyer–Peppas equation (Equation 7.1). The values obtained are presented in Table 7.2.

The values obtained for the release parameter n indicate an anomalous transport mechanism for all samples studied, which can be explained by coupling Fickian diffusion with the relaxation of the hydrogel network. In the second step, the release exponent increases in respect with that of the first step and specific rate decreases, so the release is not controlled at low Cloisite® amount.

TABLE 7.2

Kinetic Parameters of Vanillin Release from
CH/Cloisite® 15A Hydrogels

Sample	n	R	$k\ 10^{-3}$ (min^{-n})	R
CH Step 1	0.72	0.992	8.96	0.997
Step 2	0.88	0.99	2.52	0.998
CH 5 wt% Cloisite® 15A Step 1	0.68	0.985	6.85	0.99
Step 2	1.35	0.99	0.17	0.997
CH 7 wt% Cloisite® 15A	0.77	0.997	5.56	0.997
CH 9 wt% Cloisite® 15A	0.71	0.992	5.25	0.99

The clay content in the GA cross-linked chitosan samples and the values of the release rate constant are connected, so the increase in the clay content leads to a decrease in the release rate constant values from 8.96×10^{-3} for neat chitosan sample to 5.25×10^{-3} for the sample containing 9 wt% Cloisite® 15A. The relaxation of the hydrogel network at swelling equilibrium provides a better entrapment of the vanillin inside matrices, while the nanoparticles presence ensures its slower release, these systems being suitable for a sustained release of the natural antimicrobial agent, useful for antimicrobial packaging applications.

7.3.2 Vanillin Release from Xanthan/Lignin Hydrogels

Due to the its characteristics (nano-size particles, amorphous polymer, nontoxic and biodegradable polymer, antibacterial and antioxidant properties), lignin can be considered as a material providing many opportunities in various application fields, such as nanotechnology, medicine and pharmaceutical domains, agricultural, biomaterial, and packaging industries.

Research and development of antimicrobial materials for food applications such as packaging and other food contact surfaces is expected to grow in the next decade with the advent of new polymer materials and antimicrobials.

Thus, studies in materials science have been focused on the development of polymeric materials containing lignin. The incorporation of small quantities of lignin in a polymer matrix in order to stabilize materials toward photo- or thermo-oxidation reactions is a strategy that is based on antioxidant properties of lignin. The cross-linking of xanthan with lignin (GL) in the presence of EPC leads to hydrogels films that present very high swelling degree (q) and also very high swelling rate in aqueous medium. An increase of GL content in the sample led to an important decrease in the maximum swelling degree and the swelling rate of hydrogels. The highest

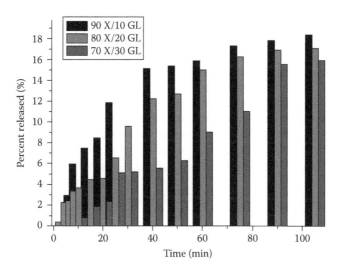

FIGURE 7.2
Release profiles of vanillin from X/GL-based hydrogels with different compositions in twice-distilled water at 37°C.

swelling degree for the xanthan/GL hydrogel was obtained for the composition 95X/5GL hydrogel, and the lowest value was observed for the 60X/30GL hydrogel.[55]

A temporary increase of inter-macromolecular spaces after swelling favors the penetration, fastening and stabilizing the polymer complex—active ingredient.

The release profiles in Figure 7.2 showed that the percentage of vanillin released decreases with increase in the amount of lignin in X/GL hydrogel composition.[56]

Thus, the 70X/30GL hydrogel released approximately 15% vanillin compared with 90 X/10GL compositions, which release about 18% within 100 min, and then the release continues at very slow rates and will be complete following polysaccharide degradation.

The kinetic parameters for vanillin released from X/GL-based hydrogels with various compositions are presented in Table 7.3.

TABLE 7.3

Kinetic Parameters of Vanillin Released from X/GL Hydrogels with Different Compositions

Hydrogels	First-Order Kinetic Model		
	n_r	R	K_r 10^{-3} (min^{-1})
90X/10GL	1	0.97	4.2
80X/20GL	1	0.98	2.23
70X/30GL	1	0.98	1.61

From the obtained release profiles, the diffusion exponent n_r was calculated according to Equation 7.1 (Table 7.3). The release of vanillin in twice-distilled water is described as a case II transport mechanism (zero-order kinetics) for all formulations, and the values of release rate constant, K_r, are decreasing with increasing GL content.

An increase in GL content in the hydrogels leads to a slower rate, and smaller percentage of vanillin is released. Lignins act as antioxidant agents; thus, by using these types of hydrogels, their compatibility and biocompatibility will be improved. By FT-IR spectroscopy, interaction forces of inter- and intramolecular nature between hydrogels and aroma compound were evidenced.

It can be concluded that the vanillin release from X/GL hydrogel matrices can be controlled with respect to both released quantity and release rate of lignin content in hydrogel formulations.

7.3.3 Polyolefin/Lignin Composites

The most common polymer matrices used in food packaging composition were polyolefins (polypropylene and polyethylene).

By blending polypropylene with lignin in various concentrations, polypropylene and lignin films have been obtained with the acceptable mechanical properties (tensile properties and modulus of elasticity).[77,78] By evaluating the mechanical properties of PP/4% lignin films, after an enzymatic treatment with white rot fungi *Phanerochaete chrysosporium*, Mikulàšová and Košíková[79] established a correlation between the decrease in the elongation at break and the amount of lignin fragments released in the medium.

Studies on the antioxidant properties of lignin in the polyolefin matrix were carried out by Pouteau et al.[24] Authors used several types of lignin from wheat (straw and bran) and kraft lignin fractions. By incorporation a low amount lignin (~1 wt.%) in polypropylene matrix, the PP/lignin films with various morphologies have been obtained. The low solubility in polyolefin matrix and high polydispersity of lignin fractions leads to the formation of lignin aggregates in the matrix. Average dimension of the aggregates has been correlated with lignin antioxidant activity. It was found that induction time decreases with increase in average size of aggregates, a fact that indicates a good compatibility between the two partners and controls the antioxidant activity. So, to improve compatibility, and consequently, antioxidant activity, the factors to be considered are low molecular weight and unexpected low OH_T (OH_{aliph} + OH_{ph}) content for these kinds of lignins.[24]

Gregorová et al.[80] reported that the increase in lignin concentration up to 5% in PP has a positive effect on the oxidative stability of the polymer.

TABLE 7.4

Values of the Processing Characteristics for PP/Lignin Blends Studied

Sample	TQ max (N-m)	TQ 1 min (N-m)	TQ 5 min (N-m)	TQ Final (N-m)
PP	64.2	22.5	11.6	10.7
PP/0.5 Irg	66.1	10.8	10.3	9.6
PP/5 AL	54.9	12.7	10.6	10.6
PP/5 AL/0.5 Irg	43.1	11.9	10.2	9.9
PP/5 AEL	48.8	13.0	11.4	11.1
PP/5 AEL/0.5 Irg	40.1	9.9	8.7	8.6
PP/5 PP-BMI/5LS	—	11.3	—	6.1
PP/15 EP-MA/5LS	—	13.4	—	3.2

Some studies established that the lignin acts as a stabilizer during polypropylene processing by using suitable mixing procedures and also it improves the thermal and surface properties of materials from polyolefins.[81,82]

Processing behavior: Cazacu et al.[59] carried out a comparative study on the effect of incorporation of three types of lignins (aspen steam explosion lignin [AEL], alkaline lignin from annual plants [AL], and ammonium sulfonate [LS]) into polypropylene matrix. Blends containing PP/5 wt% lignin with and without a commercial antioxidant agent (0.5 wt% Irganox 1079) by melt processing have been obtained in the film form.

The processing behavior was followed as: a torque-time dependence and some parameters like torque after 1 and 5 min of mixing (TQ_{1min} and TQ_5 min, respectively), maximum torque (TQ_{max}), and the final torque (TQ_{fin}) (Table 7.4).

The results presented in Table 7.4 showed that PP is sensitive to the presence of lignin and the processing parameters significantly decreased due to lignin addition. A low lignin content leads to a lower viscosity in the melt state due to the plasticizing role of lignin and also due to its low molecular weight. Also the presence of the commercial antioxidant Irg 0.5 wt% leads to decrease in the values of processing parameters in comparison with PP processing characteristics.

Mechanical properties: Blending of lignins with polypropylene resulted in materials having slightly improved tensile properties (Figure 7.3). The values of the Young modulus (Figure 7.3a) of the PP films containing 5 wt% AL are higher than value for PP. Lignins used in this study contain the nano-size particle fractions that can be uniformly dispersed in polyolefin matrix during processing. The improved dispersion increases rigidity of the studied materials; all types of lignin used impart an increased stress at break (Figure 7.3b) in respect

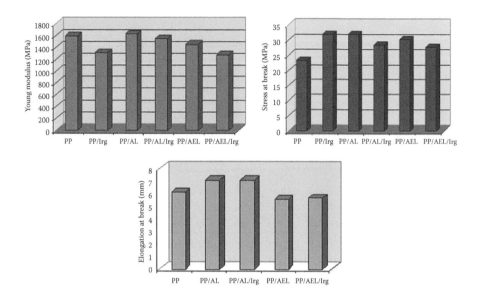

FIGURE 7.3
Tensile properties of PP/lignin blends.

with that of the neat PP. The best mechanical properties have been recorded for the system containing AL.

The values of the elongation at break (Figure 7.3c) have not been much affected by lignin incorporation; a slight increase of this tensile characteristic was registered for the PP/AL and PP/AL/Irg systems.

Antioxidant efficiency: The oxidation induction period (OIP) for the PP/lignin blends determined by isothermal DSC method[80,69] are presented in Figure 7.4 and Table 7.5.

The introduction of lignin in PP matrix leads to a long induction period of oxidation, values being different depending on the lignin type. Both lignins

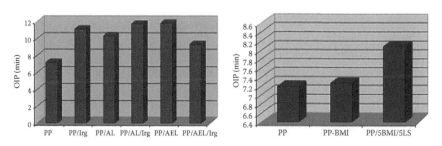

FIGURE 7.4
Values of the oxidation induction period for PP/lignin studied.

TABLE 7.5

Oxidation Induction Period (OIP), Protection Factor
(PF), and Antioxidant Effectiveness (AEX) of the
PP/Lignin Systems Determined by DSC

Sample	OIP (min)	PF	AEX
PP	7.18	—	—
PP/0.5% Irg	11.12	1.55	1.093
PP/5% AL	10.32	1.44	0.087
PP/5% AL/0.5% Irg	11.75	1.64	0.115
PP/5% AEL	9.36	1.30	0.055
PP/5% AEL/0.5% Irg	11.84	1.65	0.129
PP/5 BMI-PP/5 LS	8.08	1.12	0.012

(AL and LS) proved their efficiency as antioxidants. In case of PP/PP-BMI/LS system, the presence of LS lignin enhances the antioxidant activity. PP-BMI acts as a compatibilizing agent and improves the dispersion of lignin within PP matrix, having a positive effect on the antioxidant efficiency.

Lignin addition to PP/0.5 wt% Irg films significantly enhances its antioxidant efficiency. The system Irg/steam explosion lignin (AEL) exhibits the highest antioxidant effectiveness, the oxidation reaction of PP/AEL being initiated after 11.839 min at 240°C due to a synergetic action. On the other hand, Košiková et al.[83] supposed that lignin enhances Irganox 1079 solubility and implicit it influences the increase of Irg diffusion into polymer matrix. The protection factor (PF) and antioxidant effectiveness (AEX) vary in the same way as OIP.

Antimicrobial activity: The antimicrobial activity of PP/L with/without an antioxidant agent has been tested on *E. coli, L. monocytogenes, and S. enteritidis.*

The experimental protocol included following stages: sterilization of PP/L films; contamination with American Type Culture Collection-culture media (ATCC) culture germs of PP/L films; incubation of 24 and 48 h, at 25°C; and identification of the target germs and quantitative evaluation of bacterial colonies' growth on the PP/L films by the countering of bacteria.

From counting of colony forming units (UFC), it was observed that by incorporation of the AEL or AL a total inhibition of the growth of bacteria like *E. coli* and *L. monocytogenes* even after 24 h incubation and a partial inhibition of *S. enteritidis* bacteria (Figure 7.5 and Table 7.6) was achieved.

A different behavior was observed in the case of the incorporation of small amount of LS in PP matrix. LS have a high inhibition capacity of growth of bacteria cultures. The use of a compatibilization agent (ethylene-propylene rubber-g-maleic anhydride, EP-MA)/lignin system leads to increase of the inhibition capacity even in the case of *S. enteritidis* Figure 7.5 and Table 7.6.

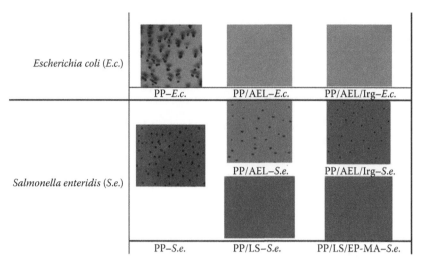

FIGURE 7.5
(See color insert.) Bacteria colonies' growth/inhibition on the PP/lignin films.

TABLE 7.6

Inhibition of Bacteria Colonies' Growth (%) on the PP/LS Films

Bacteria Type	Time of Incubation (h)	PP (%)	PP/LSA (%)	PP/LSA/EP-MA (%)
Listeria	24	32.14	78.57	85.71
monocytogenes ATCC 7644	48	39.29	83.93	100
Salmonella	24	15.63	96.88	100
enteritidis ATCC 13076	48	32.81	100	100

7.4 Conclusion

- Retarded release of vanillin aroma from chitosan nanocomposite hydrogels and xanthan/lignin hydrogels has been obtained.
- Lignin's role as antimicrobial and antioxidant agents in polyolefin composition depends on lignin type and characteristics.
- These materials from renewable resources are promising components in food packaging.

Acknowledgments

The financial support of EUREKA E!4952-BIOPACKAGING "New technologies for obtaining bioactive packaging" and COST FA0904action: Eco-sustainable Food Packaging based on Polymer Nanomaterials (PNFP) is gratefully acknowledged.

References

1. P.C. Srinivasa and R.N. Tharanathan, *Food Rev. Intern.*, 23(1) (2007) 53.
2. H.M.C. deAzeredo, *Food Res. Intern.*, 42(9) (2009) 1240.
3. D.S. Cha and M.S. Chinnan, *Crit. Rev. Food Sci.*, 44 (2004) 223.
4. J.A. Torres, M. Motoki, and M. Karel, *J. Food Process. Pres.*, 9 (1985) 75.
5. G.R. Siragusa and J.S. Dickson, *J. Food Sci.*, 57 (1992) 293.
6. F. Devlieghere, L. Vermeiren, and J. Debevere, *Food Microbiol.*, 21 (2004) 703.
7. A. Arora and G.W. Padua. *J. Food Sci.*, 75(1) (2010) R43.
8. F.S. Kittur, K.R. Kumar, and R.N. Tharanathan, *Z. Lebens. Unters. Forsch. A*, 206 (1998) 44.
9. M.R. Rinaudo and A. Domard, in *Chitin and Chitosan: Sources, Chemistry, Biochemistry, Physical Properties and Applications*, SkjakBrack G. Anthonsen T., and Stanford P. (eds.), Elsevier, New York (1989) p. 71.
10. B.L. Butler, P.J. Vergano, R.F. Testin, J.M. Bunn, and J.L. Wiles, *J. Food Sci.*, 61 (1996) 953–961.
11. P. Stossel and J.L. Leuba, *Phytopathol. Z*, 111 (1984) 82.
12. S. Yingyuad, S. Ruamsin, D. Reekprkhon, S. Douglas, S. Pongamphai, and U. Siripatrawan, *Packaging Technol. Sci.*, 19(3) (2006) 149.
13. A.L. Brody, E.R. Strupinsky, and L.R. Kline, *Active Packaging for Food Applications.* Technomic Publishing, Inc., Lancaster, PA (2001) pp. 218.
14. S. Zivanovic, S. Chi, and A.F. Draughon, *J. Food Sci.*, 70 (2005) M45.
15. Y. Pranoto, S.K. Rakshit, and V.M. Salokhe, *Food Sci. Technol. LEB*, 38 (2005) 859.
16. J. Sangsuwan, N. Rattanapanone, and P. Rachtanapun, *Postharvest Biol. Technol.*, 49 (2008) 403.
17. Y. Zheng, Y. Wu, W. Yang, C. Wang, S. Fu, and X. Shen, *J. Pharm. Sci.*, 9 (2006) 181.
18. F. Li-Fang, H. Wei, C. Yong-Zhen, X. Bai, D. Qing, W. Feng, Q. Min, and C. De-Ying, *Intern. J. Pharm.*, 375 (2009) 8.
19. J. Berger, M. Reist, J.M. Mayer, O. Felt, N.A. Peppas, and R. Gurny, *Eur. J. Pharm. Biopharm.*, 57 (2004) 19.
20. S. Harish, D. Peter Michael, A. Bensely, D. Mohan Lal, and A. Rajadurai, *Mater. Charact.*, 60(1) (2009) 44.
21. N.G. Jústiz-Smith, G.J. Virgo, and V.E. Buchanan, *Mater. Charact.*, 59(9) (2008) 1273.
22. B. Košiková, V. Demianova, and M. Kakurakova, *J. Appl. Polym. Sci.*, 47 (1993) 1065.
23. B. Košiková, K. Miklešová, and V. Demianová, *Eur. Polym. J.*, 29(11) (1993) 1495.

24. C. Pouteau, P. Dole, B. Cathala, L. Averous, and N. Boquillon, *Polym. Degrad. Stab.*, 81 (2003) 9.
25. J. Zemek, B. Košiková, J. Augustin, and D. Joniak, *Folia Microbiol.*, 24 (1979) 483.
26. M. Balat, *Energy Sour Part A Recov Util Environ Effects*, 31 (2009) 516.
27. J.L. Nelson, J.W. Alexander, L. Gianotti, C.L. Chalk, and T. Pyles, *Nutrition*, 10 (1994) 32.
28. M.A. Nada, A.I. El-Diwany, and A.M. Elshafei, *Acta Biotechnol.*, 9 (1989) 295.
29. T. Dizhbite, G. Telyesheva, V. Jurkjane, and U. Viesturs, *Bioresour. Technol.*, 95, (2004) 309.
30. F.J. Lu, L.H. Chu, and R.J. Gau, *Nutr. Cancer*, 30 (1998) 31.
31. C. Boeriu, D. Bravo, R.J.A. Gosselink, and J.E.G. van Dam, *Ind. Crops Prod.*, 20 (2004) 205.
32. V. Ugartondo, M. Mitjans, and M.P. Vinardell, *Ind. Crops Prod.*, 30 (2009) 184.
33. N. Madad, L. Chebil, C. Sanchez, and M. Ghoul, *Rasayan J. Chem.*, 4 (2011) 189.
34. K. Zhou, J. J. Yin, and L. Yu, *Food Chem.*, 95 (2006) 446.
35. L.R. Beuchat and D.A. Golden, *Food Technol.*, 43 (1989) 134.
36. M.B. Hocking, *J. Chem. Educ.*, 74 (1997) 1055.
37. R.S. Ramachandra and G.A. Ravishankar, *J. Sci. Food Agric.*, 80 (2000) 289.
38. A. Tai, T. Sawano, F. Yazama, and H. Ito, *Biochim. Biophys. Acta*, 1810 (2011) 170.
39. D.J. Fitzgerald, M. Stratford, M.J. Gasson, J. Ueckert, A. Bos, and A. Narbad, *J. Appl. Microbiol.*, 97 (2004) 104.
40. J.M. Jay and G.M. Rivers, *J. Food Saf.*, 6 (1984) 129.
41. P. Cerrutti and S.M. Alzamora, *Intern. J. Food Microbiol.*, 29 (1996) 379.
42. B. Matamoros-Leon, A. Argaiz, and A. Lopez-Malo, *J. Food Prot.*, 62 (1999) 540.
43. K.D. Moon, P. Delaquis, P. Toivonen, and K. Stanich, *Food Microbiol.*, 23 (2006) 169.
44. H.P.V. Rupasinghe, J. Boulter-Bitzer, T. Ahn, and J.A. Odumeru, *Food Res. Int.*, 39 (2006) 575.
45. P. Cerrutti, S.M. Alzamora, and S.L. Vidales, *J. Food Sci.*, 62 (1997) 608.
46. V. Penney, G. Henderson, C. Blumb, and P. Johnson-Green, *Innov. Food Sci. Emerg.*, 5 (2004) 369.
47. H.P. Vasantha Rupasinghe, J. Boulter-Bitzer, T. Ahn, and J.A. Odumeru, *Food Res. Int.*, 39 (2006) 575.
48. X. Castañón, A. Argaiz, and A. López-Malo, *Food Sci. Technol. Int.*, 5 (1999) 51.
49. D.J. Fitzgerald, M. Stratford, M.J. Gasson, and A. Narbad, *J. Food Prot.*, 67 (2004) 391.
50. A. López-Malo, S.M. Alzamora, and A. Argaiz, *J. Food Sci.*, 63 (1998) 143.
51. D.J. Fitzgerald, M. Stratford, and A. Narbad, *Intern. J. Food Microbiol.*, 86 (2003) 113.
52. A. López-Malo, S.M. Alzamora, and A. Argaiz, *Food Microbiol.*, 12 (1995) 213.
53. J. Burri, M. Graf, P. Lambelet, and J. Loliger, *J. Sci. Food Agric.*, 48 (1989) 49.
54. C.M.M. Ribeiro, M.L Beirão-da-Costa, and M. Moldão-Martins, Origanum virens L. flavor encapsulation in a spray dried starch matrix, 2nd Mercosur Congress on Chemical Engineering, ENPROMER (2005) Costa Verde, Brazil.
55. I.E. Răschip, C. Vasile, D. Ciolacu, and G. Cazacu, *High Perform. Polym.*, 19(5) (2007) 603.
56. I.E. Răschip, E.G. Hitruc, A.-M. Oprea, M.-C. Popescu, and C. Vasile, *J. Mol. Struct.*, 1003(1–3) (2011) 67.
57. Southern Clay Products, Cloisite® 15A: Typical physical properties bulletin, Product Bulletiin/Cloisite. http://www.scprod.com/product_bulletins/ PB%20Cloisite%2015A.pdf

58. G. Bhat, R.R. Hedge, M.G. Kamath, and B. Deshpande, *J. Eng. Fibers Fabrics*, 3(3) (2008) 22.
59. G. Cazacu, L. Nita, M. Pintilie, and C. Vasile, Physico-chemical characterization of lignin. Size lignin particles determination on the Zetasizer nano, COST FP0901 Meeting, Paris, France, January 25–26, 2011.
60. K. Koda, A.R. Gaspar, L. Yu, and D.S. Argyropoulos, *Holzforschung*, 59 (2005) 612.
61. X. Qiu, Q. Kong, M. Zhou, and D. Yang, *J. Phys. Chem. B*, 114 (2010) 15857.
62. R. Darie, C. Vasile, and M. Kozlowski, *Compatibilization of Complex Polymeric Systems*, LAP Lambert Academic Publishing, Saarbrucken, Germany (2011); ISBN 978-3-8443-1225-6.
63. I.M. El-Sherbiny, R.J. Lins, E.M. Abdel-Bary, and D.R.K. Harding, *Eur. Polym. J.*, 41 (2005) 2584.
64. C.N. Chiaburu, B. Stoica, A. Neamţu, and C. Vasile, *Revista Medico-Chirurgicala a Soc. Med. Nat. Iasi*, 115 (2011) 864.
65. J.M. Oh, C.S. Cho, and H.K. Choi, *J. Appl. Polym. Sci.*, 94 (2004) 327.
66. N.A. Peppas and R.W. Korsmeyer, in *Hydrogels in Medicine and Pharmacy*, Peppas N.A. (ed.), CRC Press, Boca Raton, FL (1986) pp. 109.
67. R.W. Korsmeyer and N.A. Peppas, *J. Control. Release*, 1 (1984) 89.
68. L. Serra, J. Doménechc, and N.A. Peppas, *Biomaterials*, 27(31) (2006) 5440.
69. Z. Cibulková, P. Šimon, P. Lehocký, and J. Balko, *Polym. Degrad. Stab.*, 87 (2005) 479.
70. R.A. Hule, and D.J. Pochan, *MRS Bull.*, 32 (2007) 354.
71. W. Xiaoying, D. Yumin, and L. Jiwen, *Nanotechnology*, 19 (6) (2008) 065707.
72. R. Xu, E. Manias, A.J. Snyder, and J. Runt, *J. Biomed. Mater. Res.*, 64(A) (2003) 114.
73. K. Yano, A. Usuki, and A. Okada, *J. Polym. Sci. A Polym. Chem.*, 35 (2000) 2289.
74. S.H. Cypes, W.M. Saltzman, and E.P. Giannelis, *J. Control. Release*, 90 (2003) 163.
75. W.F. Lee and Y.C. Chen, *J. Appl. Polym. Sci.*, 91 (2004) 2934.
76. A. Cojocariu, L. Profire, M. Aflori, and C. Vasile, *Appl. Clay Sci.*, 54 (2012) 1.
77. C.G. Sánchez and L.A. Espósito Alvarez, *Angew. Makromol. Chem.*, 272 (4758) (1999) 65.
78. B. Košiková and J. Lábaj, *BioResources*, 4(2) (2009) 805.
79. M. Mikulàšová and B. Košíková, *Folia Microbiol.*, 44 (1999) 669.
80. Gregorová, Y. Cibulková, B. Košiková, and P. Šimon, *Polym. Degrad. Stab.*, 89 (2005) 553.
81. G. Cazacu and V. Popa, in *Handbook of Polymer Blends and Composites*, Vasile C. and Kulshreshtha A.K. (eds.), RAPRA Technology Limited, U.K., Vol. 4B (2003) 565pp.
82. G. Cazacu, M. Pascu, L. Profire, A. Kowarski, M. Mihaes, and C. Vasile, *Ind. Crops Prod.*, 20 (2004) 261.
83. B. Košiková, A. Rvajová, and V. Demianová, *Eur. Polym. J.*, 31(10) (1995) 953.

8

Cellulose Nanowhiskers:
*Properties and Applications as Nanofillers
in Nanocomposites with Interest
in Food Biopackaging Applications*

Marta Martínez-Sanz, Amparo López-Rubio, and José María Lagarón

CONTENTS

8.1 Introduction

Cellulose is one of the most abundant biopolymers on earth. It is the major cell-wall component of plants and, therefore, it is often extracted from vegetal resources such as wood, cotton, and linter. Nevertheless, cellulose can also be extracted from some marine animals, such as tunicates, from algae, and is produced by some bacterial species. Regardless of its source, cellulose consists of a linear homopolysaccharide of poly-β(1,4)-D-glucopyranoside chains linked by β-1-4-linkages, forming rod-like crystal units that are organized parallel in a highly ordered manner.[1–3] These crystal units are held together in a paracrystalline matrix and linked along the axis by disordered amorphous domains. This is the basic structural component of cellulose and it is commonly referred to as cellulose "microfibrils," "nanofibrils," "nanocrystals," or "nanowhiskers." Cellulosic nanocrystals are increasingly being used as load-bearing constituents for new and inexpensive biodegradable materials since they present excellent mechanical properties as well as fully degradable and renewable character. Modulus values in excess of 130 GPa have been reported for cellulose nanocrystals, and mechanical strength values are close to 7–10 GPa.[4–10] In addition to these remarkable mechanical properties, cellulose nanocrystals present other interesting properties such as high sound attenuation,[11] high barrier to gases,[12] high specific surface area,[13] and low density (ca. 1500 kg/m^3),[14] which make them attractive for their use in nanocomposite materials.

The significant interest of using materials obtained from renewable sources, on which currently many efforts are being focused, is one of the reasons why the use of cellulosic materials as reinforcement agents in nanocomposites has recently gained so much attention. There is a growing worldwide interest to increment the responsible use of renewable resources in plastic commodity products in order to reduce the waste associated to their use, particularly in packaging applications.[15] The use of biodegradable materials and resources is one strategy to minimize the environmental impact of petroleum-based plastics. Despite the fact that some well-known thermoplastic biodegradable polymers, such as poly(lactic acid) (PLA) and polyhydroxyalkanoates (PHAs), have already become commercially available, they still present some lack of properties, as for example, low thermal resistance, excessive brittleness, and insufficient barrier to oxygen and/or to water as compared to other benchmark polymers such as polyolefins and Poly(ethylene terephthalate) (PET). In this sense, cellulose nanowhiskers are a good choice for reinforcing these biopolymers, provided that a good dispersion within the matrix is achieved. Since, the use of cellulose nanoparticles represents a feasible route to enhance the barrier properties of these materials while preserving their inherently good properties such as transparency and biodegradability.[16–18]

Nevertheless, the application of cellulose nanocrystals as a reinforcement agent for nanocomposites presents some drawbacks. A major disadvantage of cellulosic materials is their hydrophilic character, which makes them, in principle, incompatible with less polar or nonpolar polymers since cellulose nanowhisker agglomeration can take place to a high extent, thus leading to detrimental properties for the nanocomposite material. Another issue is related to the extraction process used to isolate cellulose nanowhiskers from the cellulosic resources. Extraction procedures need to be optimized in order to obtain a material with the desired mechanical and thermal properties and at the same time, they need to be cost efficient and provide extraction yields as high as possible.

In this chapter, we will provide a general overview about cellulosic nanofillers and will discuss the latest applications of these materials in the nanocomposite field, which could be of interest for biopackaging applications, paying special attention to the optimization of the extraction process and to approaches for enhancing the dispersion of this nanofiller within the polymeric matrix.

8.2 Morphology and Structural Characteristics of Cellulose Nanowhiskers

Cellulose nanofillers are typically micron-sized crystal units that show a whisker like rectangular cross-sectional area in the nanoscale with dimensions depending on the cellulose source. As shown in Figure 8.1, poly-β(1,4)-D-glucopyranoside chains are strongly associated through hydrogen bonds and hence packed into the microfibrils in a highly ordered way,

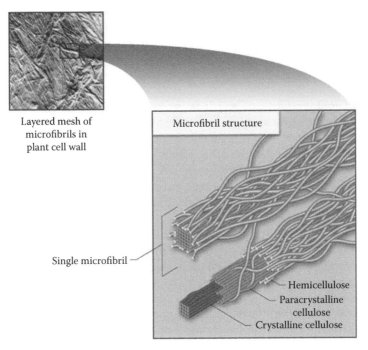

Layered mesh of microfibrils in plant cell wall

Microfibril structure

Single microfibril

Hemicellulose
Paracrystalline cellulose
Crystalline cellulose

FIGURE 8.1
(See color insert.) Schematic representation of the cellulose microfibril.[19] Each microfibril is composed of several elementary crystalline fibrils that consist in the association of β(1,4)-D-glucopyranoside chains linked into crystalline units. Some strain-distorted regions (defects) can be identified along the cellulose microfibril.

thus forming crystalline regions. However, defects that correspond to amorphous domains can be also detected along the microfibrils. The $\beta(1,4)$-D-glucopyranoside chains are aligned parallel along the longitudinal axis of the microfibril and they can be configured slightly different depending on how the chains are twisted around their axis and interact by intramolecular hydrogen bonding with neighboring chains, thereby creating different allomorphs.

Native cellulose presents a crystalline structure of cellulose I allomorph, which at the same time is a composite of two crystalline phases, I_α and I_β, which have been assigned to triclinic and monoclinic unit cells, respectively.[20–23] The ratio of these two allomorphs is different depending on the cellulose source. The cellulose I_α is an unstable phase and tends to transform into the I_β allomorph upon thermal heating,[24,25] especially in polar media such as dilute alkali solutions. Cellulose can also present the cellulose II allomorph, which is also known as regenerated cellulose and is found in this material when precipitated out of alkali solutions. Whereas the structure of cellulose I is made up of parallel chains,[26,27] the crystalline structure of cellulose II is described as antiparallel.[28,29] Cellulose II is the most thermodynamically stable form of cellulose,[28] although cellulose extracted from natural resources consists of cellulose I.

As stated previously, cellulose can be extracted from a wide variety of vegetal resources and from marine animals such as tunicates, and it can also be synthesized by some bacterial species. Plant cellulose structure and extraction process have been extensively studied and it is the most widely used source for the extraction of cellulose fibrils. Cellulose is found in the walls of vegetal resources and it is usually associated to other materials such as hemicellulose and lignin. Therefore, a first purification step is necessary in order to remove those other components and to obtain purified cellulose microfibers. On the other hand, the case of cellulose synthesized by bacteria is of special interest since it provides a method for the production of highly crystalline nearly pure cellulose. In a culture medium rich in polysaccharides, some bacterial species, such as *Gluconacetobacter xylinus*, are able to produce a layer of bacterial cellulose (BC) in the liquid/air interface. This highly hydrated pellicle consists of a random assembly of ribbon-shaped fibrils, less than 100 nm wide, which are composed of a bundle of nanofibrils.[30] Although plant-derived cellulose (PC) and BC have the same chemical structure, they have different structural organization and mechanical properties. BC shows a finer web-like network structure, higher water holding capacity, and higher crystallinity.[31,32] Furthermore, BC is practically pure cellulose, and, therefore, the purification step that is required in the case of PC is not necessary in the case of BC. The influence of the cellulose source for the extraction of cellulose nanowhiskers on the aspect ratio, crystallinity, and thermostability of the attained material will be further discussed in the next two sections.

8.3 Extraction Process of Cellulose Nanowhiskers: Optimization of BCNW Nanofabrication by Acid Hydrolysis

Cellulosic materials may be subjected to either mechanical disintegration (mechanical shearing at high pressure) for the production of microfibrillated cellulose (MFC) or to an acid hydrolysis treatment for the preparation of cellulose nanowhiskers (CNWs). For their application as nanofillers, cellulosic materials are usually subjected to hydrolysis with strong acids such as sulfuric acid or hydrochloric acid, which produce a preferential digestion of the amorphous domains of the material and cleavage of the nanofibril bundles,[33] therefore breaking down the hierarchical structure of the material into crystalline nanofibers or nanocrystals, usually referred to as cellulose nanowhiskers. The morphology of the obtained CNWs depends on the cellulose source and the hydrolysis conditions. While CNWs extracted from vegetal resources such as cotton or wood typically have a length of 100–300 nm and width of 5–20 nm,[34–36] CNWs obtained from tunicin and BC may have several micrometers in length and a width of 5–50 nm.[37–39]

Regarding the hydrolysis conditions, the acid concentration, cellulose/acid ratio, temperature, and hydrolysis time are factors that determine the CNWs' morphology. The CNWs' aspect ratio (L/D) is a crucial parameter that has a remarkable influence on the reinforcing capacity of the nanofiller when incorporating it into a polymeric matrix.[40] Materials with aspect ratios higher than 30, such as tunicin whiskers (L/d ~67), have been reported to provide a considerably higher reinforcement effect as compared to nanofillers having lower aspect ratios, such as Avicel whiskers (L/d ~10).[41] Nevertheless, it has also been reported that for aspect ratios larger than 100, the Young's modulus reaches a plateau corresponding to the maximum point of reinforcement.[40] Therefore, the acid hydrolysis conditions must be carefully studied and controlled in order to obtain a material with the desired morphology.

The most widely used procedure for the extraction of CNWs consists of sulfuric acid treatment followed by filtration or centrifugation. Sulfuric acid hydrolysis leads to stable aqueous suspensions of cellulose nanocrystals that are negatively charged and, thus, do not tend to aggregate. During the hydrolysis process, esterification of the surface hydroxyl groups from cellulose takes place, and, as a consequence, sulfate groups are introduced.[33] Despite the advantage of obtaining stable suspensions, the presence of sulfate groups in the outer surface of the material has been proven to strongly decrease the thermal stability of the material,[42] which is also a key factor when intending to use CNWs as a nanoreinforcement.

In the case of CNWs extracted from vegetal microfibrillated cellulose, the extraction procedure with sulfuric acid has already been standardized and applied in several works.[43–45] Typical hydrolysis conditions are cellulose/acid ratio of 10 g/100 mL, sulfuric acid concentration of 9.1 M, hydrolysis temperature of 44°C, and hydrolysis time of 130 min.[43–45] Higher digestion times have been observed to lead to carbonization and darkening of the product.[45] However, until recently, no extraction methods were standardized for the extraction of cellulose nanowhiskers from different sources such as BC.

Common CNW extraction methods involve centrifugation after hydrolysis with the purpose of removing the acid and the degraded material. After several centrifugation cycles, CNWs are usually obtained from the turbid liquid supernatant, while bigger cellulosic material fractions and some impurities remain in the solid precipitate. The supernatant, which usually presents a pH close to 3–3.5, is then neutralized with sodium hydroxide and subsequently subjected to dialysis. These extraction processes are usually associated to low yields, which represent a drawback of the acid hydrolysis digestion of native cellulose.

Taking into account that in the case of BC there is no hemicellulose or lignin to remove, previous studies proposed an extraction method in which bacterial cellulose nanowhiskers (BCNWs) are obtained in the centrifugation precipitate instead of the supernatant, and, thus, the yield can be as high as 89% based on the dry weight of BC[46,47] versus yields around 1%–5% when the whiskers are obtained from the liquid supernatant. In contrast with this great advantage, the highly crystalline network structure of BC requires strong hydrolysis conditions in order to break down the morphology of fibril bundles and individual nanofibrils cannot be yielded without partial carbonization and degradation of the material.[46]

In a previous study, it was found that BCNWs with a crystallinity index of ca. 86% were obtained after applying a relatively strong sulfuric acid hydrolysis. Nevertheless, the thermostability of the material was significantly diminished with respect to the untreated BC and BCNWs started to degrade at approximately 100°C,[47] which is far below the typical processing temperatures for processing most thermoplastics. Therefore, it was necessary to develop a method for the extraction of BCNWs that assured a high crystallinity but also a good thermostability and a proper morphology, which makes BCNWs suitable for nanocomposite applications. Martinez-Sanz et al. studied the effects of sulfuric acid hydrolysis time and posttreatments, such as neutralization and dialysis, on the morphology, crystallinity, and thermostability of BCNWs for the first time.[48]

The morphology of BCNWs was examined by Transmission Electron Microscopy (TEM), showing a decrease in the nanowhiskers' length when increasing hydrolysis time as expected, whereas neutralization or dialysis did not significantly affect the aspect ratio of the material (see Figure 8.2). The x-ray diffraction (XRD) patterns of the different samples showed a crystalline structure characteristic of the cellulose I allomorph. From the calculated crystallinity indexes, it was deduced that long hydrolysis times, such as 48 h, are required when intending to digest a significant fraction of amorphous material and thus obtaining a significant increase in crystallinity by comparison with the native BC.

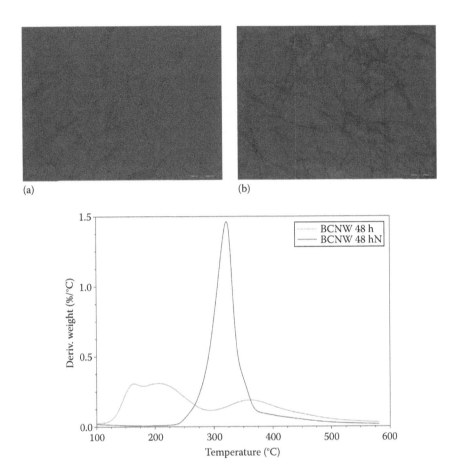

FIGURE 8.2
TEM micrographs and corresponding Derivative thermogravimetric (DTG) curves of BCNWs
subjected to sulfuric acid hydrolysis during 48 h before (a) and after neutralizing the material (b).

Nevertheless, as a consequence of this extensive acid hydrolysis treatment, the thermal stability of the material is significantly decreased, making
it unsuitable for most melt-compoundable polymer-based nanocomposite
applications. On the other hand, as observed in Figure 8.2, neutralization
led to a remarkable increase in the BCNWs' thermal stability, as determined
by thermogravimetric analyses (TGA). Furthermore, it was found out that
dialysis applied after neutralization did not present any additional improvement on the BCNWs' properties.

As a conclusion, an optimized method for the extraction of BCNW was
developed, consisting in relatively long acid hydrolysis times (longer
than 48 h), which allowed the disruption and digestion of a significant
amount of amorphous domains, followed by neutralization. This method
gave rise to the production of highly crystalline BCNW, with a high aspect
ratio and a relatively good thermostability.

8.4 Plant Cellulose Nanowhiskers vs. Bacterial Cellulose Nanowhiskers

As previously mentioned in Section 8.3, the morphology of cellulose nanowhiskers is influenced by the source from which they have been extracted. In addition, the crystallinity index of the material differs depending on its origin. The XRD patterns of BC, plant cellulose microfibers (CMF), and the corresponding nanowhiskers obtained after acid treatment are shown in Figure 8.3. In the case of BC, three major diffraction peaks are observed at

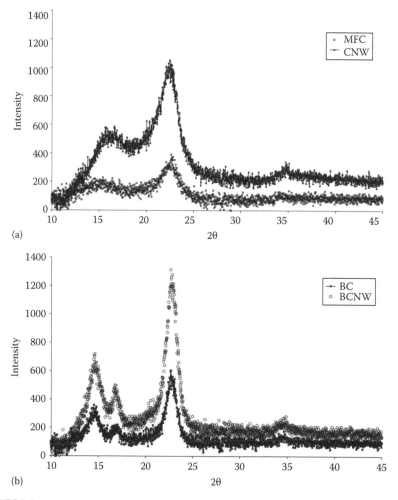

(a)

(b)

FIGURE 8.3
XRD patterns of plant cellulose microfibers (MFC) and the extracted CNWs (a) and BC and BCNWs (b).

14.5°, 16.4°, and 22.5° 2θ, which, according to the literature,[49] are ascribed to the cellulose I crystallographic planes 101, 101̄, and 002, respectively. On the other hand, the diffraction patterns corresponding to the plant-derived material correspond to those previously reported for plant derived cellulose,[50,51] showing one major diffraction peak located at 22° 2θ and a shoulder in the region 14°–17° 2θ. Crystallinity indexes determined from the XRD data are 41.4% and 60.2% for CMF and CNW, respectively, and 79.1% and 95.3% for BC and BCNW, respectively. As expected, the applied acid hydrolysis produced an increase of 16%–19% in the crystallinity index of both materials. It is worth noting that the crystallinity index of BC is noticeably higher than that of the plant-derived cellulose microfibers, hence highlighting the convenience of BC when aiming to obtain highly crystalline whiskers.

Thermal stability of cellulose nanowhiskers is also determined by the extraction resource, just as illustrated by the TGA analyses shown in Figure 8.4. As observed, BCNW present higher thermal stability than CNW. Whereas degradation starts at 186°C for the CNW, the onset temperature for BCNW is 226°C. This different behavior is probably a result of the higher length and aspect ratio of BCNW. Both materials present a degradation profile that makes them suitable for processing through melt compounding with a wide variety of biopolymers, although BCNWs permit a broader range of processing temperatures for the production of nanocomposite films.

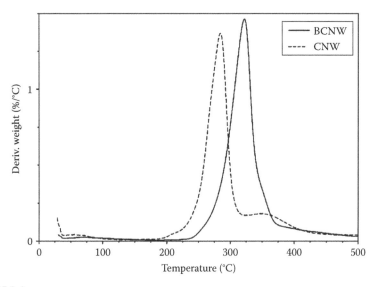

FIGURE 8.4
DTG curves of plant cellulose and bacterial cellulose nanowhiskers (CNW and BCNW, respectively).

8.5 Cellulose-Reinforced Nanocomposites

For most biodegradable polymers, it is required to improve some of their properties, such as gas barrier characteristics and mechanical properties, so that they can compete with greater advantage with petroleum-based materials. Nanoreinforcement of these biodegradable polymers to prepare nanocomposites has already been proven to be an effective way to enhance these properties concurrently. In this sense, cellulose nanowhiskers have a great potential for their use as nanofiller in bionanocomposites. While the use of cellulose nanowhiskers as the reinforcing agent presents great advantages such as renewable nature, biodegradability, low density, wide availability of sources, low energy consumption, relatively reactive surface, and relatively easy processability due to their nanoabrasive nature, which allows high filling levels,[41,52] it also has some major disadvantages, for instance, difficult compatibilization of highly hydrophilic CNWs with nonpolar matrices, moisture absorption, and limitation of processing temperature.

Reinforcement of biopolymers with dispersed cellulose nanowhiskers (CNW) has been reported to increase mechanical properties, improving the thermal stability of the materials.[14,15] However, until recently, very little was known about the effect of the CNW on the barrier properties of nanocomposites.

In the following sections, we will discuss the latest developments of our group related to the production of cellulose nanocomposites through proprietary (patent pending) processes such as via electrospinning, strategies for improving the compatibility and dispersion of cellulose nanowhiskers within different polymeric matrices, and the investigation of the effect of the addition of cellulose nanowhiskers on the barrier properties of polymeric matrices.

8.5.1 Casting

The most widely used method for the preparation of cellulose nanocomposites is the solvent casting method, where CNWs are mixed with a polymer solution and cast. The film is then formed after evaporation of the solvent. This technique can be used either with water-soluble polymers, in which case, never-dried CNW suspensions are mixed with the polymeric solution, or with non-hydrosoluble systems, in which case, CNWs may be subjected to solvent exchange procedures with the purpose of maintaining the CNWs in their non-agglomerated wet state in the organic solvent, thereby omitting the drying of the nanomaterial.

As an example of hydrosoluble system, we will discuss the effects of the addition of CNWs into κ/ι-hybrid carrageenan,[53] while for non-hydrosoluble systems, biocomposites of thermoplastic PLA[45,54] will be analyzed.

8.5.1.1 Nanobiocomposites of κ/ι-Hybrid Carrageenan and Cellulose Nanowhiskers

Carrageenans are structural polysaccharides from seaweed, which have been extensively used in foods, cosmetics, and pharmaceuticals.[55] As a polysaccharide, their films present high moisture permeability and low oxygen and lipid permeability at lower relative humidities and compromised barrier and mechanical properties at high relative humidities.[56] With the aim of improving the properties of these materials, especially the water resistance, both CNWs and the vegetal cellulose microfibers from where they were extracted, with concentrations ranging from 1 to 5 wt.-%, were incorporated into carrageenan by solution casting.

Morphological characterization showed a good dispersion of CNWs within the matrix for relatively low filler contents; however, for loadings higher than 3 wt.-%, CNW tended to agglomerate due to hydrogen bonding–induced self-association. When the plasticizer glycerol was added to the biopolymer, it was seen to form large segregated domains. Whereas cellulose microfibers were not homogeneously dispersed and they segregated to the matrix fraction, when cellulose nanowhiskers were used the glycerol domains became smaller and more homogeneously dispersed across the matrix.[53]

Figure 8.5 summarizes the results of water permeability measurements of carrageenan and its biocomposites with both CNWs and cellulose microfibers for various filler contents with and without glycerol. As deduced from Figure 8.5a, for the films without glycerol reductions of water vapor permeability of ca. 68%, 70%, and 58% were obtained in films containing 1, 3, and 5 wt.-% CNWs, respectively, whereas decreases of ca. 40%, 56%, and 8% were observed with the addition of 1, 3, and 5 wt.-% cellulose microfibers. Therefore, the highest permeability drop took place for a 3 wt.-% filler loading, while higher contents led to increased permeability as a consequence of the nanofiller agglomeration. The results also point out the higher efficiency of CNWs in reducing the water permeability as compared to the microfibers. When glycerol was added to the system (see Figure 8.5b), reductions of ca. 50%, 60%, and 63% were attained for 1, 3, and 5 wt.-% CNW loadings, respectively. In contrast, the incorporation of microfibers resulted in increased permeability due to most likely segregation and agglomeration of the microfibers outside glycerol domains.

Evaluation of water solubility by measuring water uptake at several relative humidities revealed that the permeability drop was a consequence of a strong reduction in water uptake rather than a diffusion-driven tortuosity effect. CNWs reduced the water sorption more efficiently than microfibers due to both their higher crystallinity and the higher dispersion within the matrix due to the nanosize.

It was proven therefore that it is possible to enhance the water barrier and resistance of carrageenan by the incorporation of CNWs.

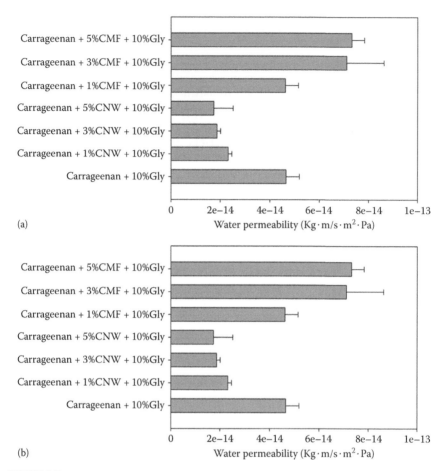

FIGURE 8.5

Water permeability of carrageenan and its nanocomposites containing 1, 3, and 5 wt.-% of CNW and of cellulose microfibers (CMF) (a). Water permeability of carrageenan with 10 wt.-% glycerol and its nanocomposites containing 1, 3, and 5 wt.-% of CNW and of cellulose microfibers (CMF) (b). (From Sanchez-Garcia, M.D. et al., *J. Agric. Food Chem.*, 58, 12847, 2010. With permission.)

8.5.1.2 PLA Nanocomposites Reinforced with Cellulose Nanowhiskers

As previously commented, one of the main drawbacks of thermoplastic bio-polymers such as PLA is the insufficient barrier to oxygen and/or to water as compared to other packaging polymers such as PET. The incorporation of CNWs prepared from highly purified alpha-cellulose microfibers into a PLA matrix was evaluated as a strategy for enhancing the barrier properties. Nanocomposites were prepared by the addition of CNWs with contents ranging from 1 to 5 wt.-% by a solvent casting method. Both freeze-dried and chloroform solvent–exchanged CNWs were incorporated into the PLA matrix.[45]

Morphological characterization of the prepared films by means of TEM indicated that a higher dispersion was achieved when using freeze-dried

CNWs instead of those subjected to solvent exchange with chloroform. This result is contrary to what was expected and can be explained by the fact that some water may still be retained during the solvent exchange procedure and/or agglomeration may take place during the centrifugation process. An additional sonication or homogenization step after the solvent exchange process might prevent CNWs agglomeration.

Thermal properties of the nanocomposites were evaluated though DSC analyses. It was observed that melting point and crystallinity increased with the CNW content, hence suggesting a filler-induced crystallinity development. It was previously reported that CNWs induced crystal nucleation in Polyhydroxybutyrate-co-hydroxyvalerate (PHBV) nanocomposites[43] and it was explained by the assumption that CNWs act as nucleating agents, which enhance the crystallization rate of the matrix molecules.[57] In contrast, the opposite effect has also been reported for polyvinyl alcohol (PVA) copolymers[58] and PLA[59] and it was attributed to positive interactions between the cellulosic surface and the polymeric matrix, thus restricting the capability of the matrix chains to grow bigger crystalline domains.

Regarding the mass transport properties of the nanocomposites, it was concluded that higher reductions in both water and oxygen permeability were attained when using freeze-dried CNWs instead of the ones subjected to solvent exchange, in accordance with the higher dispersion observed for the freeze-dried material. As shown in Figure 8.6, nanocomposites containing between 2 and 3 wt.-% freeze-dried CNWs exhibit the highest water and oxygen barrier, reducing the water permeability by up to 82% and the oxygen permeability by up to 90% with 3 wt.-% of nanofiller. The presence of highly crystalline cellulose nanoshields, PLA crystallinity development, and sorbed moisture filling the free volume were put forward as the most likely factors behind this behavior. From a previous work,[54] the water barrier properties of PLA biocomposites containing cellulose microfibers were seen to be only reduced (by ca. 10%) for a loading of 1 wt.-% of the filler. Thus, CNWs were seen to enhance more efficiently the barrier properties of PLA as compared to cellulose microfibers.

Moisture sorption was on the other hand responsible for a filler-induced plasticization effect, which, together with the fact that filler loadings were below the percolation threshold, resulted in lower mechanical performance of the nanocomposites as compared to neat PLA.

In conclusion, this work shows the adequacy of CNWs for significantly improving the barrier properties to gases and vapors of PLA, therefore allowing the production of fully renewable biocomposites with interest in biopackaging applications.

8.5.2 Electrospinning

Electrospinning is a versatile method to produce continuous polymer fibers with diameters in the sub-micron range through the action of an external

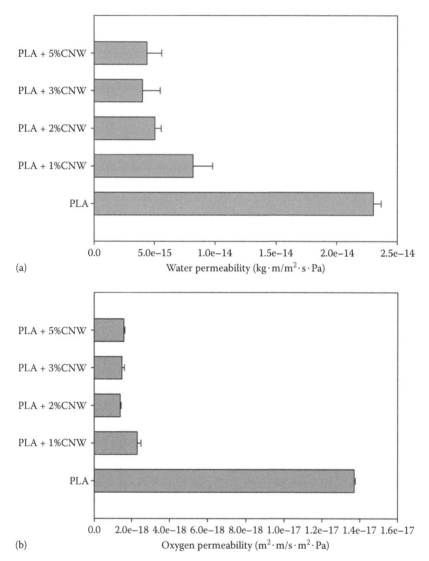

FIGURE 8.6
Mass transport properties of PLA and their nanocomposites with 1, 2, 3, and 5 wt.-% freeze-dried CNW. (a) Water permeability and (b) oxygen permeability. (From Sanchez-Garcia, M.D. and Lagaron, J.M., *Cellulose*, 17, 987, 2010. With permission.)

electric field applied between two electrodes and imposed on a polymer solution or melt. When the electrical force at the interface of the polymer solution exceeds the surface tension, a charged jet is formed. The jet initially extends in a straight line, then undergoes a vigorous whipping motion caused by the electrohydrodynamic instabilities. As the solvent in the jet solution evaporates, the polymer is collected onto a grounded substrate to form a nonwoven mat with significant large surface-to-volume ratio.[60] Since recently, this technique has received substantial attention for the fabrication of polymer nanofibers in a wide range of applications that demand high-performance fibers. Electrospinning has recently been applied to obtain cellulose fibers[60,61] as well as hybrid fibers reinforced by cellulose whiskers[46,62]. In our group, electrospinning has been recently applied for the production of hybrid fibers reinforced with BCNW. As the matrix material, both polar polymers, such as ethylene-vinyl alcohol copolymers (EVOH),[47,63] and non-polar polymers, such as poly(methyl methacrylate) (PMMA)[46] and PLA,[64] were used. These hybrid fibers were seen to incorporate dispersed BCNW, presenting interesting properties and also providing a method for incorporating well-dispersed nanowhiskers into polymeric materials by means of techniques, for example, melt compounding, which we will discuss in the next section.

8.5.2.1 Electrospun EVOH Fibers Reinforced with BCNW

As previously explained in this chapter, some bacterial species are able to synthesize cellulose. The most efficient production is carried out by the bacterial species *Gluconacetobacter xylinus* in a rich saccharide medium under static condition at around 28°C–30°C.[31] We obtained BC pellicles by following the procedure described elsewhere,[47] and after cleaning with NaOH and cutting the pellicles into pieces, a material like the one shown in Figure 8.7 was attained. For the production of BCNW, this material was subjected to sulfuric acid hydrolysis. In this case, the attained BCNW were not neutralized, and therefore, just as discussed in Section 8.3, they presented a relatively

FIGURE 8.7
Visual aspect of BC mats after cleaning with water and NaOH solution.

low thermal stability. On the other hand, the acid digestion produced an increase in the crystallinity of the material from 73% (BC) to 82% (BCNW), as estimated by x-ray diffraction.

Hybrid BCNW and EVOH fibers were generated having a more uniform morphology than the pure polymer electrospun fibers. A method was developed for improving the incorporation of BCNW into the EVOH electrospun fibers, consisting of the addition of the BCNW in the form of a centrifuged precipitate, versus the most conventionally employed freeze-dried nanowhiskers. The degree of nanofiller incorporation into the fibers, estimated by means of transmission Fourier transform infrared spectroscopy (FT-IR) was higher when using the partially hydrated material, whereas the strong network created by hydrogen bonding by subjecting the material to freeze-drying led to agglomeration and low incorporation of BCNW into the fibers. Sonication was found to be efficient for improving the incorporation of freeze-dried BCNW, and it was also seen to enhance interfacial interaction but to reduce the incorporation of the filler in the matrix in the case of the centrifuged material.

By applying the optimized method, it was possible to incorporate BCNWs concentrations up to ca. 24 wt.-%, although a complete incorporation of the nanofiller into the fibers was only achieved with solutions containing up to 20 wt.-% of the filler. For all the electrospun materials, morphologies containing some beads were obtained, and it was confirmed by polarized light optical microscopy that these fibers presented a highly homogeneous dispersion of highly crystalline BCNW, since fibers of pure EVOH did not show birefringence under crossed polarizers.

Differential scanning calorimetry (DSC) analyses suggested that the incorporation of the nanofiller reduced the crystallinity of the as-obtained EVOH fibers and produced an increase in the glass transition temperature of these during the second heating run.

Thermogravimetric analyses showed that even though EVOH protects the nanowhiskers from thermal degradation, the electrospun hybrid fibers present a relatively lower thermal stability than the pure EVOH fibers. FT-IR analyses of the samples subjected to different thermal treatments confirmed that the stiffening effect observed by DSC only occurs after melting of the EVOH phase and is cooperative with a partial acid chemical development in the BCNWs, which promotes strong chemical interactions between the polymeric matrix and the nanofiller. Therefore, there is a compromise between the stiffening effect induced by heating of the material and the thermal stability of the fibers.

These results confirmed the suitability of the electrospinning technique as a way to incorporate BCNW into a polymeric matrix and at the same time pointed out the need for an optimized hydrolysis procedure, which was discussed in Section 8.3, with the aim of obtaining a more thermally stable material that can be processed with EVOH by melt compounding without suffering thermal degradation.

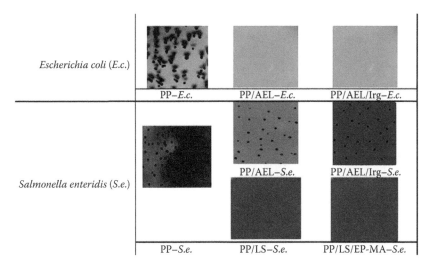

FIGURE 7.5
Bacteria colonies' growth/inhibition on the PP/lignin films.

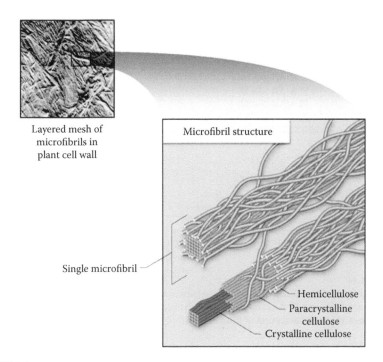

FIGURE 8.1
Schematic representation of the cellulose microfibril.[19] Each microfibril is composed of several elementary crystalline fibrils that consist in the association of β(1,4)-D-glucopyranoside chains linked into crystalline units. Some strain-distorted regions (defects) can be identified along the cellulose microfibril.

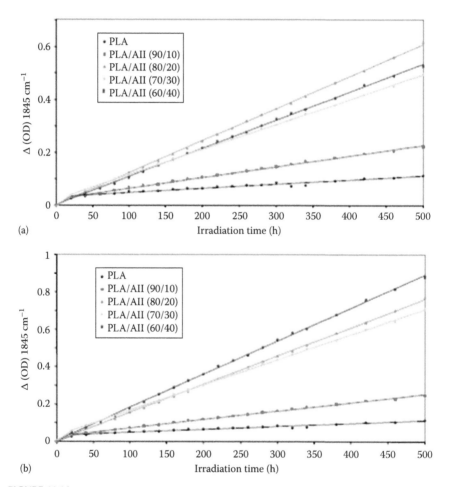

FIGURE 11.14

Evolution of anhydride absorption band intensity at $\lambda_{max} = 1845\,cm^{-1}$ as a function of UV exposure time for PLA and PLA composites with different filler ratios. (a) Without correction of filler amount. (b) With correction of filler amount.

8.5.2.2 Electrospinning of Anisotropic Biohybrid Fiber Yarns Containing BCNW

The mechanical properties of nonwoven materials generated by electrospinning may be enhanced by cross-linking or annealing; however, this leads to an undesired reduced interactive surface area of the fibers, and intrinsic reinforcement of the fiber cores is preferred.[65–67] It has been suggested that elongated nanoparticles can act as reinforcement agents and electrospinning was demonstrated to be efficient for aligning rod-like particles within fibers, parallel to the main fiber axis.[68–71] The anisotropic particle arrangement may allow for unidirectional stress transfer to the elongated particles provided that good interfacial adhesion between the polymer and the particles is achieved and that uniform particle distribution and uniform fibers are attained.[72,73]

The electrospinning technique has been applied for the production of uniform PMMA fibers incorporating BCNWs.[46] A solvent exchange procedure was used in this case in order to obtain suspensions of BCNWs in an organic solvent suitable for the polymeric matrix, which in this case was DMF/THF. It was possible to incorporate BCNWs, which were aligned parallel with the axis of the electrospun fibers, in concentrations as high as 20 wt.-% BCNW.

BCNWs were seen homogeneously distributed over the fracture surface of the hybrid fibers without signs of outer surface protrusion with loadings up to 7 wt.-% BCNW. However, higher BCNW concentrations led to agglomeration, which was related to thermodynamic instability of the electrospinning solutions resulting in BCNW phase separation before the nanofiller was rapidly sealed inside the PMMA matrix at solidification. Both TGA and DSC results suggested that the dispersion was efficient only at relatively low BCNW concentrations. TGA results showed that at BCNW loadings of 7 wt.-% and above, the onset of degradation shifted to lower temperatures as compared to the pure PMMA, suggesting that the cellulose formed domains of sufficient size to approach the degradation behavior found in the absence of the polymeric matrix. Additionally, the T_g determined from DSC analyses was seen to slightly increase when incorporating BCNW up to 3.5 wt.-% whereas higher fractions of BCNW showed a gradual decrease in the T_g.

The cellulose hybrid fibers were then aligned into anisotropic nanocomposite yarns by running the fibers over a hollow spool rotating at 15 rpm/min after the fibers were collected from a water surface (see Figure 8.8a). The obtained yarn of fibers was not a perfectly aligned continuous system and, as observed in Figure 8.8b, some kinks could be detected after examination of the material by Scanning Electron Microscopy (SEM). Nevertheless, the strategy developed in this work appears to be a useful method for aligning fibers on a large scale and it may allow for new mechanically robust nonwoven fiber systems, or be used as implemented on existing electrospun formulations that are lacking mechanical integrity.

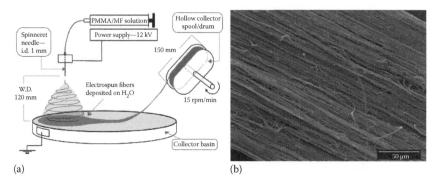

(a)   (b)

FIGURE 8.8

(a) Electrospinning setup with a spinneret–water surface working distance of 120 mm and a spool running at 15 rpm/min. The spool (hollow) was designed to collect fibers with minimal spool-fiber yarn contact. (b) Aligned electrospun PMMA-BCNW hybrid fibers containing 7 wt.-% BCNW. (From Olsson, R.T. et al., *Macromolecules*, 43, 4201, 2010. With permission.)

8.5.2.3 Electrospun PLA Fibers Reinforced with BCNW

In addition to relatively hydrophilic polymers such as EVOH, it has been demonstrated in a recent work that it is possible to obtain electrospun fibers consisting of a hydrophobic matrix such as PLA and incorporating well-dispersed BCNW in concentrations up to 15 wt.-%, presenting a good thermostability and stiffer behavior than pure PLA fibers.

In first place, the adequacy of three different solvents, such as 1,1,1,3,3,3-Hexafluoro-2-propanol (HFP), acetone–chloroform, and chloroform (incorporating PEG), was evaluated and both pure and hybrid fibers containing 5 wt.-% BCNW were generated. Although the morphology of fibers was influenced by the electrospinning solvent, TEM analyses showed that partially hydrated BCNW were well dispersed along the fibers for the three solvent systems. However, once again the convenience of manipulating partially hydrated BCNW versus the freeze-dried product was proven. When using freeze-dried BCNW, the fibers presented an heterogeneous morphology with beads and BCNW appeared as small aggregates due to the strong intramolecular bonding promoted by the freeze-drying process, which hindered the stretching of the nanowhiskers.

The incorporation of PEG into the fibers was required when chloroform was the solvent, but it was shown to be detrimental for the thermal properties of the fibers. On the one hand, it resulted in a significant reduction of the T_g due to a plasticization effect, and on the other hand, they showed decreased thermal stability when incorporating BCNW due to a weak level of interaction between the matrix and the nanofiller.

By selecting the optimum solvents, i.e., HFP and acetone–chloroform, fibers with BCNW loadings up to 15 wt.-% were produced in the next stage of the study. It was observed that the viscosity of the solutions markedly increased with the addition of BCNW and, hence, solid concentrations lower than in the previous stage were also used, facilitating the electrospinning process and also resulting in a higher degree of BCNW incorporation.

The solvent HFP was found to be more compatible with the BCNW, hence resulting in stiffer and more thermally stable fibers as compared to the ones obtained from acetone–chloroform solutions. An increase in the T_g of fibers attained from HFP solutions, detected by DSC, suggested that strong interactions between the PLA and the BCNW were developed in this case. Finally, it was observed that the thermal stability of the HFP fibers increased by incorporating BCNW, whereas it decreased in the case of the acetone–chloroform fibers due to poor adhesion between the matrix and the nanofiller. In any case, the attained biohybrid fibers could be subjected to temperatures typical for polymer processing techniques, and, therefore, it allows for new processing routes in the production of PLA reinforced with cellulose nanowhiskers.

8.5.3 Melt Compounding

Solution casting has been the most widely used technique for the incorporation of CNW. Nevertheless, very few reports exist on the production of nanocomposites reinforced with CNWs via industrial directed techniques, such as melt extrusion methods. When aiming to develop nanocomposite materials through melt compounding, one of the main issues is to achieve a good dispersion of the nanofiller within the matrix. An attempt to prepare nanocomposites of PLA reinforced with CNW by pumping a suspension of CNW in DMAc/LiCl into the polymer melt during the extrusion process was reported.[74] Nevertheless, aggregation and thermal degradation took place to a certain extent. Subsequently, the possibility of using PVA as compatibilizer was investigated, but a bad dispersion in the PLA matrix was observed.[75]

As previously discussed in Section 8.4.2, the electrospinning technique provides a method for incorporating well-dispersed CNWs. Therefore, a procedure in which electrospinning was used as a vehicle for the incorporation of highly dispersed BCNW into a polymeric matrix by melt compounding was developed.[76] The adequacy of the electrospinning technique, as well as a novel solution-precipitation method, was evaluated as a way to obtain nanocomposites of EVOH with highly dispersed CNWs through melt mixing. Both plant CNWs and BCNWs were used as the nanofiller.

The method for incorporating the nanowhiskers prior to the melt compounding step was found to be a key parameter that determined the morphology of the obtained nanocomposites. Except for the case of direct melt mixing of the polymer with the freeze-dried nanomaterial, which resulted in agglomeration of the filler, a general morphology of homogeneously dispersed nanowhiskers was observed in the rest of the nanocomposite materials by cryo-SEM and TEM. For a fixed concentration of 2 wt.-% BCNW, the optimum dispersion was achieved when applying the electrospinning technique and the precipitation method for preincorporating partially hydrated and freeze-dried BCNW, respectively, into EVOH. When incorporating the nanofiller by means of the precipitation method, the optimal loading in terms of a good dispersion was 3 wt.-% for BCNW, whereas it corresponded to 1 wt.-% for plant CNW since higher concentrations induced agglomeration of the nanofiller.

Water permeability was strongly influenced by the incorporation method. For a loading of 2 wt.-% BCNW, the water permeability increased by 69% as compared to pure EVOH for the direct melt mixing of freeze-dried BCNW with the matrix, whereas no significant effect was observed for the preincorporation of BCNW through precipitation or electrospinning. The optimal concentration in terms of water permeability seemed to be 4 wt.-% BCNW, which resulted in a 22% permeability drop.

Regarding the mechanical properties, as shown in Figure 8.9, the nanofiller loading was an important parameter since the formation of a strong nanowhiskers' network is required in order to attain optimum mechanical properties. It has been previously reported that the high polymer melt viscosity that occurs during extrusion limits the random movement and consequently hinders the interconnection between cellulose nanowhiskers. As a result, the percolating structure is not easily formed and the reinforcing efficiency is lower than for nanocomposites obtained by solvent casting.[77] Nevertheless, reaching the percolation threshold, which corresponded to 3 wt.-% BCNW, allowed the production of nanocomposites with increased elastic modulus and tensile strength but still maintaining a relatively ductile behavior.

The incorporation of BCNW through electrospun fibers resulted in nanocomposites with a high dispersion level, but, on the other hand, it prevented the formation of the nanofiller percolation network. As a result, the material did not show any significant improvement on the mechanical and barrier properties. On the contrary, the agglomeration produced for the freeze-dried BCNW was detrimental for both the barrier and mechanical properties.

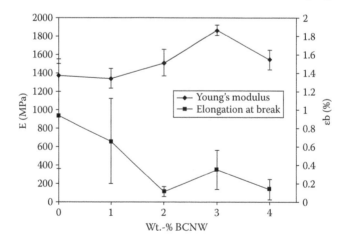

FIGURE 8.9
Mechanical properties of EVOH nanocomposite films with different BCNW loadings.

Furthermore, it was observed that plant-derived CNW did not produce significant improvements on the permeability and mechanical properties, probably due to the lower aspect ratio, which does not lead to increased tortuosity and requires higher loadings for the formation of the percolation threshold. Nevertheless, the incorporation of freeze-dried material seemed to favor the blocking capacity of CNW, which was also higher than for freeze-dried BCNW.

Therefore, this work demonstrated that it is possible to obtain nanocomposites with highly dispersed nanowhiskers by using the electrospinning technique. It is also possible to enhance the barrier and mechanical properties of EVOH even though the melt compounding method hinders the formation of the nanofiller percolation network.

8.6 Future Perspectives and Conclusions

Through this chapter, it has been shown that CNWs present a great potential for their use as nanofillers in biocomposites. Provided that a good dispersion of the nanowhiskers is achieved, they are able to make suitable the application of biopolymers in the field of packaging applications by enhancing the gas and vapor barrier properties of the polymeric matrices. The renewable and biodegradable character of CNWs makes them even more attractive since it is possible to obtain high performance fully biodegradable packaging materials. Therefore, special interest is being currently focused on the production of biopolyesters such as PLA and PHAs loaded with CNWs.

One of the most important issues when developing biocomposites incorporating CNWs is to attain a dispersion level high enough in order to improve the properties of the base material. Along this chapter, it has been demonstrated how CNWs can be incorporated into the polymeric matrix with a high dispersion level not only by lab-scale techniques such as solution casting but also by more industrially oriented processes such as melt compounding. In this sense, innovative techniques such as electrospinning have been proven to be efficient for the incorporation of CNWs into polymeric polar and nonpolar matrices, in relatively high amounts. The hybrid electrospun fibers have interesting properties and may be used as a vehicle for the incorporation of highly dispersed nanowhiskers into a polymeric matrix by a subsequent melt compounding step.

A different alternative that is being currently studied for improving the dispersion of CNWs into nonpolar matrices such as biopolyesters consists in the surface modification of nanocrystals. For that purpose, methods such as reaction with silanes or silylation,[78,79] acetylation by reaction with acid or anhydride groups,[80–84] and the adsorption of cationic surfactants on the surface of cellulose nanocrystals[85] have been developed and work is in progress for the optimization of these methods. Furthermore, the route of incorporating CNWs into the polymeric chain directly during the polymerization process, i.e., in situ polymerization, which has not been extensively studied yet, seems to be a promising method for the incorporation of highly dispersed nanowhiskers into polymers such as PLA.[86–89]

As related to the viability of the production of CNWs, it has been shown that it is possible to extract this material both from vegetal resources and from bacterial synthesized cellulose. BC presents the advantage of high purity and does not need any step for removing impurities such as hemicellulose and lignin. On the other hand, BC is typically produced from expensive culture media, containing glucose as carbon source and other nutrient sources, resulting in high production costs. An interesting strategy to overcome this issue consists in the use of cheap carbon and nutrient sources, such as agro-forestry industrial residues. Several industrial wastes or by-products, such as corn steep liquor,[90] fruit juices,[91] and coconut water,[92] as well as agro-forestry residues,[93] have been already successfully used as carbon sources for the production of BC. Hence, optimization of the BC production process by using industrial or agricultural wastes is one of the future trends that could enable the low-cost production of a high-value material by decreasing the environmental problems associated with the disposal of these wastes.

References

1. D.P. Delmer and Y. Amor, *Plant Cell*, 7 (1995) 987.
2. M. Mutwil, S. Debolt, and S. Persson, *Current Opinion in Plant Biology*, 11 (2008) 252.
3. A.C. Neville and S. Levy, *Planta*, 162 (1984) 370.
4. Y. Nishiyama, *Journal of Wood Science*, 55 (2009) 241.
5. A.N. Nakagaito, S. Iwamoto, and H. Yano, *Applied Physics A: Materials Science and Processing*, 80 (2005) 93.
6. L.M.J. Kroon-Batenburg, J. Kroon, and M.G. Northolt, *Polymer Communications Guildford*, 27 (1986) 290.
7. S. Iwamoto, W. Kai, A. Isogai, and T. Iwata, *Biomacromolecules*, 10 (2009) 2571.
8. Y.C. Hsieh, H. Yano, M. Nogi, and S.J. Eichhorn, *Cellulose*, 15 (2008) 507.
9. D.G. Hepworth and D.M. Bruce, *Journal of Materials Science*, 35 (2000) 5861.
10. W. Helbert, J.Y. Cavaille, and A. Dufresne, *Polymer Composites*, 17 (1996) 604.
11. J. George, M.S. Sreekala, and S. Thomas, *Polymer Engineering and Science*, 41(2001) 1471.
12. H. Fukuzumi, T. Saito, T. Iwata, Y. Kumamoto, and A. Isogai, *Biomacromolecules*, 10 (2009) 162.
13. R.T. Olsson, M.A.S. Azizi Samir, G. Salazar-Alvarez, L. Belova, V. Ström, L.A. Berglund, O. Ikkala, J. Nogués, and U.W. Gedde, *Nature Nanotechnology*, 5 (2010) 584.
14. M. Henriksson and L.A. Berglund, *Journal of Applied Polymer Science*, 106 (2007) 2817.
15. K. Petersen, P.V. Nielsen, and M.B. Olsen, *Starch/Staerke*, 53 (2001) 356.

16. M.F. Koenig and S.J. Huang, *Polymer*, 36 (1995) 1877.
17. E.S. Park, M.N. Kim, and J.S. Yoon, *Journal of Polymer Science, Part B: Polymer Physics*, 40 (2002) 2561.
18. H. Tsuji and T. Yamada, *Journal of Applied Polymer Science*, 87 (2003) 412.
19. U.S. Department of Energy. http://genomicsgtl.energy.gov/roadmap/ (July 5, 2011).
20. D.L. VanderHart and R.H. Atalla, *Macromolecules*, 17 (1984) 1465.
21. J. Sugiyama, R. Vuong, and H. Chanzy, *Macromolecules*, 24 (1991) 4168.
22. F. Horii, A. Hirai, and R. Kitamaru, *Macromolecules*, 20 (1987) 2117.
23. R.H. Atalla and D.L. VanderHart, *Science*, 223 (1984) 283.
24. H. Yamamoto and F. Horii, *Macromolecules*, 26 (1993) 1313.
25. F. Horii, H. Yamamoto, R. Kitamaru, M. Tanahashi, and T. Higuchi, *Macromolecules*, 20 (1987) 2946.
26. J. Sugiyama, J. Persson, and H. Chanzy, *Macromolecules*, 24 (1991) 2461.
27. E. Dinand, M. Vignon, H. Chanzy, and L. Heux, *Cellulose*, 9 (2002) 7.
28. I.M. Saxena and R.M. Brown Jr, *Annals of Botany*, 96 (2005) 9.
29. A.D. French, N.R. Bertoniere, R.M. Brown, H. Chanzy, D. Gray, K. Hattori, and W. Glasser, *Encyclopedia of Polymer Science and Technology*, 5 (2003) 473.
30. S. Yamanaka, K. Watanabe, N. Kitamura, M. Iguchi, S. Mitsuhashi, Y. Nishi, and M. Uryu, *Journal of Materials Science*, 24 (1989) 3141.
31. M. Iguchi, S. Yamanaka, and A. Budhiono, *Journal of Materials Science*, 35 (2000) 261.
32. Y.Z. Wan, Y. Huang, C.D. Yuan, S. Raman, Y. Zhu, H.J. Jiang, F. He, and C. Gao, *Materials Science and Engineering C*, 27 (2007) 855.
33. B.G. Rånby, *Acta Chemica Scandinavica*, 3 (1949) 649.
34. J. Araki, M. Wada, S. Kuga, and T. Okano, *Colloids and Surfaces A: Physicochemical and Engineering Aspects*, 142 (1998) 75.
35. V. Favier, H. Chanzy, and J.Y. Cavaille, *Macromolecules*, 28 (1995) 6365.
36. G. Siqueira, J. Bras, and A. Dufresne, *Biomacromolecules*, 10 (2009) 425.
37. A. Hirai, O. Inui, F. Horii, and M. Tsuji, *Langmuir*, 25 (2009) 497.
38. M.M. De Souza Lima, and R. Borsali, *Macromolecular Rapid Communications*, 25 (2004) 771.
39. J. Araki and S. Kuga, *Langmuir*, 17 (2001) 4493.
40. S.J. Eichhorn, A. Dufresne, M. Aranguren, N.E. Marcovich, J.R. Capadona, S.J. Rowan, C. Weder, W. Thielemans, M. Roman, S. Renneckar, W. Gindl, S. Veigel, J. Keckes, H. Yano, K. Abe, M. Nogi, A.N. Nakagaito, A. Mangalam, J. Simonsen, A.S. Benight, A. Bismarck, L.A. Berglund, and T. Peijs, *Journal of Materials Science*, 45 (2010) 1.
41. M.A.S. Azizi Samir, F. Alloin, and A. Dufresne, *Biomacromolecules*, 6 (2005) 612.
42. M. Roman and W.T. Winter, *Biomacromolecules*, 5 (2004) 1671.
43. L. Jiang, E. Morelius, J. Zhang, M. Wolcott, and J. Holbery, *Journal of Composite Materials*, 42 (2008) 2629.
44. L. Petersson, I. Kvien, and K. Oksman, *Composites Science and Technology*, 67 (2007) 2535.
45. M.D. Sanchez-Garcia and J.M. Lagaron, *Cellulose*, 17 (2010) 987.
46. R.T. Olsson, R. Kraemer, A. Lopez-Rubio, S. Torres-Giner, M.J. Ocio, and J.M. Lagaron, *Macromolecules*, 43 (2010) 4201.
47. M. Martínez-Sanz, R.T. Olsson, A. Lopez-Rubio, and J.M. Lagaron, *Cellulose*, 18 (2011) 335.
48. M. Martínez-Sanz, A. Lopez-Rubio, and J. Lagaron, *Carbohydrate Polymers*, 85 (2011) 228.

49. M.A. Moharram and O.M. Mahmoud, *Journal of Applied Polymer Science*, 105 (2007) 2978.
50. D. Bondeson, A. Mathew, and K. Oksman, *Cellulose*, 13 (2006) 171.
51. S. Elanthikkal, U. Gopalakrishnapanicker, S. Varghese, and J.T. Guthrie, *Carbohydrate Polymers*, 80 (2010) 852.
52. P. Podsiadlo, S.Y. Choi, B. Shim, J. Lee, M. Cuddihy, and N.A. Kotov, *Biomacromolecules*, 6 (2005) 2914.
53. M.D. Sanchez-Garcia, L. Hilliou, and J.M. Lagaron, *Journal of Agricultural and Food Chemistry*, 58 (2010) 12847.
54. M.D. Sanchez-Garcia, E. Gimenez, and J.M. Lagaron, *Carbohydrate Polymers*, 71 (2008) 235.
55. G.A. De Ruiter and B. Rudolph, *Trends in Food Science and Technology*, 8 (1997) 389.
56. A.L. Brody, *Food Technology*, 59 (2005) 65.
57. G. Zhang and D. Yan, *Journal of Applied Polymer Science*, 88 (2003) 2181.
58. M. Roohani, Y. Habibi, N.M. Belgacem, G. Ebrahim, A.N. Karimi, and A. Dufresne, *European Polymer Journal*, 44 (2008) 2489.
59. L. Petersson and K. Oksman, *Composites Science and Technology*, 66 (2006) 2187.
60. C.W. Kim, D.S. Kim, S.Y. Kang, M. Marquez, and Y.L. Joo, *Polymer*, 47 (2006) 5097.
61. C.W. Kim, M.W. Frey, M. Marquez, and Y.L. Joo, *Journal of Polymer Science, Part B: Polymer Physics*, 43 (2005) 1673.
62. M.S. Peresin, Y. Habibi, J.O. Zoppe, J.J. Pawlak, and O.J. Rojas, *Biomacromolecules*, 11 (2010) 674.
63. M. Martínez-Sanz, R. Olsson, A. Lopez-Rubio, and J. Lagaron, *Journal of Applied Polymer Science*, 124 (2012) 1398.
64. M. Martínez-Sanz, A. Lopez-Rubio, and J.M. Lagaron, *Journal of Polymers and the Environment*, under review.
65. J. Fang, T. Lin, W. Tian, A. Sharma, and X. Wang, *Journal of Applied Polymer Science*, 105 (2007) 2321.
66. X. Wang, K. Zhang, M. Zhu, B.S. Hsiao, and B. Chu, *Macromolecular Rapid Communications*, 29 (2008) 826.
67. Y.Z. Zhang, J. Venugopal, Z.M. Huang, C.T. Lim, and S. Ramakrishna, *Polymer*, 47 (2006) 2911.
68. D. Chen, T. Liu, X. Zhou, W.C. Tjiu, and H. Hou, *Journal of Physical Chemistry B*, 113 (2009) 9741.
69. Y. Dror, W. Salalha, R.L. Khalfin, Y. Cohen, A.L. Yarin, and E. Zussman, *Langmuir*, 19 (2003) 7012.
70. M.T. Hunley, P. Pötschke, and T.E. Long, *Macromolecular Rapid Communications*, 30 (2009) 2102.
71. E. Katz, A.L. Yarin, W. Salalha, and E. Zussman, *Journal of Applied Physics*, 100 (2006).
72. J. Amiran, V. Nicolosi, S.D. Bergin, U. Khan, P.E. Lyons, and J.N. Coleman, *Journal of Physical Chemistry C*, 112 (2008) 3519.
73. D. Blond, W. Walshe, K. Young, F.M. Blighe, U. Khan, D. Almecija, L. Carpenter, J. McCauley, W.J. Blau, and J.N. Coleman, *Advanced Functional Materials*, 18 (2008) 2618.
74. K. Oksman, A.P. Mathew, D. Bondeson, and I. Kvien, *Composites Science and Technology*, 66 (2006) 2776.
75. D. Bondeson and K. Oksman, *Composites Part A: Applied Science and Manufacturing*, 38 (2007) 2486.

76. J.M. Lagaron, M. Martinez-Sanz, and A. Lopez-Rubio, Procedure to obtain nanocomposite materials. Patent Application P201030663.
77. P. Hajji, J.Y. Cavaille, V. Favier, C. Gauthier, and G. Vigier, *Polymer Composites*, 17(1996) 612.
78. M. Andresen, L.S. Johansson, B.S. Tanem, and P. Stenius, *Cellulose*, 13 (2006) 665.
79. C. Goussé, H. Chanzy, G. Excoffier, L. Soubeyrand, and E. Fleury, *Polymer*, 43 (2002) 2645.
80. B. Braun and J.R. Dorgan, *Biomacromolecules*, 10 (2009) 334.
81. D.Y. Kim, Y. Nishiyama, and S. Kuga, *Cellulose*, 9 (2002) 361.
82. G. Rodionova, M. Lenes, Ø. Eriksen, and Ø. Gregersen, *Cellulose*, 18 (2011) 127.
83. J.F. Sassi and H. Chanzy, *Cellulose*, 2 (1995) 111.
84. H. Yuan, Y. Nishiyama, M. Wada, and S. Kuga, *Biomacromolecules*, 7 (2006) 696.
85. K. Xhanari, K. Syverud, G. Chinga-Carrasco, K. Paso, and P. Stenius, *Cellulose*, 18 (2011) 257.
86. J. Chen, F. Qin and R.J. Yang, *Cailiao Gongcheng/Journal of Materials Engineering*, 1 (2011) 28.
87. V. Katiyar, N. Gerds, C.B. Koch, J. Risbo, H.C.B. Hansen, and D. Plackett, *Polymer Degradation and Stability*, 95 (2010) 2563.
88. Y. Li, C. Chen, J. Li and X.S. Sun, *Polymer*, 52 (2011) 2367.
89. I. Moura, R. Nogueira, V. Bounor-Legare, and A.V. MacHado, *Reactive and Functional Polymers*, 71 (2011) 694.
90. H. El-Saied, A.I. El-Diwany, A.H. Basta, N.A. Atwa, and D.E. El-Ghwas, *BioResources*, 3 (2008) 1196.
91. A. Kurosumi, C. Sasaki, Y. Yamashita, and Y. Nakamura, *Carbohydrate Polymers*, 76 (2009) 333.
92. S. Kongruang, *Applied Biochemistry and Biotechnology*, 148 (2008) 245.
93. P. Carreira, J.A.S. Mendes, E. Trovatti, L.S. Serafim, C.S.R. Freire, A.J.D. Silvestre, and C.P. Neto, *Bioresource Technology*, 102 (2011) 7354.

9

Edible Nano-Laminate Coatings for Food Applications

M.A. Cerqueira, A.I. Bourbon, A.C. Pinheiro, H.D. Silva,
M.A.C. Quintas, and Antonio A. Vicente

CONTENTS

9.1 Introduction

Food industry seeks for new products and methodologies that lead to sustainability programs. In this context, food technology plays an important role, with main research areas such as chemistry, biochemistry, engineering, and physics, in seeking new know-how for the improvement of food quality during processing, storage, and handling. The creation of new methodologies and materials that increase the storability of foods without affecting their quality is one of the focuses of today's researchers.[1,2] In this context, nanotechnology appears in the past years as an emerging technology that holds potential to change the food industry.[3,4] Nanotechnology involves the control of nanostructures within sizes from 1 to 100 nm,[5] which, with the adequate development of technology, can be used to manipulate or assemble materials providing commercial, technological, and scientific opportunities for the industry.[4]

The utilization of edible coatings has increased on the food packaging market and has been accompanied by a vast amount of knowledge acquired through research and product development work, as well as by advances in materials science and processing technology. Based on this, in the past years, a great interest has developed in the replacement of commercial synthetic plastics by materials obtained from renewable resources (e.g., polysaccharides, proteins, and lipids).[6–9] Over the past years, many works have been developed and it has been shown that coatings of biopolymers present a promising and increasing potential for application in food and biomedical industries.[10–13]

The application of nanotechnology to the coating technology may allow the modification of many of their macroscale characteristics, increasing and improving their applications in the food industry (e.g., water solubility, thermal stability, barrier, and mechanical properties).[4,14–17] The application of different materials as a single or multiple layer has been proved to be an efficient methodology to improve physical properties of these materials,[18] which is in agreement with recent literature that suggests that biopolymers can present improved functionality when used at nanoscale.[19]

Coatings with functional properties, such as antimicrobial and antioxidant properties, have also been developed for many purposes. The control of microbial growth in food products is one of the applications of the coatings in the packaging industry, while the incorporation of functional agents in the coatings and their controlled release in order to prevent physical, chemical, or biological degradation of foods is another application of these coatings, preserving at the same time moisture, flavor, and quality.[20]

The application of nanotechnology to coating science lead to the formation of nano-laminate coatings that can be produced by successive adsorption of oppositely charged polyelectrolytes on a solid carrier providing potential applications of such coatings for food preservation and implant devices with enhanced physical properties.[20,21] A nano-laminate coating can be described as two or more layers of material with nanometer dimension that are physically or chemically bonded to each other. One of the most powerful methods for their construction is the layer-by-layer (LbL) deposition technique, in which a charged surface is coated with multiple nano-laminate layers of different materials. These materials include either polycations or polyanions, depending on their functional group type.[22] The nano-laminate coating construction will depend on the single layer or multilayer building. The sequential electrostatic LbL adsorption of charged polymers with opposite charges will allow the production of thin layers with different properties. These different charges will allow the construction of multilayers through electrostatic interactions. The number of layers, the polymers used, and the conditions of the construction process will allow the control of the thickness and properties of the coating. Initially applied to flat and macroscopic[22] objects, the LbL technique was later also applied to colloidal systems (e.g., nanoparticles).[23]

This chapter will allow readers to understand the applicability of the edible nano-laminate coatings to food systems by dealing with materials and methodologies that can be used in the construction and characterization of these nano-laminate coatings, templates and shapes used, and finally some of the examples where they have already been applied.

9.2 Biobased Materials

9.2.1 Polysaccharides

Polysaccharides play a major role in the food industry. They can exhibit a wide range of chemical structures and different functions, offering a large variety of applications. Natural polysaccharides can be derived from plant biomass, including algae, from microbial fermentation, or from animals. Some polysaccharides exhibit polyelectrolyte properties, that is, they contain a large number of ionizable groups, which under appropriate conditions dissociate, leaving charges on the chain (polyion) and counterions in the solution.[24]

TABLE 9.1

Polyelectrolytes Derived from Natural Polysaccharides

Sources	Polysaccharides	Main Properties
Botanical	Pectins	Gelling or thickening polymer depending on the cations and temperature
		Anionic polysaccharide
	Cellulose ionic derivatives	Gelling or thickening polymers
	Starch ionic derivatives	Thickening and emulsifying polymers
	Galactomannans ionic derivatives	Thickening agents
Algae	Alginates	Gelling polymer in the presence of divalent counterions
		Anionic polysaccharides
	Carrageenans	Gelling or thickening polymers, depending on the cations and temperature
		Anionic polysaccharides
	Fucoidans	Wide range of biological activities
		Anionic polysaccharides
Microbial	Xanthan	Thickening agents
	Succinoglycan	Anionic polysaccharides
Animal	Chitosan	Antimicrobial activity
		Cationic polysaccharide

Some of the polyelectrolytes derived from natural polysaccharides that are most relevant to the food industry, their origins, and their main properties are presented in Table 9.1. Their principal characteristics are discussed later.

9.2.1.1 Botanical Sources

Pectins are present in the primary cell walls and middle lamella of plant tissues.[25] They are heteropolysaccharides mainly composed of 1,4-linked α-D-galacturonic acid units.[26] Some carboxylic groups of galacturonic acid units are methyl esterified and the percentage of esterified groups is expressed as the degree of esterification (DE). Depending on the DE, pectins are divided into two major groups: high methoxyl pectins (HMP, DE > 50%) that form thermoreversible gels in acid conditions and in the presence of sucrose and low methoxyl pectins (LMP, DE < 50%) that form gels in the presence of calcium ions.[27] Pectins are negatively charged at pH > 3.5,[28] and due to their polyelectrolyte nature, pectin gels are sensitive to variations of pH and to the nature and quantity of cations present in the system.[29]

Carboxymethylcelluloses (CMCs) are the most important cellulose derivatives among the water-soluble products. CMC, like other cellulose ethers, can be obtained by the substitution of hydroxyl groups of the cellulose molecule with ether groups. These ionic polymers become soluble in water when the degree of substitution is higher than 0.5. Also, as a consequence of their polyelectrolyte properties, the viscosity of CMC aqueous solutions varies with salt concentration.[30]

Starch phosphates are anionic compounds that can be obtained by treating starch with phosphoric acid. They exhibit improved properties compared with unmodified starch, such as higher viscosity and the production of more stable dispersions. In the food industry, low substituted starch phosphates are applied as thickening, emulsifier, and stabilizer agents in a wide range of products.[31] Also, due to their anionic character, starch phosphates can also be used as flocculating agents and ion exchangers.[32] Cationic starches can be obtained by the reaction of starch with reagents containing amino, imino, ammonium, or sulfonium groups.

Galactomannans are neutral polysaccharides isolated from various albuminous or endospermic seeds. They are composed of linear β (1 → 4) D-mannan (M) chains with varying amounts of single D-galactose substituents linked to the main backbone by (1 → 6)-α-glycosidic bonds.[33] Galactomannans have been modified in order to extend their range of applications.[34,35] Sierakowski et al. (2000) performed specific chemical modifications at C-6 position of a galactomannan and obtained an anionic polyelectrolyte having a greater solubility in aqueous solvent than the original material.[35]

9.2.1.2 Algal Sources

Alginates are anionic polysaccharides extracted from brown algae (mainly from the orders *Laminariales* and *Fucales*), where they are structure-forming components, giving the plant both mechanical strength and flexibility. Alginates are linear block copolymers composed of varying proportions of 1,4-linked β-D mannuronic acid (M) and α-L-guluronic acid (G), forming regions of M-blocks, G-blocks, and blocks of alternating sequence (MG-blocks). The alginates' physical properties depend on the M/G ratio.[36] Alginates exhibit thickening properties and have the ability to form gel in the presence of multivalent counterions, which is a direct consequence of the fact that alginates are polyelectrolytes. Alginates show also interesting ion-binding properties, which represent the basis for their gelling properties. Sodium alginate presents a zeta potential value of −62.13 ± 4.01 mV at pH 7 and at a concentration of 0.2%.[16]

Carrageenans are sulfated polysaccharides extracted from red seaweeds (Rhodophyceae). Their linear polymers are composed of repeating galactose units and 3,6-anhydrogalactose, both sulfated and nonsulfated, joined by alternating α (1–3) and β (1–4) glycosidic links. The various carrageenans structures differ in 3,6-anhydrogalactose and ester sulfate content and these variations influence the gel strength, texture, solubility, melting and setting temperatures, syneresis and synergistic properties.[37] The main carrageenan types, lambda, kappa, and iota, differ in their thickening and gelling properties. Lambda carrageenan is a thickening polymer that is not affected by the presence of salts and kappa forms strong and rigid gels with potassium ions, while iota interacts with calcium ions to give soft, elastic gels.[27] Carrageenans are anionic polyelectrolytes: a solution of 0.2% of kappa-carrageenan at pH 7 exhibits a zeta-potential value of -60.53 ± 0.15 mV.[38]

Fucoidans are heterogeneous sulfated polysaccharides isolated from brown algae. Each brown algae species synthesizes highly branched polysaccharides with specific sugar composition, originating diverse fucoidan structures. Moreover, every algae specie forms several types of fucoidans.[39] Fucoidans exhibit highly complex structures due to the presence of branching, random distribution of the sulfate units, different types of glycosidic linkages, and also the presence of acetylation, methylation, and pyruvilation.[40] Recently, new fucoidans have been fully characterized, and it was shown that fucoidans can have an α-(1–3) backbone or repeating disaccharide units of α-(1–3) and α-(1–4) linked fucose residues with branching attached at C2 positions and may be sulfated at C4, C2, or at both positions of fucose units.[41–43] Due to the presence of sulfate groups, fucoidans exhibit negative charge at neutral pHs.[44] Fucoidans exhibit a wide range of biological activities such as anticoagulant,[45] antitumor,[46] anti-inflammatory,[47] antiviral,[48] antibacterial,[49] and antioxidant.[50]

9.2.1.3 Microbial Sources

Xanthan gum is an anionic extracellular polysaccharide secreted by the microorganism *Xanthomonas campestris*.[51] It is a heteropolysaccharide composed of a linear (1 → 4)-linked β-D-glucose with a trisaccharide side chain on every other glucose at C-3, containing a glucuronic acid residue (1 → 4)-linked to a terminal mannose unit and (1 → 2)-linked to a second mannose that connects to the backbone.[52] Xanthan gum is soluble in cold water and their solutions exhibit a high shear-thinning behavior, which results from the ability of the xanthan molecules, in solution, to form intermolecular aggregates through hydrogen bonding and polymer entanglements.

Succinoglycans are anionic polysaccharides produced by bacteria of the species *Pseudomonas*, *Rhizobium*, *Agrobacterium*, and *Alcaligenes*. Succinoglycans are composed of a repeating unit of a branched octasaccharide of seven β-linked glucose residues and one β-linked galactose residue, with noncarbohydrate substitutions of acetate on the backbone and pyruvate and succinate on the branch.[53] The high molecular weight (MW) of succinoglycan and the presence of an ordered conformation induce a high viscosity in aqueous solutions, which is stable in a highly acidic medium.[54] This property makes succinoglycans an attractive thickening agent for food industry.

9.2.1.4 Animal Sources

Chitosan is a linear polysaccharide obtained by deacetylation of chitin, which is a polysaccharide widely distributed in nature (e.g., crustaceans, insects, and certain fungi).[55] Chitosan is composed of N-acetyl-2-amino-2-deoxy-d-glucopyranose and 2-amino-2-deoxy-d-glucopyranose units linked by $(1 \rightarrow 4)$-β-glycosidic bonds.[56] It is a polycationic polymer that has one amino group and two hydroxyl groups in the repeating glucosidic residue.[57] Chitosan is a very interesting polymer for numerous applications due to its biocompatibility, biodegradability and nontoxicity and due its intrinsic antimicrobial activity.[58,59] As a cationic polymer (zeta-potential of +58.28 ± 4.18 mV for 0.2% solution at pH 3), chitosan may be associated with anionic polyelectrolytes, leading to the formation of a polyelectrolyte multilayer.[16]

9.2.2 Proteins

Proteins are widely applied in the food industry due to their ability to add unique functional properties such as emulsifying, foaming, gelling, and solubility attributes.[60,61] The selection of suitable protein or a combination of proteins to fabricate biopolymer-based systems is based on their properties, which depends on both intrinsic (e.g., molecular structure, composition) and extrinsic factors (e.g., temperature, chemical environment, pH).[62] The type, number, and particular sequence of amino acids in a protein determines the MW, conformation, electrical charge, hydrophobicity, physical interactions, and functionalities of proteins. Food proteins such as soy proteins, milk proteins (caseins and whey proteins), and egg proteins, are frequently used in food matrices (Table 9.2).

TABLE 9.2

Functional Roles of Food Proteins

Source	Proteins	Functional Properties
Milk proteins	β-lactoglobulin	Emulsifying, foaming, and gelling properties
	Bovine serum albumin	Solubility, foaming, and emulsifying properties
	α-lactalbumin	Gelling properties and fat and flavor binding
	Casein	Emulsifying, foaming, and gelling properties
	Lactoferrin	Gelling properties, antimicrobial, anti-inflammatory, antitumor
Milk or egg proteins	Lysozyme	Antimicrobial
Egg protein	Ovalbumin	Gelling and emulsifying properties and foam stability
Animal or vegetable proteins	Gelatin	Emulsifying and gelling properties
Vegetable protein	Soy protein	Gelling properties and thermal stability

Food proteins may serve as an effective transporter of bioactive molecules because of their ligand binding properties. They are widely used in formula food due their high nutritional value. However, an important handicap of the use of proteins is their allergenicity. There are some proteins such as milk, soy, or egg proteins that may cause an allergic or intolerant reaction to some of the consumers.[63] Some of the food protein characteristics are discussed in the following sections.

9.2.2.1 Soy Protein

Soy proteins are abundant, renewable, and biodegradable, making it attractive to be applied in food products.[64] They are mainly globular and have an MW ranging from 8 to 600 kDa.[65] Numerous studies have shown that soy proteins are quite heterogeneous. The major components are classified according to their sedimentation properties in which a larger Svedberg number (S) indicates a larger protein. The storage proteins, 7 S (conglycinin) and 11 S (glycinin), are the principal components of soy protein.[66] Soy protein functionality is partly dependent on the conglycinin-to-glycinin ratio, which can vary among genotypes. Glycinin represents around 30% of the total proteins in soybeans. It is a quaternary structure composed of three acidic and three basic subunits of ca. 35,000 and 20,000 Da, respectively.[67] The isoelectric points of the basic subunits range between 8.0 and 8.5 and of the acidic subunits from 4.7 to 5.4. This may account for the limited solubility of 11S globulins at low ionic strength, around pH 6.0.[66]

9.2.2.2 Milk Proteins

Milk proteins are widely available, inexpensive, natural, and GRAS (generally recognized as safe) raw materials with high nutritional value and good sensory properties. Their structural properties and functionalities make them highly suitable as delivery systems in food industry.[68] In milk, there are two major protein types, which in bovine milk are defined by acid precipitation: the caseins, which precipitate as a group at pH 4.6, and the whey proteins, which can be subdivided into the major mammary synthesized proteins and the minor, usually blood, proteins.[69]

Caseins are conjugated proteins, most of them with phosphate group(s) esterified to serine residues. These proteins are phosphoproteins precipitated from raw milk at pH 4.6 at 20°C and comprise approximately 80% of the total protein content in milk. The principal proteins of this group are classified according to the homology of their primary structures into $\alpha s1$-, $\alpha s2$-, β-, and κ-caseins.[68] The conformation of caseins is similar to denatured globular proteins. Within the group of caseins, there are several distinguishing features, based on their charge distribution and sensitivity to calcium precipitation.[70] The casein micelle is indeed a remarkable example of natural nanovehicle for nutrient delivery.[68]

The following proteins belong to whey protein group: in higher quantities β-lactoglobulin and α-lactalbumin and in less amounts serum albumin, lactoferrin, immunoglobulins, and proteose peptones. Whey proteins give 20% of total protein content in bovine milk. They are globular and are present in milk as discrete molecules with varying numbers of disulfide cross-links. These proteins are more heat sensitive and less sensitive to calcium than caseins.[71]

β-lactoglobulin (β-Lg) is a globular protein with an MW of 18 Da and is very acid stable.[72] It is generally in dimer form at the isoelectric pH of 5.2 and alkaline pH range. Temperature affects the three-dimensional structure of β-Lg. Although β-Lg is found mainly in the dimer form in milk, monomers appear when temperature is increased up to 65°C. β-Lg is one of those milk proteins that are responsible for milk protein intolerance or allergy in humans.[46,47,72,73] Bovine α-lactalbumin (α-La) is a small globular protein that is relatively stable with MW around 14 Da. The molecule has an ellipsoid shape with a deep cleft dividing the protein in two parts. Four disulfide bonds make this protein relatively heat stable. α-La was found to be a cofactor in lactose synthesis and the concentrations of this protein and of lactose in milk are correlated. It is a strong binder of calcium and other ions, including Zn(II), Mn(II), Cd(II), Cu(II), and Al(III), and changes conformation markedly on calcium binding.[70] One interesting feature of α-La is that it seems to exist in three different structures: the calcium-bound, the calcium-free, and the low pH or A form.[69] Bovine serum albumin (BSA) is a globular protein having an MW of 66 kDa, 580 amino acid residues, and 17 intrachain disulfide bonds. BSA contains three domains specified for metal-ion binding, lipid binding, and nucleotide binding.[74] This protein is also involved in the transport of insoluble free fatty acids in the blood. Glycomacropeptide (GMP) is an acidic peptide with isoelectric point between 4 and 5, highly soluble, and heat stable.[75] GMP has various bioactive properties: it has antibacterial activity, modulates immune system responses, and regulates blood circulation.[75]

Lysozyme is a protein secreted in milk and its MW is 15 kDa. This protein is of interest for use in food systems since it is a naturally occurring enzyme with antimicrobial activity. Lysozyme is active against a number of gram-positive bacteria including *Listeria monocytogenes*.[76]

Lactoferrin is a basic, positively charged iron-binding glycoprotein of the transferrin family with an MW of 80 kDa and an isoelectric point around 8.0–8.5, present in various external secretions of mammals such as milk.[77,78] One of the main interests in lactoferrin resides in its various biological activities, such as antimicrobial, anti-inflammatory, antitumor, immuno-modulatory, and enzymatic activities.[77,79]

9.2.2.3 Other Proteins

Ovalbumin (OVA) is a highly functional food protein that is frequently used in food matrix design. It has an MW of 47,000 Da and isoelectric point (pI) of 4.8.[80] OVA exhibits several interesting functionalities such as its ability to form gel networks and stabilization of emulsions and foams. Moreover, OVA was chosen as a carrier for drug delivery owing to its availability and low cost, compared with other proteins. Due to its pH- and temperature-sensitive properties, it has a high potential for use as a carrier for controlled drug release.[80,81]

Collagen is the structural building material of vertebrates and the most abundant mammalian protein that accounts for 20%–30% of total body proteins. Collagen has a unique structure, size, and amino acid sequence, which results in the formation of triple helix fiber. Collagen is abundant, renewable, and a biodegradable protein making it attractive to be applied in food products.[82]

Gelatin is obtained by controlled hydrolysis of fibrous, insoluble protein, collagen, which is widely found as the major component of skin, bones, and connective tissue.[83] Gelatin is considered as an interesting biodegradable material, which is nontoxic and inexpensive and has therefore an immense potential to be used for the preparation of colloidal drug delivery systems.[84]

9.3 Development of Nano-Laminate Coatings

9.3.1 LbL Technique

In the last 60 years, various methods have been developed and implemented for the construction of thin films. In 1917, Lagmuir[85] started the work on liquid monolayers that was complemented later by Blodgett.[86] In this method, the films from the liquid surface were transferred to solid supports through dipping or removing the solid support from the aqueous subphase that intercepted the monolayer. This process when repeated lead to the formation of a multilayer film.[87]

In 1991, Decher and coworkers changed the area of thin film assembly by the introduction of the LbL technique.[88] LbL technique involves the adsorption of charged molecules in an aqueous solution to a support. The method involves the sequential adsorption of oppositely charged species onto a solid support and involves as forces ionic interactions. It does not involve covalent bonding; however, secondary forces such as hydrogen bonding may act. This technique can be used for the development of nanocoatings in solid supports allowing nanoscale fabrication of structures.[22] During the multilayer construction process, the removal of the polymer excess is of extreme importance to avoid the aggregation and formation of complexes.

A great number of compounds can be used for film formation, which will depend finally on the functionality of the system (e.g., sensors, membranes, nanocapsules, liposomes, and materials for tissue engineering).[89–92] These compounds can include charged polymers (polyelectrolytes), functionalized nanoparticles and nanotubes, phthalocyanines, and biomolecules.[93–97]

The main advantages of using the LbL technique for film preparation is that it is a simple method where water is the main solvent used, allowing the control of thickness, composition, and structure of the films at nanometer scale. LbL technique besides the control of thickness also allows the control of parameters such as the surface density and roughness[98] and the improvement of mechanical properties and the permeability of the systems.[99] It also allows the preparation of nanocomposites with a high level of dispersion and at the same time with a higher filler load.[100]

The layers of the nano-laminate systems could present thickness values on the order of 1 nm per layer, being the obtained value dependent on the charge density, MW, and ionic strength of the adsorbing materials. LbL technique depends neither on the surface area of the support nor on its shape, the charge properties of the surface and the assembling species characteristics being the most influent factors.[20,22,88]

The LbL technology allows precise control over the thickness and properties of the interfacial films, enabling the creation of thin films (1–100 nm per layer).[19] The simplicity of the technique and the availability of established characterization methods render nanolayered films extremely versatile in a wide range of applications, the food industry being one of the areas that can benefit from these nanostructures, once the nanolayered films could be created entirely from food-grade ingredients (proteins, polysaccharides, and lipids).

9.3.2 Substrates, Templates, and Shapes Used in Nano-Laminate Coatings

The LbL technique can be applied to modify the surface properties of a substrate of interest[16,101] or the substrate can be used just as a support to allow the formation and the characterization of the multilayer structure.[38] The LbL technique can be used on surfaces of almost any kind and shape; however, the prerequisite for successful LbL coating is the presence of a minimal surface charge, which is one of the few disadvantages of the technique.[21]

The LbL technique has been applied in two main classes of substrates: planar and colloidal. The LbL assembly on nonporous planar substrates typically involves solid supports for the production of nanolayered films. The most commonly used substrates are glass, quartz, synthetic polymers (e.g., polyethylene terephthalate, PET), mica, and gold-coated substrates. The use of porous planar substrates for LbL assembly allows the production of nanotubes. The most common colloidal substrates are spherical colloids (e.g., solid spheres, emulsions) and high-aspect-ratio colloids (such as rods and fibers).[23] Nanoparticles produced from edible biopolymers (such as proteins, polysaccharides, and lipids) may also be important vehicles to encapsulate, protect, and deliver bioactive or functional components in food matrices. These different-shaped particles may have applications as different kinds of functional ingredients within the food industry, the application of nano-laminate coatings, through the LbL assembly method, being considered an easy standard method for the surface functionalization of a broad range of nanoparticles.[102] An advantage of LbL technique is the preparation of nanoparticles of core-independent but shell-dependent characteristics.[103] This would result in nanoparticles with controlled size and suitable for targeting different sites in the body depending on the site-specific needs.[104,105] Hollow capsules with wall thicknesses at the nanoscale can also be obtained using the LbL technique removing the core through chemical decomposition or dissolution.[106] Being so, free-standing nanostructured materials with different morphologies and functions can be obtained using a template with the required shape for the LbL deposition, followed by the removal of the template. This approach is known as the LbL templating technique.[107]

9.4 Nano-Laminate Coating Characterization

In order to better understand the nano-laminate systems, the analytical techniques that can be used for the characterization of these nanosystems are described in this section. The analytical approaches have been subdivided into two groups: characterization techniques and imaging techniques, where some are more appropriate for planar supports and other for colloidal systems.

9.4.1 Physicochemical Characterization Techniques

In this section, techniques involving essentially physical determinations (e.g., size, size distribution, zeta potential, and crystallinity of the nanosystems) are described.

9.4.1.1 Dynamic Light Scattering

Dynamic light scattering (DLS) is a technique that measures Brownian motion and relates this to the size of the particles through Stokes–Einstein equation, allowing to determine the size of nanosystems being used to evaluate the size distribution of nanoemulsion multilayers[108,109] and nanoparticles.[110]

9.4.1.2 Zeta Potential

Zeta potential is a scientific term for electrokinetic potential in colloidal systems, and it measures the magnitude of the repulsion or attraction between particles, this being a physical property exhibited by any particle in suspension. A value of 30 mV (positive or negative) can be taken as the arbitrary value that separates low-charged surfaces from highly charged surfaces.[110] Briefly, zeta potentials from 0 to ±30 mV indicate instability; on the other hand, zeta potentials higher than ±30 mV indicate stability.[111] For instance, Preetz et al. evaluated the zeta potential of nanocapsules and Li et al. evaluated the zeta potential of nanoemulsion multilayers: the information given by the zeta potential allows stating that the nanoemulsions with highly charged surfaces are stable and will resist droplet aggregation.[109,110]

9.4.1.3 Quartz Crystal Microbalance

Quartz crystal microbalance (QCM) is based on the piezoelectric effect, where the mass is deposited onto the crystal surface, and the piezoelectric properties of the quartz crystal change its oscillation frequency.[112] It can detect monolayer surface coverage by small molecules or polymer films, providing information about the energy dissipating properties. QCM can characterize surface film mass, thickness, viscoelasticity, hydrophobicity, roughness, and energy dissipative or viscoelastic behavior,[17,112,113] allowing a direct observation of the adsorption process in situ.[44,114]

Channasanon et al.[114] used QCM as a tool to follow the assembly process of chitosan-poly(sodium styrenesulfonate) nano-laminate films by monitoring the frequency change as a function of the number of depositions.

Indest et al.[44] observed in situ the adsorption behavior of fucoidan to a chitosan layer, using the same process. Recently, Pinheiro et al.[17] confirmed through QCM measurements that the alternating deposition of k-carrageenan and chitosan resulted in the formation of a stable multilayer structure.

9.4.1.4 Ultraviolet–Visible Spectroscopy

Ultraviolet–visible (UV/Vis) spectroscopy refers to absorption spectroscopy or reflectance spectroscopy in the ultraviolet–visible spectral region. UV/Vis spectroscopy is being used as a technique to verify if the adsorption of polysaccharides in nano-laminate films is occurring.[16,100] Carneiro-da-Cunha et al.[16] monitored layer deposition by measuring the increase in absorbance at 260 nm for nanomultilayer films (up to five layers), this being in agreement with other works where chitosan deposition was by LbL technique.[16] Pinheiro et al.[17] followed the multilayer film growth of k-carrageenan/chitosan nanolayered film on A/C PET surface by UV/VIS spectroscopy.

9.4.1.5 Contact Angle

Contact angle measurement is a method for surface analysis (in terms of surface energy and tension). The measurement provides information regarding the bonding energy of the solid surface and surface tension of the droplet. Because of its simplicity, contact angle measurement has been accepted for material surface analysis related to wetting, adhesion, and absorption.[115]

Contact angle measurements revealed that the deposited layers in a nano-layered coating are typically highly interpenetrated and that the deposition process is a surface charge dominated adsorption process.[16,114,115]

9.4.1.6 Fourier Transform Infrared

Fourier transform infrared (FTIR) technique is extremely accurate and reproducible, and the resulting spectrum represents the molecular absorption and transmission, creating a molecular fingerprint of the sample.[116] Lawrie et al.[117] studied the interactions between alginate and chitosan in the alginate–chitosan polyelectrolyte complexes in the form of a film through FTIR. Channasanon et al.[114] verified by Attenuated Total Reflectance (ATR)-FTIR analysis the coverage of the multilayer film on the treated PET substrate.

9.4.1.7 Differential Scanning Calorimetry

Differential scanning calorimetry (DSC) is a thermoanalytical technique that directly measures heat changes that occur in biomolecules during controlled increase or decrease in temperature and can be used to determine the enthalpy change during, for example, denaturation to identify phase transitions including the melting of crystalline regions and to analyze, for example, the proportion of solid fat or the proportion of ice crystals in emulsions.[118–120] Carneiro-da-Cunha et al.[16] showed through DSC analyses an increase in the melting energy, when comparing a nanolayered film with the PET film used as support.

9.4.1.8 X-Ray Photoelectron Spectroscopy

X-ray photoelectron spectroscopy (XPS) is a quantitative spectroscopic technique that measures the elemental composition, empirical formula, chemical state, and electronic state of the elements that exist within a material.[44,117] Indest et al.[44] studied the surface chemistry and morphology of the chitosan/fucoidan or chitosan/chitosan sulfate coated hydrolyzed PET (PET-H) films using XPS and showed that a small amount of sulfur was present in the fucoidan layer and that the fucoidan layer was stable on the film surface. Lawrie et al.[117] studied the interactions between alginate and chitosan in alginate–chitosan polyelectrolyte complexes in the form of a film through XPS, observing that when chitosan is the outermost layer in an LbL assembly, the amine groups become deprotonated upon washing and drying with air. Nevertheless, when alginate is the outermost layer, the amine groups of the underlying chitosan layer will be protonated to a larger degree due to interaction with the deprotonated carboxylate groups of alginate, and this gives a major stability to the LbL assembly.[117] A low protonation degree means a low charge density with a small repulsive electrostatic force between its chains, resulting in a less stable form.[121]

9.4.2 Imaging Techniques

Imaging techniques can be applied in order to enable information and visualization of the shape and aggregation state and to confirm the size of the nanosystems. Some of the imaging methods that are used to characterize nanosystems are presented in the following sections.

9.4.2.1 Transmission Electron Microscopy

Transmission electron microscopy (TEM) is a technique capable of a resolution on the order of 0.2 nm.[3,122] TEM is very frequently used in materials and biological sciences; samples must be thin and able to resist the high vacuum present in the instrument. Nonetheless, TEM has some drawbacks, requiring extensive sample preparation (very thin samples) for some materials, being a relatively time-consuming technique, revealing low "productivity."[3,122] The structures of the samples may be changed in the preparation process, the field of view is small and the electron beam may damage the sample.[3,122] Preetz et al.[110] used TEM for the investigation of the shape, morphology, and surface structure of polyelectrolyte nanocapsules and for estimation of the shell thickness (3–10 nm).

9.4.2.2 Scanning Electron Microscopy

Scanning electron microscopy (SEM) is able to produce high-resolution images of a sample surface.[3,123] SEM images have a characteristic three-dimensional appearance and are useful for the visualization of the surface's structure. SEM allows a large amount of sample to be in focus at one time, the higher magnification produces high-resolution images, and this combined with the larger depth of focus, greater resolution, and easy sample observation makes SEM one of the most used techniques in nanosystem characterization. Despite of this, SEM requires a high vacuum and sample conductivity,[3,122] thus rendering it inappropriate for certain types of samples. De Mesquita et al.[100] and Carneiro-da-Cunha et al.[16] showed that it is possible to characterize the thickness of assembled nanolayered films using SEM.

9.4.2.3 Atomic Force Microscopy

Atomic force microscopy (AFM) allows biomolecules to be imaged not only under physiological conditions but also during biological processes. The high resolution (±0.1 nm) attained by AFM has been utilized to directly view single atoms or molecules that have dimensions of a few nanometers. Due to a nanometer-sized probe, AFM can obtain high-resolution three-dimensional profile images of the surface under study. AFM can image nearly any type of surface, including polymers, ceramics, composites, glass, and biological samples. Direct observation of biomolecular systems allows analyzing structural and functional properties at the submolecular level. AFM can be used for structural characterization of, for example, proteins, polysaccharides, liposomes, and multilayered structures.

AFM is complementary to other image techniques but is not the best option when the tip is in direct contact with a surface that is soft, sticky, or has loose particles floating on it.[3,124]

Preetz et al. showed the differences between nanoemulsions and nanocapsules by studying the shape, morphology, and mechanical properties of the emulsion and capsule shell through AFM and demonstrated that the shell around an oil droplet solidified with increasing amounts of polyelectrolytes.[110] Indest et al.[44] studied the topography of modified PET films using AFM.

9.4.2.4 Confocal Laser Scanning Microscopy

Confocal laser scanning microscopy (CLSM) allows obtaining high-resolution optical images with depth selectivity. The key feature of confocal microscopy is the ability to acquire in-focus images from selected depths, a process known as optical sectioning, allowing three-dimensional reconstructions of topologically complex objects, this being useful for surface profiling in opaque specimens. In nonopaque specimens, it is possible to see interior structures with enhanced quality of image over simple microscopy. Confocal microscopy provides the capacity for direct, noninvasive, serial optical sectioning of intact, thick, living specimens with a minimum of sample preparation as well as a marginal improvement in lateral resolution. Biological samples are often treated with fluorescent dyes to make selected objects visible. However, the actual dye concentration can be low to minimize the disturbance of biological systems: some instruments can track single fluorescent molecules.[125] Using CSLM and fluorescent-labeled chitosan (FITC-chitosan), Richert et al.[125] studied the diffusion ability of chitosan and saw that the FITC-chitosan added at the outermost layer of the film diffused through the whole film up to the substrate.

9.4.3 Transport and Mechanical Properties

In food systems, the application of edible coatings has as main objective the control of gas and water vapor exchanges between the food system and its surrounding atmosphere, leading to the extension of the products' shelf life. This control may slow down the typical degradation reactions that occur in food systems. Moreover, they can protect from physical damage throughout the distribution chain. Although coatings have been applied for a large number of years to food products, the use of these materials at the nanoscale may improve their functionality.[19] It is then of utmost importance to characterize the transport of gases and water through the nano-laminate coating and its mechanical properties.

9.4.3.1 Transport Properties

Due to nanoscale nature of these systems, experimental determination of a standalone nano-coating transport properties is not an easy task. To solve this issue, experiments can be done using a well-characterized support in which the coating is constructed using any technique mentioned in a previous section. The transport property alone can then be determined by[126]

$$TP_{nano\ coating} = \frac{L_{nano\ coating}}{(L_{total}/TP_{total}) - (L_{support}/TP_{support})} \quad (9.1)$$

where
  TP refers to the transport property in study
  L is the thickness

This approach has been successfully used in nanolayered systems like κ-carrageenan/lysozyme,[38] chitosan/alginate,[16] and κ-carrageenan/chitosan.[17]

9.4.3.2 Mechanical Properties

As for the determination of mechanical properties of the film, standard properties measurements on coatings, like tensile strength (TS) and elongation at break (E), cannot be performed. In this case, the use of a support to assemble the nanocoating will affect the results in an immeasurable way, thus impairing any conclusion.

Nanoindentation can be used for testing the mechanical properties of these materials. Small loads and tip sizes are used, so the indentation area may only be a few square micrometers or even nanometers. The recorded data can be interpreted to obtain hardness, modulus, and other mechanical properties with very high accuracy and precision.[16]

Nanoindentation was used to assess the mechanical properties of a nano-layered chitosan/alginate coating.[16] The hardness of the support (PET) alone and for the nano-laminate coating assembled over the support was measured. The hardness of the nanocoating was determined as 0.245 GPa. This is higher than previous reports on coating of multilayered chitosan with other materials.[127]

9.4.4 Controlled Release

In the past years, the controlled release from nanosystems has been extensively studied. Most of these studies are devoted to developing drug delivery systems. In foods, the main applications of controlled release from nanosystems are related with human gastrointestinal environment when developing particles to deliver nutrients in the intestine.[128] Also, research on "intelligent" packaging of foods shows strong potential on extending shelf life, such as the release of antimicrobial agents when a packed food is exposed to a temperature abuse.[8,129]

Release of an active compound from a nanosystem can be due to the collapse of the nanosystem itself. However, in cases like nanocoatings aimed at extending shelf life of perishable foods, the release may be due to transport phenomena within the nanosystem itself. In these cases, understanding the transport properties and how they are related with the system's structural rearrangement will be crucial. Moreover, understanding the transport of bioactive compounds in edible nanocoatings can provide valuable information for the incorporation of compounds able to extend products' shelf life, such as antioxidants or antimicrobials.

Release of bioactive compounds from bio-polymeric matrices may occur due to mechanisms of Fick's diffusion, polymer matrix swelling, polymer erosion, and degradation[128,130] and a different mechanism may prevail, depending on the system and environmental conditions.

These mechanisms may be generally classified into three different types: ideal Fickian diffusion (Brownian transport), case II transport (polymer relaxation-driven), and anomalous behavior (ranging from Fickian to case II transport).

In recent years, several works concerning the release of functional compounds from conventional edible films have been published.[131–133] Comparatively, few works can be found on release mechanisms involved at the nanoscale. Literature suggests that in the case of nanolayered coatings, the release behavior depends on the permeability and on the disassembly or erosion of the multilayer structure and on other experimental variables.[134] Literature suggests that, depending on nanolayered coatings' composition and on the released compound, release may follow a Fickian[135] or an anomalous transport behavior.[136] Also, some authors attribute the observed behavior to a Fickian transport (Equation 9.2) through the nanolayers, followed by release due to polymer dissolution.[137] However, there are still a lot of unknown variables that certainly play a role during the release of compounds from nanocoatings, especially regarding the effect of environmental conditions.

$$\frac{M_t}{M_\infty} = 1 - \frac{8}{\pi^2} \cdot \exp(-k_F \cdot t) \tag{9.2}$$

where
M_t is the total mass released from the polymer at time t
M_∞ is the mass of the compound release at equilibrium
k_F is Fickian diffusion rate constant

Recently, the transport mechanism of a model compound (methylene blue) release from a K-carrageenan/chitosan nanolayered coating in acid and neutral solutions was studied. For that, the linear superimposition model was used (Equation 9.3), which accounts for both Fickian and case II transport effects[133,138,139] and is then a suitable approach to investigate transport mechanisms in these systems. This approach assumes that the observed transport of molecules within the polymer can be described by the sum of the molecules transported due to Brownian motion (Fick's transport), with the molecules transported due to polymer relaxation:

$$\frac{M_t}{M_\infty} = X\left[1 - \frac{8}{\pi^2}\sum_{n=1}^{\infty}\frac{1}{n^2}\exp(-n^2 k_F t)\right] + (1-X)[1-\exp(-k_R t)] \qquad (9.3)$$

where

M_t is the total mass released from the polymer at time t
M_∞ is the mass of the compound released at equilibrium
k_F is the Fickian diffusion rate constant
k_R is the relaxation rate constant
X is the fraction of compound released by Fickian transport

For the case of experiments under acidic conditions, the model was able to adequately describe the experimental results (Figure 9.1). However, this model was not accurate for the same system under pH = 7, neither simple Fick's diffusion was able to describe it appropriately (Figure 9.2).

Finally, it is important to stress that these and the rest of literature results are related with release to liquid systems. The future of this area of research

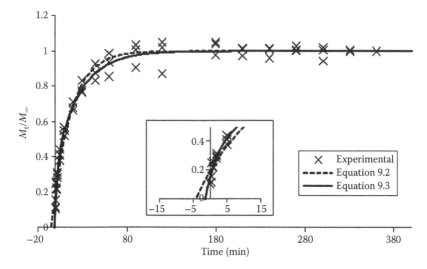

FIGURE 9.1
Example of Fick's (Equation 9.2) and LSM (Equation 9.3) description of methylene blue release at 37°C from the fourth layer at pH = 2. Insets show the detail of the model fitting to the initial experimental data.[140] (Reprinted from *Innovative Food Science & Emerging Technologies*, A.C. Pinheiro, A.I. Bourbon, M.A.C. Quintas, M.A. Coimbra, A.A. Vicente, K-carrageenan/chitosan nanolayered coating for controlled release of a model bioactive compound, in press, Copyright 2012, with permission from Elsevier.)

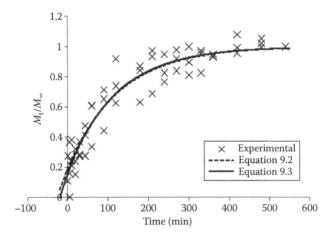

FIGURE 9.2

Fick's (Equation 9.2) and LSM (Equation 9.3) description of methylene blue release at pH 7. Example is for the fourth layer at 4°C.[140] (Reprinted from *Innovative Food Science & Emerging Technologies*, A.C. Pinheiro, A.I. Bourbon, M.A.C. Quintas, M.A. Coimbra, A.A. Vicente, K-carrageenan/chitosan nanolayered coating for controlled release of a model bioactive compound, in press, Copyright 2012, with permission from Elsevier.)

should focus on simulating release to solid foods, which are the main application of edible nanocoatings and where other variables (e.g., water activity of the food, respiration rate) must be taken into account.

9.5 Nano-Laminate Coating Applications

The formation of nano-laminate coatings using the LbL assembly technique has been extensively explored over the last years.[16,20,101,141] Nanolaminates can provide to food scientists interesting advantages for the preparation of edible coatings and films over conventional technologies, such as the simplicity of their preparation (simple processing operations such as dipping and washing procedures), the low concentrations of polymers needed, and the possibility of activating the release of functional agents in response to environmental changes, such as dilution, pH, ionic strength, or temperature.[19]

The choice of the type of adsorbent substances used to form each one of the layers, the total number of layers incorporated in the film, the sequence of different layers, and the conditions of preparation used to form each layer will determine the functionality of the final film (e.g., permeability to gases, mechanical properties, and sensitivity to environmental conditions).[19]

Recently, some works have been developed in order to understand how the application of bio-based materials using LbL could affect the substrates' properties and their suitability to be applied as a functional material (Table 9.3).

TABLE 9.3

Nano-Laminate Coating Applications

Nanostructure	Polyelectrolytes	Properties	References
Nanofilms	Chitosan and hyaluronan	Antimicrobial	[142]
	Chitosan and heparin	Antiadhesive and antibacterial	[101]
	K-carrageenan and lysozyme	Good gas-barrier properties	[38]
	Chitosan and sodium alginate	Highly functional properties	[16]
	Montmorillonite clay and cationic polyacrylamide	Good gas-barrier properties	[18]
	Hen egg white lysozyme and poly(L-glutamic acid)	Antimicrobial	[20]
	Poly-L-lysine and poly-L-glutamic acid	Bioactive compound release system	[134]
Nanoparticles	Sodium alginate and calcium phosphate	Bioactive compound release system	[58]
	Dextran-hydroxyethyl methacrylate and poly-L-arginine	Bioactive compound release system	[99]
	Poly-L-arginine and dextran sulfate	Bioactive compound release system	[55]
Nanoemulsions	β-lactoglobulin, chitosan, alginate, and pectin	Higher physical stability	[109]
	Chitosan and dextran sulfate	Bioactive compound release system	[143]
	Chitosan and κ-carrageenan	Good mechanical resistance	[110]

Richert et al.[142] studied the formation of polysaccharide films based on the alternate deposition of chitosan and hyaluronan. The deposition process was characterized by an exponential increase of the mass and thickness of the film with the number of deposition steps, which according to the authors is related to the ability of the polycation (chitosan) to diffuse "in" and "out" of the whole film at each deposition step. The authors also analyzed the effect of the MW of the chitosan on the buildup process and observed a faster growth for low MW chitosan. The influence of the salt concentration during buildup was also investigated. It was observed that whereas the chitosan/hyaluronan films grow rapidly at high salt concentrations (0.15 M NaCl) with the formation of a uniform film after only a few deposition steps, it is very difficult to build the film at salt concentrations of 10^{-4} M. The chitosan/hyaluronan films built at high salt concentrations were found to be chondrocyte resistant and, more interestingly, bacterial resistant (bacterial adhesion was reduced by 80% as compared to that on a glass substrate). Therefore, the authors concluded that chitosan/hyaluronan multilayers may be used as antimicrobial coatings.[142]

Also the alternating deposition of pectin and chitosan at a pH under which both polymers carry a charge resulted in the formation of a multilayer structure.[144] This deposition was irreversible over practical timescales and the thickness of an individual layer showed a dependence on biopolymer concentration; this effect was particularly marked for pectin.

Martins et al.[145] reported the design of a polyelectrolyte multilayer of chitosan and alginate at neutral pH. The authors showed that the density charge of chitosan has a profound influence on the film formation and on its structural properties and that the pH-dependent character of polyelectrolytes enables the production of multilayered systems with different viscoelastic behaviors. This work showed the possibility of processing multilayered films containing chitosan at physiological conditions that could enable the introduction of pH-sensitive molecules.[145]

Also using chitosan as polycation and in this case heparin as the negative charge polymer, Fu et al.[101] developed antiadhesive and antibacterial multilayer films through the alternate deposition of the polymers onto an aminolyzed PET film. The chitosan/heparin multilayer film depressed *Escherichia coli* adhesion dramatically and was effective in reducing the numbers of viable bacterial cells adhered onto the PET surface. The authors also showed that pH influences both surface properties of the films (i.e., composition, roughness, wettability) and their antiadhesive and antibacterial properties.[101]

In a similar work, a nanolayered coating composed of κ-carrageenan and lysozyme was assembled on an aminolyzed/charged PET support, by alternate five-layer deposition.[38] The κ-carrageenan/lysozyme nanolayered coating exhibited low water vapor and oxygen permeabilities. Also Carneiro-da-Cunha et al.[16] developed a nanolayered film composed of chitosan and sodium alginate on to an aminolyzed/charged PET and characterized its surface, gas barrier, thermal, and mechanical properties. The alginate/chitosan nanolayered film exhibited highly functional properties (such as good gas barrier properties) and presented promising applications onto different types of substrates aiming at, for example, the preparation of multilayer coatings for biomedical applications or multilayer edible coatings for food applications with enhanced mass transfer and mechanical properties.

The adsorption behavior of fucoidan and chitosan derivatives (chitosan sulfate) on PET film surface was studied by Indest et al.[44] and it was found that chitosan/fucoidan films were thinner and more compressed, while in the case of chitosan/chitosan sulfate films large amounts of chitosan sulfate were adsorbed, indicating a loose and thick adsorbed film.[44]

Other authors[18] produced thin films composed of negatively charged montmorillonite clay and a cationic polyacrylamide using a PET substrate. The film exhibited a low oxygen transmission rate (OTR) and its high barrier behavior was explained based on the presence of a brick wall nanostructure within the thin film that probably produces an extensive tortuous path for a diffusing oxygen molecule. Also, the thin film showed an optical transparency greater than 90% and potential for microwaveability and was reported as a good candidate for foil replacement in food packaging while being possibly also useful for flexible electronics packaging.

Rudra et al.[20] developed an antimicrobial multilayer coating through LbL deposition of hen egg white lysozyme (HEWL) and poly(L-glutamic acid) (PLGA). The PLGA/HEWL nanolayered films inhibited the growth of the model microorganism *Micrococcus luteus* in the surrounding liquid medium and therefore were mentioned as having promising applications in food preservation.[20]

LbL nano-laminate films have been also investigated as bioactive compound release systems[134,146,147] due to the possibility of controlling the release of molecules through manipulating the film's properties and to incorporate a wide range of functional biomolecules without substantial loss of their biological functions.[136]

Functional LbL multilayer structures can be obtained by chemical grafting of polyelectrolytes by functional moieties,[148] alternate deposition of polyelectrolytes and functional molecules,[149] or by postdiffusion of the functional molecules into the multilayers.[134,147] A diversity of nanolayered films has been developed to control bioactive compounds' release by deploying different triggers such as pH, ionic strength, temperature, and enzymes.[150–152]

Jiang and Li[134] developed polypeptide multilayer nanofilms composed of poly-L-lysine and poly-L-glutamic acid on quartz slides or stainless steel discs using the LbL technique and studied the loading and release behavior of small charged bioactive molecules (e.g., cefazolin, gentamicin, and methylene blue). Their loading and release were found to be pH dependent and could also be controlled by changing the number of film layers and drug incubation time and applying heat treatment after film formation. Antibiotic-loaded polypeptide multilayer nanofilms showed controllable antibacterial properties against *Staphylococcus aureus*. The authors concluded that the developed biodegradable polypeptide multilayer nanofilms are capable of loading both positively and negatively charged drug molecules and promise to serve as drug delivery systems.

Polypeptide multilayer nanofilms have been prepared on quartz microscope slides by LbL self-assembly to study the postfabrication loading and release of a model therapeutic, methylene blue (MB).[147] The results showed that the amount of MB loaded and the rate of MB release have been influenced by the choice of polypeptide film species, pH, number of capping layers, and release medium.

In the late 1990s, Caruso et al. extensively investigated the adsorption of polyelectrolytes onto colloidal particles using the LbL technique. Electrostatic interactions between the particles and polymer were used to build up the multilayers onto colloidal templates and to obtain hollow spheres with well-defined physical and chemical properties.[39,58] This structure provides one of the most versatile techniques for encapsulating bioactive compounds with a high loading of the latter due to high surface area of the porous particle and is applicable to a wide range of materials of different sizes, from proteins to low MW drugs.[153]

LbL assembly of polyelectrolytes can also be applied on soft and porous templates such as nanogels. The physical–chemical properties of nanogel capsules can be tailored or their stability improved, by depositing a shell of polyelectrolytes. Wong et al.[154] applied LbL assembly on thermoresponsive nanogels and demonstrated that the layers were not stripped off during the phase transition of the nanogels. Wong and Richtering[153] reviewed the application of the LbL technique as an attractive process to modify and tune the properties of microgel particles' surface. These authors observed that the choice of the parameters of polyelectrolyte pairs may govern the transport of substances into and out of the gel capsule.[153,154]

The encapsulation of lipophilic molecules inside the oily core of lipid nanoemulsions, using the LbL adsorption technique, is also one of the alternatives that can be used to protect lipophilic functional components within lipid droplets from chemical degradation.[155–157]

The application of nano-laminate coatings to nanoemulsions can be used to overtake some of the nanoemulsions' limitations such as limited stability to pH, salt, heating, freezing, dehydration, and chilling.[4,109,158] The research in the development of nano-laminate coatings applied to nanoemulsions is in an early stage of development; however, some works already performed seem to show that their successful development may be a reality. Li et al.[109] have shown that it is possible to prepare multilayer nanoemulsions by the LbL technique, using a globular protein (β-lactoglobulin) to form the inner layer of the multilayer (first layer), using a cationic polysaccharide (chitosan) as an intermediate layer (second layer), and an anionic polysaccharide (pectin or alginate) as an outer layer (third layer). The nanoemulsions with the first layer alone were unstable to droplet aggregation at intermediate pH values due to the low net surface charge. The two-layer nanoemulsions were stable from pH 3 to 6 but were highly unstable at higher pH values. Nanoemulsions that had a third outer layer (alginate or pectin) were unstable to droplet aggregation at pH 3, but stable at higher pH values. The addition of polysaccharides through the LbL technique to triacylglycerol nanoemulsions can improve their physical stability, while reducing the rate of lipid digestion by restricting the access of lipases to the emulsified triacylglycerols.[109]

Also through LbL technique Madrigal-Carballo et al.[143] developed a novel delivery system for ellagic acid by deposition of biopolymers (chitosan and dextran sulfate) onto soybean lecithin liposomes. Release studies revealed that with the biopolymers' addition a better thermal and pH stability was achieved, as well as improved release properties.[143]

Preetz et al.[110] developed an LbL-based nanocapsule system starting from a nanoemulsion. Octenyl succinic anhydride starch (OSA starch) was used as an emulsifier and chitosan and k-carrageenan were the polysaccharides chosen. The addition of the polysaccharides promoted a higher stiffness of the wall with the increasing number of shell layers observed using AFM. A stiffer capsule might be advantageous during storage due to better mechanical resistance together with the potential to provide a stronger barrier to molecules' migration; nanocapsules developed by LbL were thus expected to prolong drug delivery time as compared to nanoemulsions.[110]

Aoki et al.[156] have used a three-step process based on LbL technique to produce multilayer nanoemulsions. Nanoemulsions were prepared using high-pressure homogenization, stabilized by sodium dodecyl sulfate. Then, a secondary layer was formed by chitosan and as tertiary layer pectin was used. This work shows that nanoemulsion stability could be improved only by coating the droplets with two or more layers.[159]

Medeiros et al.[38] applied a nano-laminate coating (κ-carrageenan/lysozyme), through the LbL technique, on 'Rocha' (*Pyrus communis* L.) fresh-cut and whole pears; their results showed that the nanolayered coating assembled on the fruits' surface had a positive effect on fruit quality and contributed to extend their shelf life.

The aforementioned examples, resummed in Table 9.3, clearly demonstrate that nano-laminate coatings hold a very interesting and yet unexplored potential to be successfully used by the food industry in a great range of applications.

Acknowledgments

The authors Miguel A. Cerqueira, Ana I. Bourbon, Ana C. Pinheiro, Hélder D. Silva, and Mafalda A.C. Quintas are recipients, respectively, of the fellowships: SFRH/BPD/72753/2010, SFRH/BD/73178/2010, SFRH/BD/48120/2008, SFRH/BD/81288/2011, and SFRH/BPD/41715/2007, supported by Fundação para a Ciência e Tecnologia, POPH-QREN and FSE (FCT, Portugal).

References

1. P. Laurienzo, M. Malinconico, G. Mazzarella, F. Petitto, N. Piciocchi, R. Stefanile, and M.G. Volpe, *Journal of Dairy Science*, 91 (2008) 1317.
2. M.S. Rahman, *Handbook of Food Preservation*, 2nd edn. Boca Raton, FL: CRC Press (2007).
3. D.M.A.M. Luykx, R.J.B. Peters, S.M. van Ruth, and H. Bouwmeester, *Journal of Agricultural and Food Chemistry*, 56 (2008) 8231.
4. Q. Huang, H. Yu, and Q. Ru, *Journal of Food Science*, 75 (2010) R50.
5. M. Quintanilla-Carvajal, B. Camacho-Díaz, L. Meraz-Torres, J. Chanona-Pérez, L. Alamilla-Beltrán, A. Jimenéz-Aparicio, and G. Gutiérrez-López, *Food Engineering Reviews*, 2 (2010) 39.

6. A.E. Pavlath and W. Orts, Edible films and coatings: Why, what, and how?, in *Edible Films and Coatings for Food Applications*, K.C. Huber and M.E. Embuscado, Eds. New York: Springer (2009), p. 57.
7. R.N. Tharanathan, *Trends in Food Science and Technology*, 14 (2003) 71.
8. J.H. Han, C.H.L. Ho, and E.T. Rodrigues, Intelligent packaging, in *Innovations in Food Packaging*, J. Han, Ed. Baltimore, MD: Elsevier Science & Technology Books (2005), p. 138.
9. M.A. Cerqueira, B.W.S. Souza, J.A. Teixeira, and A.A. Vicente. *Food Hydrocolloids*, 27 (2012) 175.
10. B. Medeiros, A. Pinheiro, M. Carneiro-da-Cunha, and A. Vicente, *Journal of Food Engineering*, 110 (2012) 457–464.
11. Y. Zhong, B. Li, and D.T. Haynie, *Nanotechnology*, 17 (2006) 5726.
12. P. Kujawa, G. Schmauch, T. Viitala, A. Badia, and F.M Winnik, *Biomacromolecules*, 8 (2007) 3169.
13. M.A. Cerqueira, A.I. Bourbon, A.C. Pinheiro, J.T. Martins, B.W.S. Souza, J.A. Teixeira, and A.A. Vicente, *Trends in Food Science and Technology*, 22 (2011) 662.
14. D.J. McClements, E.A. Decker, and J. Weiss, *Journal of Food Science*, 72 (2007) R109.
15. D.J. McClements, E.A. Decker, Y. Park, and J. Weiss, *Critical Reviews in Food Science and Nutrition*, 49 (2009) 577.
16. M.G. Carneiro-da-Cunha, M.A. Cerqueira, B.W.S. Souza, S. Carvalho, M.A.C. Quintas, J.A. Teixeira, and A.A. Vicente, *Carbohydrate Polymers*, 82 (2010) 153.
17. A.C. Pinheiro, A.I. Bourbon, B.G.S. Medeiros, L.H.M. da Silva, M.C.H. Silva, M.G. Carneiro-da-Cunha, M.A. Coimbra, and A.A. Vicente, *Carbohydrate Polymers*, 87 (2012) 1081.
18. W.-S. Jang, I. Rawson, and J.C. Grunlan, *Thin Solid Films*, 516 (2008) 4819.
19. J. Weiss, P. Takhistov, and D.J. McClements, *Journal of Food Science*, 71 (2006) R107.
20. J.S. Rudra, K. Dave, and D.T. Haynie, *Journal of Biomaterials Science*, 17 (2006) 1301.
21. M.M. de Villiers, D.P. Otto, S.J. Strydom, and Y.M. Lvov, *Advanced Drug Delivery Reviews*, 63 (2011) 701.
22. G. Decher, Polyelectrolyte multilayers, an overview, in *Multilayer Thin Films*. Wiley-VCH Verlag GmbH & Co. KGaA (2003), p. 1.
23. G.B. Sukhorukov, E. Donath, H. Lichtenfeld, E. Knippel, M. Knippel, A. Budde, and H. Möhwald, *Colloids and Surfaces A: Physicochemical and Engineering Aspects*, 137 (1998) p. 253.
24. B. Nyström, A.-L. Kjøniksen, N. Beheshti, A. Maleki, K. Zhu, K.D. Knudsen, R. Pamies, J.G. Hernández Cifre, and J. García de la Torre, *Advances in Colloid and Interface Science*, 158 (2010) 108.
25. L. Liu, J. Cao, J. Huang, Y. Cai, and J. Yao, *Bioresource Technology*, 101 (2010) 3268.
26. B.R. Thakur, R.K. Singh, A.K. Handa, and M.A. Rao, Chemistry and uses of pectin – A review. *Critical Reviews in Food Science and Nutrition*, 37(1) (1997) 47–73.
27. M. Rinaudo, Polyelectrolytes derived from natural polysaccharides, in *Monomers, Polymers and Composites from Renewable Resources*, M.N. Belgacem and A. Gandini, Eds. Oxford, UK: Elsevier (2008), p. 495.
28. J. Liu, E. Verespej, M. Alexander, and M. Corredig, Comparison on the effect of high-methoxyl pectin or soybean-soluble polysaccharide on the stability of sodium caseinate-stabilized oil/water emulsions. *Journal of Agricultural and Food Chemistry*, 55(15) (2007) 6270–6278.

29. H.-U. Endreβ and S.H. Christensen, Pectins, in *Handbook of hydrocolloids*, G.O. Phillips and P.A. Williams, Eds. Cambridge, UK: Woodhead Publishing Limited (2009), p. 274.
30. X. Yang and W. Zhu, *Cellulose*, 14 (2007) 409.
31. L. Passauer, H. Bender, and S. Fischer, *Carbohydrate Polymers*, 82 (2010) 809.
32. M. Meiczinger, J. Dencs, G. Marton, and B. Dencs, *Industrial & Engineering Chemistry Research*, 44 (2005) 9581.
33. W.C. Wielinga, Galactomannans, in *Handbook of hydrocolloids*, G.O. Phillips and P.A. Williams, Eds. Cambridge, UK: Woodhead Publishing Limited (2009), p. 228.
34. E. Frollini, W.F. Reed, M. Milas, and M. Rinaudo, Polyelectrolytes from poly-saccharides: Selective oxidation of guar gum – a revisited reaction. *Carbohydrate Polymers*, 27(2) (1995) 129–135.
35. M.R. Sierakowski, M. Milas, J. Desbrières, and M. Rinaudo, *Carbohydrate Polymers*, 42 (2000) 51–57.
36. E.-S. Chan, B.-B. Lee, P. Ravindra, and D. Poncelet, *Journal of Colloid and Interface Science*, 338 (2009) 63.
37. A.P. Imeson, Carrageenan and furcellaran, in *Handbook of hydrocolloids*, G.O. Phillips and P.A. Williams, Eds. Cambridge, UK: Published by Woodhead Publishing Limited (2009), p. 164.
38. B. Medeiros, A. Pinheiro, J. Teixeira, A. Vicente, and M. Carneiro-da-Cunha, *Food and Bioprocess Technology*, 5 (2012) 2435–2445.
39. M. Kusaykin, I. Bakunina, V. Sova, S. Ermakova, T. Kuznetsova, N. Besednova, T. Zaporozhets, and T. Zvyagintseva, Structure, biological activity, and enzymatic transformation of fucoidans from the brown seaweeds. *Biotechnology Journal*, 3(7) (2008) 904–915.
40. V.H. Pomin and P.A.S. Mourão, Structure, biology, evolution, and medical importance of sulfated fucans and galactans. *Glycobiology*, 18(12) (2008) 1016–1027.
41. M.I. Bilan, A.A. Grachev, N.E. Ustuzhanina, A.S. Shashkov, N.E. Nifantiev, and A.I. Usov, A highly regular fraction of a fucoidan from the brown seaweed *Fucus distichus* L. *Carbohydrate Research*, 339(3) (2004) 511–517.
42. L. Chevolot, B. Mulloy, J. Ratiskol, A. Foucault, and S. Colliec-Jouault, A disaccharide repeat unit is the major structure in fucoidans from two species of brown algae. *Carbohydrate Research*, 330(4) (2001) 529–535.
43. T.N. Zvyagintseva, N.M. Shevchenko, A.O. Chizhov, T.N. Krupnova, E.V. Sundukova, and V.V. Isakov, Water-soluble polysaccharides of some far-eastern brown seaweeds. Distribution, structure, and their dependence on the developmental conditions. *Journal of Experimental Marine Biology and Ecology*, 294(1) (2003) 1–13.
44. T. Indest, J. Laine, L.S. Johansson, K. Stana-Kleinschek, S. Strnad, R. Dworczak, and V. Ribitsch, *Biomacromolecules*, 10 (2009) 630.
45. T. Nishino, G. Yokoyama, K. Dobashi, M. Fujihara, and T. Nagumo, Isolation, purification, and characterization of fucose-containing sulfated polysaccharides from the brown seaweed ecklonia kurome and their blood-anticoagulant activities. *Carbohydrate Research*, 186(1) (1989) 119–129.
46. Y. Aisa, Y. Miyakawa, T. Nakazato, H. Shibata, K. Saito, Y. Ikeda, and M. Kizaki, Fucoidan induces apoptosis of human HS-Sultan cells accompanied by activation of caspase-3 and down-regulation of ERK Pathways. *American Journal of Hematology*, 78(1) (2005) 7–14.

47. J.W. Yang, S.Y. Yoon, S.J. Oh, S.K. Kim, and K.W. Kang, Bifunctional effects of fucoidan on the expression of inducible nitric oxide synthase. *Biochemical and Biophysical Research Communications*, 346(1) (2006) 345–350.

48. N.M.A. Ponce, C.A. Pujol, E.B. Damonte, M.L. Flores, and C.A. Stortz, Fucoidans from the brown seaweed Adenocystis utricularis: extraction methods, antiviral activity and structural studies. *Carbohydrate Research*, 338(2) (2003) 153–165.

49. S. Hirmo, M. Utt, M. Ringner, and T. Wadström, Inhibition of heparan sulphate and other glycosaminoglycans binding to Helicobacter pylori by various poly-sulphated carbohydrates. *FEMS Immunology And Medical Microbiology*, 10(3–4) (1995) 301–306.

50. X. Zhao, C.-H. Xue, and B.-F. Li, Study of antioxidant activities of sulfated poly-saccharides from *Laminaria japonica*. *Journal of Applied Phycology*, 20(4) (2008) 431–436.

51. G. Sworn, Xanthan gum, in *Handbook of Hydrocolloids*, G.O. Phillips and P.A. Williams, Eds. Cambridge, U.K.: Woodhead Publishing Limited (2009).

52. P-e. Jansson, L. Kenne, and B. Lindberg, *Carbohydrate Research*, 45 (1975) 275.

53. S. Simsek, B. Mert, O.H. Campanella, and B. Reuhs, *Carbohydrate Polymers*, 76 (2009) 320.

54. A. Boutebba, M. Milas, and M. Rinaudo, *International Journal of Biological Macromolecules*, 24 (1999) 319.

55. C. Peniche, Chitin and Chitosan: Major Sources, Properties and Applications, in *Monomers, Polymers and Composites from Renewable Resources*, M.N. Belgacem and A. Gandini, Eds. Oxford, UK: Elsevier (2008).

56. M. Dash, F. Chiellini, R.M. Ottenbrite, and E. Chiellini, *Progress in Polymer Science*, 36 (2011) 981.

57. P. Agrawal, G.J. Strijkers, and K. Nicolay, *Advanced Drug Delivery Reviews*, 62 (2010) 42.

58. F. Shahidi, J.K.V. Arachchi, and Y.J. Jeon, *Trends in Food Science and Technology*, 10 (1999) 37.

59. I.M. Helander, E.L. Nurmiaho-Lassila, R. Ahvenainen, J. Rhoades, and S. Roller, *International Journal of Food Microbiology*, 71 (2001) 235.

60. C.M. Bryant and D.J. McClements, *Trends in Food Science & Technology*, 9 (1998) 143.

61. L. Chen, G.E Remondetto, and M. Subirade, *Trends in Food Science & Technology*, 17 (2006) 272.

62. C.M. Oliver, L.D. Melton, and R.A. Stanley, *Critical Reviews in Food Science and Nutrition*, 46 (2006) 337.

63. A.A. Vicente, M.A. Cerqueira, L. Hilliou, and C. Rocha, Protein-based res-ins for food packaging, in *Multifunctional and Nanoreinforced Polymers for Food Packaging*, J.M. Lagaron, Ed. Cambridge, U.K.: Woodhead Publishing Limited (2011), p. 610.

64. J.W. Rhim, J.H. Lee, and P.K.W. Ng, *LWT—Food Science and Technology*, 40 (2007) 232.

65. W.J. Wolf, *Journal of Agricultural and Food Chemistry*, 18 (1970) 969.

66. J. Kinsella, *Journal of the American Oil Chemists' Society*, 56 (1979) 242–258.

67. C.M.M. Lakemond, H.H.J. de Jongh, M. Hessing, H. Gruppen, and A.G.J. Voragen, *Journal of Agricultural and Food Chemistry*, 48 (2000) 1991.

68. Y.D. Livney, *Current opinion in colloid & interface science*, 15 (2010) 73.

69. L.K. Creamer and A.K.H. MacGibbon, *International Dairy Journal*, 6 (1996) 539.

70. D.W.S. Wong, W.M. Camirand, A.E. Pavlath, N. Parris and M. Friedman, *Critical Reviews in Food Science and Nutrition*, 36 (1996) 807.
71. P.F. Fox, *International Journal of Dairy Technology*, 54 (2001) 41.
72. G. Kontopidis, C. Holt, and L. Sawyer, *Journal of Dairy Science*, 87 (2004) 785.
73. S. Prabakaran and S. Damodaran, *Journal of Agricultural and Food Chemistry*, 45 (1997) 4303.
74. J. Li and P. Yao, *Langmuir*, 25 (2009) 6385.
75. C. Thomä-Worringer, J. Sørensen, and R. López-Fandiño, *International Dairy Journal*, 16 (2006) 1324.
76. S.I. Park, M.A. Daeschel, and Y. Zhao, *Journal of Food Science*, 69 (2004) M215.
77. G. Brisson, M. Britten, and Y. Pouliot, *International Dairy Journal*, 17 (2007) 617.
78. B.W.A. Van der Strate, L. Beljaars, G. Molema, M.C. Harmsen, and D.K.F. Meijer, *Antiviral Research*, 52 (2001) 225.
79. S.A. González-Chávez, S. Arévalo-Gallegos, and Q. Rascón-Cruz, *International Journal of Antimicrobial Agents*, 33 (2009) 301.e1.
80. J. Hu, S. Yu, and P. Yao, *Langmuir*, 23 (2007) 6358.
81. S. Yu, J. Hu, X. Pan, P. Yao, and M. Jiang, *Langmuir*, 22 (2006) 2754.
82. S. Sundar, J. Kundu, and S.C. Kundu, *Science and Technology of Advanced Materials* 11 (2010) 014104.
83. C. Coester, P. Nayyar, and J. Samuel, *European Journal of Pharmaceutics and Biopharmaceutics*, 62 (2006) 306.
84. M. Jahanshahi, G. Najafpour, and M. Rahimnejad, *African Journal of Biotechnology*, 7 (2008) 362.
85. I. Langmuir, *Journal of the American Chemical Society*, 39 (1917) 1848.
86. K.B. Blodgett, *Journal of the American Chemical Society*, 56 (1934) 495.
87. F.J. Pavinatto, L. Caseli, and O.N. Oliveira, *Biomacromolecules*, 11 (2010) 1897.
88. G. Decher and J.-D. Hong, *Makromolekulare Chemie. Macromolecular Symposia*, 46 (1991) 321.
89. C. Wang, S. Ye, L. Dai, X. Liu, and Z. Tong, *Biomacromolecules*, 8 (2007) 1739.
90. S.T. Dubas, T.R. Farhat, and J.B. Schlenoff, *Journal of the American Chemical Society*, 123 (2001) 5368.
91. V. Zucolotto, K.R.P. Daghastanli, C.O. Hayasaka, A. Riul, P. Ciancaglini, and O.N. Oliveira, *Analytical Chemistry*, 79 (2007) 2163.
92. C.R. Wittmer, J.A. Phelps, C.M. Lepus, W.M. Saltzman, M.J. Harding, and P.R. Van Tassel, *Biomaterials*, 29 (2008) 4082.
93. S.-H. Lee, S. Balasubramanian, D.Y. Kim, N.K. Viswanathan, S. Bian, J. Kumar, and S.K. Tripathy, *Macromolecules*, 33 (2000) 6534.
94. J.H. Fendler, *Chemistry of Materials*, 8 (1996) 1616.
95. S.N. Kim, J.F. Rusling, and F. Papadimitrakopoulos, *Advanced Materials*, 19 (2007) 3214.
96. J.R. Siqueira, L.H.S. Gasparotto, F.N. Crespilho, A.J.F. Carvalho, V. Zucolotto, and O.N. Oliveira, *The Journal of Physical Chemistry B*, 110 (2006) 22690.
97. J.A. He, L. Samuelson, L. Li, J. Kumar, and S.K. Tripathy, *Langmuir*, 14 (1998) 1674.
98. P. Bertrand, A. Jonas, A. Laschewsky, and R. Legras, *Macromolecular Rapid Communications*, 21 (2000) 319.
99. B.G. De Geest, C. Déjugnat, E. Verhoeven, G.B. Sukhorukov, A.M. Jonas, J. Plain, J. Demeester, and S.C. De Smedt, *Journal of Controlled Release*, 116 (2006) 159.
100. J.P. de Mesquita, C.L. Donnici, and F.V. Pereira, *Biomacromolecules*, 11 (2010) 473.

101. J. Fu, J. Ji, W. Yuan, and J. Shen, *Biomaterials*, 26 (2005) 6684.
102. J.W. Ostrander, A.A. Mamedov, and N.A. Kotov, *Journal of the American Chemical Society*, 123 (2001) 1101.
103. H.I. Labouta and M. Schneider, *International Journal of Pharmaceutics*, 395 (2010) 236.
104. K. Sato, K. Yoshida, S. Takahashi, and J.-I. Anzai, *Advanced Drug Delivery Reviews*, 63 (2011) 809.
105. J. Kreuter, *European Journal of Drug Metabolism and Pharmacokinetics*, 19 (1994) 253.
106. F. Caruso and J.F. Quinn, *Chemistry in Australia*, 71 (2004) 18.
107. Y. Wang, A.S. Angelatos, and F. Caruso, *Chemistry of Materials*, 20 (2007) 848.
108. R. Pongsawatmanit, T. Harnsilawat, and D.J. McClements, *Colloids and Surfaces A: Physicochemical and Engineering Aspects*, 287 (2006) 59.
109. Y. Li, M. Hu, H. Xiao, Y. Du, E.A. Decker, and D.J. McClements, *European Journal of Pharmaceutics and Biopharmaceutics*, 76 (2010) 38.
110. C. Preetz, A. Hauser, G. Hause, A. Kramer, and K. Mäder, *European Journal of Pharmaceutical Sciences*, 39 (2010) 141.
111. ASTM, American Society for Testing and Materials (1985) D 4187.
112. K.A. Marx, *Biomacromolecules*, 4 (2003) 1099.
113. P.J. Molino, O.M. Hodson, J.F. Quinn, and R. Wetherbee, *Langmuir*, 24 (2008) 6730.
114. S. Channasanon, W. Graisuwan, S. Kiatkamjornwong, and V.P. Hoven, *Journal of Colloid and Interface Science*, 316 (2007) 331–343.
115. D. Yoo, S.S. Shiratori, and M.F. Rubner, *Macromolecules*, 31 (1998) 4309.
116. T. Nicolet. (2001). Introduction to Fourier transform infrared spectrometry. http://mmrc.caltech.edu/FTIR/FTIRintro.pdf (accessed on January 11, 2011).
117. G. Lawrie, I. Keen, B. Drew, A. Chandler-Temple, L. Rintoul, P. Fredericks, and L. Grøndahl, *Biomacromolecules*, 8 (2007) 2533–2541.
118. P. Thanasukarn, R. Pongsawatmanit, and D.J. McClements, *Food Hydrocolloids*, 18 (2004) 1033–1043.
119. H. Salmah, H. Ismail, and A.A. Bakar, *Journal of Applied Polymer Science*, 107 (2008) 2266.
120. C.G. Venturini, E. Jäger, C.P. Oliveira, A. Bernardi, A.M.O. Battastini, S.S. Guterres, and A.R. Pohlmann, *Colloids and Surfaces A: Physicochemical and Engineering Aspects*, 375 (2011) 200.
121. J.-H. Zhu, X.-W. Wang, S. Ng, C.-H. Quek, H.-T. Ho, X.-J. Lao, and H. Yu, *Journal of Biotechnology*, 117 (2005) 355.
122. Z.L. Wang, *The Journal of Physical Chemistry B*, 104 (2000) 1153.
123. L. Reimer, *Measurement Science and Technology*, 11 (2000) 1826.
124. C.I. Moraru, C.P. Panchapakesan, Q. Huang, P. Takhistov, S. Liu, and J.L. Kokini, *Food Technology*, 57 (2003) 24.
125. L. Richert, P. Lavalle, E. Payan, X.Z. Shu, G.D. Prestwich, J.-Fo Stoltz, P. Schaaf, J.-C. Voegel, and C. Picart, *Langmuir*, 20 (2003) 448.
126. K. Cooksey, K.S. Marsh, and L.H. Doar, *Food Technology*, 53 (1999) 60.
127. H.J. Martin, K.H. Schulz, J.D. Bumgardner, and J.A. Schneider, *Thin Solid Films*, 516 (2008) 6277.
128. Q. Li, C.-G. Liu, Z.-H. Huang, and F.-F. Xue, *Journal of Agricultural and Food Chemistry*, 59 (2011) 1962.

129. J.H. Han and A. Gennadios, Edible films and coatings: a review, in *Innovations in Food Packaging*, J.H. Han, Ed. New York: Academic Press (2005), p. 239.
130. R.A. Jain, *Biomaterials*, 21 (2000) 2475.
131. N. Faisant, J. Siepmann, and J.P. Benoit, *European Journal of Pharmaceutical Sciences*, 15 (2002) 355.
132. M.A. Del Nobile, A. Conte, A.L. Incoronato, and O. Panza, *Journal of Food Engineering*, 89 (2008) 57.
133. S. Flores, A. Conte, C. Campos, L. Gerschenson, and M. Del Nobile, *Journal of Food Engineering*, 81 (2007) 580.
134. B. Jiang and B. Li, *International Journal of Nanomedicine*, 4 (2009) 37.
135. M. Polakovic, T. Görner, R. Gref, and E. Dellacherie, *Journal of Controlled Release*, 60 (1999) 169.
136. X.Y. Wang, X. Hu, A. Daley, O. Rabotyagova, P. Cebe, and D.L. Kaplan, *Journal of Controlled Release*, 121 (2007) 190.
137. A.A. Antipov, G.B. Sukhorukov, E. Donath, and H. Möhwald, *The Journal of Physical Chemistry B*, 105 (2001) 2281.
138. A.R. Berens and H.B. Hopfenberg, *Polymer*, 19 (1978) 489.
139. J. Crank, *The Mathematics of Diffusion*, 2nd ed. Oxford, U.K.: Clarendon Press (1975).
140. A.C. Pinheiro, A.I. Bourbon, M.A.C. Quintas, M.A. Coimbra, and A.A. Vicente, K-carrageenan/chitosan nanolayered coating for controlled release of a model bioactive compound, (2012), Reprinted from Innovative Food Science & Emerging Technologies, in press.
141. P. Lavalle, C. Gergely, F.J.G. Cuisinier, G. Decher, P. Schaaf, J.C. Voegel, and C. Picart, *Macromolecules*, 35 (2002) 4458.
142. L. Richert, P. Lavalle, E. Payan, X.Z. Shu, G.D. Prestwich, J.-F. Stoltz, P. Schaaf, J.-C. Voegel, and C. Picart, *Langmuir*, 20 (2004) 448.
143. S. Madrigal-Carballo, S. Lim, G. Rodriguez, A.O. Vila, C.G. Krueger, S. Gunasekaran, and J.D. Reed, *Journal of Functional Foods*, 2 (2010) 99.
144. M. Marudova, S. Lang, G.J. Brownsey, and S.G. Ring, *Carbohydrate Research*, 340 (2005) 2144.
145. G.V. Martins, J.F. Mano, and N.M. Alves, *Carbohydrate Polymers*, 80 (2010) 570.
146. A.J. Chung and M.F. Rubner, *Langmuir*, 18 (2002) 1176.
147. Y. Zhong, C.F. Whittington, L. Zhang, and D.T. Haynie, *Nanomedicine: Nanotechnology, Biology and Medicine*, 3 (2007) 154.
148. D.M. Kaschak, J.T. Lean, C.C. Waraksa, G.B. Saupe, H. Usami, and T.E. Mallouk, *Journal of the American Chemical Society*, 121 (1999) 3435.
149. E. Rousseau, M. Van der Auweraer, and F.C. De Schryver, *Langmuir*, 16 (2000) 8865.
150. J.F. Quinn and F. Caruso, *Langmuir*, 20 (2004) 20.
151. T. Serizawa, M. Yamaguchi, and M. Akashi, *Macromolecules*, 35 (2002) 8656.
152. K.C. Wood, J.Q. Boedicker, D.M. Lynn, and P.T. Hammond, *Langmuir*, 21 (2005) 1603.
153. J.E. Wong and W. Richtering, *Current Opinion in Colloid & Interface Science*, 13 (2008) 403.
154. J.E. Wong, C.B. Müller, A.M. Díez-Pascual, and W Richtering, *The Journal of Physical Chemistry B*, 113 (2009) 15907.

155. S. Hirsjärvi, Y. Qiao, A. Royere, J. Bibette, and J.-P. Benoit, *European Journal of Pharmaceutics and Biopharmaceutics*, 76 (2010) 200.
156. L. Moreau, H.-J. Kim, E.A. Decker, and D.J. McClements, *Journal of Agricultural and Food Chemistry*, 51 (2003) 6612.
157. S. Mun, E.A. Decker, and D.J. McClements, *Langmuir*, 21 (2005) 6228.
158. D. Guzey and D.J. McClements, *Advances in Colloid and Interface Science*, 128–130 (2006) 227.
159. T. Aoki, E.A. Decker, and D.J. McClements, *Food Hydrocolloids*, 19 (2005) 209.

10

Potential Application of Nanomaterials in Food Packaging and Interactions of Nanomaterials with Food

Zehra Ayhan

CONTENTS

10.1 Introduction

Nanotechnology has potential applications in all aspects of food chain including food processing, food quality monitoring, food packaging, and storage.[1] Major areas of food industry that could benefit from nanotechnology are[1] development of new functional materials for food packaging,[2] microscale and nanoscale processing,[3] product development,[4] and methods and instrumentation design for improved food safety and biosecurity.[2]

Application of nanotechnology in food packaging is considered highly promising since this technology could improve safety and quality of food while reducing the use of valuable raw materials and the generation of packaging waste. Nanotechnology is applicable in food packaging to improve packaging performances such as gas, moisture, UV and volatile barriers, mechanical strength, heat resistance and flame retardancy, and weight.[3,4] Nanotechnology can provide shelf life extension via active packaging, product condition monitoring through intelligent packaging, and delivery and controlled release of nutraceuticals.

Barrier properties, mechanical and oxidation stability, and biodegradability of conventional polymeric materials could be improved by nanotechnology. Poor barrier and mechanical properties of edible and biodegradable films could be improved by using nanoparticles so that the use of these materials in food industry could be expanded. The use of biodegradable nanocomposites will help to reduce packaging waste while extending shelf life of processed foods. Bioactive compounds nanoencapsulated into the packaging are a promising approach due to controlled release of these compounds into the food product. Another potential application of nanotechnology in smart packaging is the use of nanosensors embedded in the packaging to monitor product condition, detect food spoilage, and alert the consumer when food is spoiled.[5,6]

This chapter reviews the potential use of nanomaterials such as nanocomposites, bionanopolymers, and active nanomaterials in food packaging applications. Interactions of such nanomaterials with food components and migration from nanomaterials into food are evaluated considering food quality and safety. Commercial applications of nanomaterials and nanosensors in food packaging are documented. Consumer perception of nanotechnology used in food packaging is also briefly addressed. Since potential effects of nanoparticles on health and environment and regulatory issues are already documented in the recent literature, they are not covered in this chapter.

10.2 Nanomaterials in Food Packaging Applications

10.2.1 Nanocomposites in Food Packaging Applications

Nowadays, multilayered packaging materials are extensively used in ready-to-eat meat, dairy, and other products to assure food quality, safety, and required shelf life due to their gas barrier properties. It is important to select packaging material with low oxygen (OTR) and water vapor transmission rates (WVTR). For this purpose, barrier materials such as polyvinylidene chloride (PVdC) or ethylene vinyl alcohol (EVOH)-containing multilayer materials are preferred. However, laminated materials especially with PVdC and EVOH are highly expensive and these materials are not eco-friendly. Due to complex production techniques, high cost, and nonrecyclable nature of the multilayered materials, nanomaterials with improved barrier properties could be a good alternative to multilayered materials.

Scientific studies showed that nanofillers especially nanoclays used in the polymer matrix provided marked increase in the barrier properties of materials especially to gas and water. There are many studies showing the effectiveness of nanoclays in reducing oxygen and water vapor permeabilities of different polymers.[7-16] Nanoclay particles incorporated into polymer used in food packaging can reduce oxygen permeability as much as 75%. The lowest OTR was reported for nano-polyethyleneterephthalate (PET).[17,18] A recent study reported almost 100% reduction in the oxygen permeability for a transparent clay-polymer material.[19] Pererira de Abreu et al. reported that the incorporation of nanoparticles into most common polyolefin films (polypropylene [PP] and polyethylene [PE] films) used in food packaging provided good exfoliation and barrier and mechanical properties. The OTR in nano-PP was 22% lower than the control PP. The OTR was reduced 12.5% in nano-PE compared to control-PE.[20]

Many studies have focused on preparation, material properties, and characterization of nanocomposites. The polymers used for clay-based nanocomposites are polyamides, nylons, polyolefins, polystyrene (PS), ethylene-vinylacetate (EVA), polyimides, and PET. Improvements in the development of polymer nanocomposites increased the potential use of these materials for food packaging applications such as processed meats, cheese, confectionery, cereal, boil-in-the-bag foods, extrusion coating applications for fruit juices and dairy products, or co-extrusion processes for beer and carbonated drink bottles.[21] However, there has been limited research on the application of nanomaterials in real food systems and interactions between nanomaterials and foods including minimally processed ready-to-eat foods.

PE with 21% of a nanoagglomerate containing 30% of a nanoparticulate powder was tested for green tea. Compared to the normal packaging material (low-density PE [LDPE]), oxygen and water vapor transmission rates of the nanomaterial decreased by 2.1% and 28%, respectively. The vitamin C, chlorophyll, polyphenols, and amino acid contents of green tea in nanopackaging were higher than those of standard packaging by 7.7%, 6.9%, 10%, and 2%, respectively, after 240 days, indicating better preservation of green tea by nanopackaging.[22] A study showed that isotactic PP filled with innovative calcium carbonate nanoparticles was able to preserve minimally processed apple slices for up to 10 days, limiting oxidation and microbiological growth.[23]

Nanostructured calcium silicate–alkane phase change composite material was incorporated in bubble wrap liners inside paperboard packages to provide thermal buffering during transport and temporary storage of chilled perishable food. Two kilograms of asparagus as a model perishable food sealed in a plastic bag was placed inside the paperboard container. The composites provided sufficient thermal buffering capacity to maintain the temperature at 10°C for 5 h inside the container when outside temperature increased to 23°C.[24] The freshness of the perishable products such as fruits and vegetables could be maintained during transportation from the supplier to the market effectively.

Yang et al.[25] tested PE blended with nanopowder (Ag, TiO$_2$, and kaolin) for preservation of fresh strawberry at 4°C for 12 days of storage. Nano-PE had lower OTR and WVTR in comparison with regular PE. Results showed that nano-PE showed better performance in maintaining sensory, physico-chemical, and physiological quality of strawberry fruits compared to regular PE. Packaging with nanomaterials reduced fruit decay rates, controlled total soluble solids and malondialdehyde (MDA) production, maintained ascorbic acid, and inhibited activity of enzymes such as polyphenoloxi-dase (PPO) and pyrogallol peroxidase (POD).[25] However, this study lacks in measuring oxygen and carbon dioxide inside the package during stor-age to comment on product respiration considering the gas permeability of materials tested.

Polyvinyl chloride (PVC) film coated with nano-ZnO powder was used to package fresh-cut "Fuji" apple at 4°C for 12 days, and the results were compared with the uncoated PVC. Nanocoated PVC film reduced fruit decay rate, slowed down ethylene production, maintained °Brix and titrat-able acidity, and inhibited the enzyme activity (PPO and POD). Authors claimed that nanopackaging resulted in more oxygen and less carbon dioxide inside the package compared to uncoated PVC, indicating low respiration in the nanopackages. However, considering the same level of gas concentrations at the beginning and similar OTRs of both type of packages, it is not easy to understand why significant change in gas con-centration (rapid decrease in oxygen and increase in carbon dioxide) was observed in uncoated PVC.[26]

It might be very risky to use nanomaterials with low gas permeability for respiring products such as fruits and vegetables, which require materials with high permeability. Nanomaterials could be an alternative to regular synthetic materials if they are produced with higher gas permeability for respiring products. The material permeability and product respiration rate should match to obtain gas equilibrium inside the package for successful modified atmosphere package (MAP) applications.

Nanocomposites were mostly tested in different foods in the form of active nanocomposites and bionanocomposites to increase the effectiveness on food quality and safety (Table 10.1). That is why the applications of nano-composites for food are evaluated under subtitles of active/bioactive nano-materials and bionanopolymers.

10.2.2 Bionanopolymers in Food Packaging Applications

Most of the food packaging materials are oil based and nondegradable increasing environmental pollution. Biodegradable materials have been con-sidered as alternatives to synthetic materials; however, the use of biodegrad-able packaging is limited due to poor barrier and mechanical properties. If barrier, mechanical, and thermal properties of biodegradable films are improved by introduction of inorganic particles such as nanoclays into

TABLE 10.1

Effect of Different Types of Nanomaterials/Nanocoatings on Food Quality, Safety, and Shelf Life

Type of Nanomaterial	Food/Food Spoilage or Pathogenic Microorganisms	Effect of Nanomaterial on Food Quality, Safety, and Shelf Life	Reference
iPP (calcium carbonate nanoparticles)	Apple slices	Limited oxidation and microbial growth, and shelf life of 10 days	[23]
Nanosilver coating	Asparagus	Decreased microbial growth and increased shelf life	[60]
LDPE nanocomposite	Fresh orange juice	Shelf life up to 28 days	[63]
LDPE w/5% nanosilver	Orange juice	Antimicrobial activity against *Lactobacillus plantarum*	[68]
Agar hydrogel w/silver nanoparticle	Fior di Latte cheese	Controlled microbial spoilage, no adverse effects on sensory properties and prolonged the shelf life	[64]
Nanosilver containing polyethylene oxide–like coating on PE film	Apple juice	Inhibited *Alicyclobacillus acidoterrestris* (spoilage microorganism in acidic beverages)	[69]
Nanosilver coating	Microwave-freeze-dried sea cucumber	Reduced microbial count (99% of *Bacillus subtilis*)	[71]
Silver nanoparticles-PVP (polyvinylpyrolidone) coating	Green asparagus	Lower weight loss, greener color, firmer texture, better microbial quality and increased shelf life	[72]
TiO_2-coated OPP bags	Fresh-cut lettuce	Reduced microbial contamination (*E. coli*)	[76]
Silver-coated refrigerators in contact w/food	Meat, cheese, vegetables	No contamination of food surfaces, high sensory quality, no suppression of antibacterial effect of silver due to proteins of meat and cheese	[74]
PE with nanopowder (Ag, TiO_2, kaolin)	Chinese jujube	Positive effects on physicochemical and sensory quality (significant reduction in fruit softening, weight loss, browning and climatic evolution) and increased shelf life	[77]
Absorbent pads with 1% $AgNO_3$	Poultry breast (modified atmosphere packaged)	Very effective for mesophilic and lactic acid bacteria	[65]

(continued)

TABLE 10.1 (continued)

Effect of Different Types of Nanomaterials/Nanocoatings on Food Quality, Safety, and Shelf Life

Type of Nanomaterial	Food/Food Spoilage or Pathogenic Microorganisms	Effect of Nanomaterial on Food Quality, Safety, and Shelf Life	Reference
Cellulose-based absorbent pads w/nanosilver	Beef meat (modified atmosphere packaged)	Significant reduction of total aerobic bacteria and *Pseudomonas* spp. and no effect on lactic acid bacteria	[66]
Cellulose-based absorbent pads w/nanosilver	Fresh-cut melons (passive MAP)	Controlled spoilage microorganisms, retarded senescence, reduced °Brix	[67]
PE-nano particle powder	Green tea	Higher content of vitamin C, chlorophyll and polyphenols during 240 days of storage	[22]
PE w/nano powder (Ag, TiO_2, kaolin)	Fresh strawberry	Reduced fruit decay, controlled Brix and MDA, and inhibited enzyme activity, kept vitamin C	[25]
PVC coated w/nano ZnO	Fresh-cut 'Fuji' apple	Reduced fruit decay rate, slowed down respiration rate and ethylene production, inhibited enzyme activity	[26]
Chitosan w/nano $CaCO_3$	Fresh-cut yam	Higher vitamin C, L value, and titratable acidity, less weight loss and total phenolic content	[52]
Chitosan/nano SiOx coating	Fresh-cut bamboo shoots	Reduction in respiration rate and ethylene production, less enzymatic activity, lower browning index	[53]
Silver-coated silicone rubber surfaces	Tested against *Listeria monocytogenes*	4–5 log reduction, no bacteria after 12–18 h on the surfaces	[73]
Ag-MMT embedded in agar, zein, polycaprolactone (PCL)	Tested against *Pseudomonas* spp	Antimicrobial activity of agar, no antimicrobial effect of zein and PCL	[70]

polymer matrix, then they could contribute to the longer shelf life with better food quality and reduce the packaging waste.

The application of nanotechnology to biopolymers can improve both material properties and cost-price efficiency.[27] Research in this area showed that material properties specifically barrier, thermal, mechanical, and physicochemical properties were improved compared to starting polymers and conventional microscale composites.[9,10,28] The incorporation of nanoparticles into biopolymeric matrix not only improves the barrier, mechanical, and thermal properties of natural polymers but also enhances the biodegradability. The biodegradability could be controlled by selecting organically modified layered silicates.[29]

Among biodegradable nanocomposites, starch and derivatives, polylactic acid (PLA), polybutylene succinate (PBS), polyhydroxybutyrate (PHB), and polycaprolactone (PCL) are the ones suitable for food packaging applications.[27]

Starch is one of the important biopolymer that is extruded into thermoplastic starch with low mechanical resistance and high permeability of oxygen and water.[30,31] The nanoclay addition to thermoplastic starch improved the tensile strength and lowered water vapor permeability in comparison with native starch.[32] Potato starch/clay nanocomposite films produced by polymer melt processing technique showed a good intercalation of the polymeric phase into clay interlayer galleries. The mechanical properties such as modulus and tensile strength were improved. The migration results done with vegetables and simulants showed that the dried films satisfied overall migration limit of 60 mk/kg required for plastics.[33]

Electrospinning of zein, the major storage protein of corn, forms white nanofiber-based structures with improved thermal properties and stability. This nanofiber could be used as reinforcing fiber in plastic food packaging applications such as bioactive packaging, and in edible carriers for encapsulation of food additives and modification of food properties.[34,35] Kriegel et al.[36] used a method based on electrospinning to produce nanoslim fibers from chitin. These strong and naturally antimicrobial nanofibers were used to develop green biodegradable food packaging materials.[36]

Biopolyesters such as PLA, PHB, and PCL are biodegradable polymers but their use in food packaging is limited due to poor gas barrier properties and brittleness. These properties could be overcome by adding nanoclays in a biopolyester matrix to form nanocomposite structure. The addition of nanoclays in the form of kaolinite nanofillers to PLA and PLA/PCL blends improved gas barrier, thermal stability, and mechanical properties compared with the polymer and blends without clay.[37,38] The tensile strength and Young's modulus of PLA/bacterial cellulose nanocomposite increased by 203% and 146%, respectively, in comparison with PLA. Due to the size effect of nanofibrillar bacterial cellulose, the PLA/bacterial cellulose nanocomposite had very high light transmission rate.[39] Using nanoparticles in PLA will solve the weakness of PLA such as brittleness.[3] PLA and PHB could be produced as films or trays and have the potential to be used for foods such as dairy products, beverage, fresh meat, and ready meals.[27] Incorporation of nano-sized montmorillonite into PLA increased the potential use of PLA in food packaging applications such as processed meats, cheese, confectionery, cereals, and boil-in-the-bag foods.[40]

Mechanical and barrier properties of chitosan films were improved by using different types of nanofillers such as unmodified montmorillonite, organically modified montmorillonite, nanosilver, and silver zeolite. Tensile strength increased by 7%–16%, whereas water vapor permeability decreased by 25%–30% depending on the nanoparticle tested. Chitosan-based nanocomposites, especially silver-containing ones, showed antimicrobial activity in a promising range.[41]

Chitosan nanoparticles incorporated into hydroxypropyl methylcellulose (HPMC) significantly improved mechanical (tensile strength) and barrier properties (water vapor and oxygen permeabilities).[42] HPMC, a water-soluble material, is a promising material as edible coating and film for food packaging, but it has poor mechanical and water vapor barrier properties.[43–45] The reason for particular interest in chitosan is due to its biodegradable, bioabsorbable, and bactericidal properties.[46,47] Chitosan nanoparticles of 60 nm were found to be nontoxic.[48] The use of chitosan nanoparticles can increase the potential use of cellulose-based materials in biodegradable food packaging.

The physical properties of edible films and coatings could also be improved by incorporation of nanoparticles.[2,27] Addition of clay montmorillonite into pectin lowered oxygen diffusion.[11] The physical properties of gelatin were improved by using montmorillonite.[49] Another study reported that stability of chitosan/layered nanocomposites increased.[50] The reinforcement of cellulose nanofibers (CNFs) into edible films produced from mango puree improved mechanical properties (tensile strength and Young's modulus) except elongation and decreased water vapor permeability of the films. However, the influence of CNF on Tg was low.[51]

Fresh-cut yam was coated with chitosan containing nano-$CaCO_3$ and stored at 10°C. The results indicated that titratable acidity, vitamin C content, and L value of the fresh-cut yam coated with chitosan–nano-$CaCO_3$ were higher than the product coated with only chitosan. The weight loss and total phenolic content of the product coated with chitosan–nano-$CaCO_3$ were less than that of chitosan coated. Chitosan modified with nano-$CaCO_3$ was suggested for use in fresh-cut yam to prolong the shelf life.[52]

Fresh-cut bamboo shoots were coated with 1% of chitosan/nano-SiOx complex and stored at 4°C for 8 days. The effects of nanocoating on the quality and physiology of fresh-cut bamboo shoots were investigated. The coating reduced the respiratory rate and ethylene production rate, delayed the increase in enzymatic activity, and maintained high L value and low level of browning index.[53]

The nanoparticles used in edible coatings and films can be used as carrier of additives such as antimicrobials or antioxidants to improve product quality and control microbial spoilage during long-term storage. Although the properties of the edible films and coating could be improved by nanoparticles, the consumer perception has to be considered when nanoparticles become the edible part of food.

The applications of bionanocomposites can reduce the packaging waste and improve the preservation period of foods. The potential applications of nanotechnology would be in the development of biodegradable packaging materials with additional bioactive functions.

10.2.3 Active/Bioactive Nanomaterials in Food Packaging Applications

Materials having substances such as antimicrobials, oxygen scavengers, ethylene removers, and carbon dioxide absorbers/removers are promising applications of active packaging, which can release these substances into packaged food or absorb from the packaged food or the environment. Mostly metal nanoparticles, metal oxide nanomaterials, and carbon nanotubes are used to produce antimicrobial nanomaterials. Silver, gold, and zinc nanoparticles having antimicrobial function are extensively studied, and silver was already commercialized in some applications.[54] Silver with high temperature stability and low volatility is known to be a very effective antimicrobial at the nanoscale being effective against 150 different bacteria.[55–59]

Materials at nanoscale have a higher surface-to-volume ratio when compared to microscale materials. Higher surface allows nanomaterials to attach more active molecules with greater efficiency. For example, nanocomposites with low silver content showed a better increased efficacy against *Escherichia coli* than microcomposites with a much higher silver content.[60] Antifungal active paper incorporating cinnamon oil with solid wax paraffin using nanotechnology as an active coating was proved to be effective packaging material for bakery products.[61]

Nanocomposites including antimicrobials could be useful for products susceptible to surface spoilage such as cheese, bakery goods, and sliced meat products which are vacuum packed or film wrapped where most of the product surface is in contact with the packaging.[62] Materials containing nanosilver as antimicrobial agent could be used in fruits and vegetables to extend the shelf life since silver absorbs and decomposes ethylene. It was reported that a coating containing nanosilver was effective in decreasing microbial growth and increasing shelf life of asparagus.[60]

LDPE nanocomposite films containing Ag and ZnO nanoparticles were evaluated to determine the shelf life of fresh orange juice. Packages including nanomaterials, except 1% nano-ZnO, kept the microbial load of fresh juice below the limit of microbial shelf life up to 28 days.[63] Agar hydrogel with silver nanoparticles as an antimicrobial packaging system was evaluated to prolong the shelf life of Fior di Latte cheese. Antimicrobial packaging system including nanosilver increased the shelf life of the product by controlling microbial spoilage without affecting sensory properties.[64]

Fernandez et al.[65] reported that there is a satisfactory correlation between silver ion release and antimicrobial effect against *E coli* and *Staphylococcus aureus* in vitro studies. The same researchers used absorbent pads including 1% $AgNO_3$ in modified atmosphere package of poultry breast at refrigeration conditions. The exudates were absorbed in silver-loaded absorbent pads, which were very effective for mesophilic and lactic acid bacteria.[65] Nanosilver adsorbed cellulose fibers were used as absorbent pads in the modified atmosphere packages of beef meat. The levels of total aerobic bacteria and *Pseudomonas* spp. were significantly reduced and *Enterobacteriaceae* was under the detection limit in the presence of silver; however, lactic acid bacteria were not significantly affected.[66] The antimicrobial activity of cellulose-based absorbent pads including silver nanoparticles was investigated for fresh-cut melons under passive modified atmosphere during storage at 4°C for

10 days. It is reported that the silver-loaded absorbent pads in contact with fresh-cut melon controlled the spoilage microorganisms with lower yeast counts and lower°Brix values, and retarded the senescence.[67] These studies showed that silver-based nanotechnology has potential to control microbial growth in unsanitary absorbent pads at specific food packaging applications specifically in meat packaging.

Nanocomposite LDPE films containing Ag (LDPE + 1.5% nanosilver and LDPE + 5% nanosilver) and ZnO (LDPE + 0.25% nano-ZnO and LDPE + 1% nano-ZnO) were tested against *Lactobacillus plantarum* in orange juice for 112 days of storage at 4°C. LDPE packages containing 5% nanosilver had highest antimicrobial activity on *Lactobacillus plantarum* among all applications. Decrease in antimicrobial activity of the film containing 1% nano-ZnO compared to the film with 0.25% ZnO was attributed to agglomeration of nanoparticles during film processing.[68]

The effectiveness of Ag-containing polyethylene oxide–like coating on a PE film was tested against *Alicyclobacillus acidoterrestris,* cause of spoilage in acidic beverages, in acidified model system and in apple juice. Results showed that the active film inhibited the growth of gram-positive and spore-forming *Alicyclobacillus acidoterrestris* in both model system and real food system.[69]

The antimicrobial effect of silver–montmorillonite (Ag-MMT) embedded in agar, zein, and PCL polymer matrices was tested against cocktail of three strains of *Pseudomonas* spp. Although agar has antimicrobial activity, zein and PCL did not show any antimicrobial effect. The antimicrobial effect of the active films was attributed to the water content of polymeric matrix.[70]

Nanosilver coating is tested for microbial control on microwave-freeze-dried sea cucumber. Nanosilver coating was quite effective to reduce microbial count (99% of *Bacillus subtilis*) with no significant effect on drying efficiency.[71] The application of silver nanoparticle–polyvinylpyrolidone (PVP) coating including 0.06 mg L^{-1} silver increased the shelf life of green asparagus spears to 25 days at 2°C. The coated asparagus had lower weight loss, greener color, firmer texture, and better microbial quality compared to uncoated samples.[72]

Silver nanoparticles were coated onto medical and food grade silicone rubber, stainless steel, and paper surfaces, and the antimicrobial effect of silver was tested against *Listeria monocytogenes*. There was 4–5 log reduction in bacterial count and no bacteria detected after 12–18 h on silver-coated silicone rubber surfaces. This study showed that silver is one of the effective bactericide.[73]

Another interesting study was performed to determine antimicrobial effect of sterile silver-coated refrigerators in contact with food. Foods such as meat, cheese, and vegetables were stored in silver-coated and control refrigerators without silver at room temperature and 5°C for different time intervals. The products were analyzed in terms of sensory and microbial quality. All food stored in silver-treated refrigerator tend to have higher sensory quality than the food stored in untreated refrigerators. This study also showed that the silver-based material used in refrigerators protects the inner walls of the refrigerator and prevents contamination upon contact with the food surface.[74] The authors also claimed that the antibacterial effect of silver was not suppressed by proteins of meat and cheese products as it was already reported in the literature.[55,75]

Titanium dioxide (TiO_2)-coated oriented PP (OPP) bags were tested on the fresh-cut lettuce inoculated with *E. coli*. The number of *E. coli* cells of the cut lettuce packaged in TiO_2-coated bags exposed to UV light decreased from 6.4 log CFU/g to 4.9 log CFU/g on the first day of storage; however, there was only 0.3 log CFU/g reduction in the product packaged in uncoated bags. The results showed that TiO_2-coated film could reduce the microbial contamination on the surface of fresh-cut products with reducing the risk for microbial growth.[76]

Effect of nanomaterials synthesized by blending PE with nanopowder (nano-Ag, nano-TiO_2, and kaolin) was tested for preservation of Chinese jujube. The nanomaterial had positive effects on physicochemical and sensory quality of the product compared with normal PE with no nanopowder. Fruit softening, weight loss, browning, and climactic evolution of the product were significantly inhibited by nanomaterial after 12 days of storage. The nanomaterial is suggested for Chinese jujube to increase shelf life and improve the quality.[77]

The antibacterial and antifungal activity and mechanisms of nanoscale chitosan were reported.[78–80] The antibacterial properties of a novel chitosan–Ag nanoparticle were studied using recombinant *Escherichia coli* with green fluorescent protein (GFP). The presence of metal nanoparticles (2.15% w/w) in the composite significantly inactivated *E. coli* compared to unaltered chitosan.[81]

Although most studies are focused on nanosilver in antimicrobial packaging, one of the problems with nanosilver is its high cost. The use of nanoparticles of zinc, calcium, magnesium, and titanium oxides could be an alternative for antimicrobial packaging applications since these materials are much cheaper.[62] Titanium oxide, zinc oxide, silicon oxide, and magnesium oxide can also function as UV blockers and photocatalytic disinfecting agents.[54]

Most of the studies in active packaging area are dominated in antimicrobial packaging materials. Oxygen scavenging films using nanotechnology are another active packaging application in order to control oxidation reaction of food, which is one of the big problems in food deterioration. Incorporation of oxygen scavengers into food packaging materials can maintain very low oxygen levels, which is desired in food preservation.[1,60] Active packaging films for selective control of oxygen transmission were developed based on the nanotechnology approach.[82] Oxygen scavenger films were developed by adding titania nanoparticles into different polymers. The mechanism is based on the photocatalytic activity of nanocrystalline titania (TiO_2) under UV exposure, which is the major drawback.[83,84]

Bioactive compounds nanoencapsulated into the biodegradable/edible or sustainable packaging systems are another promising approach for many applications in food technology since nanoparticles allow better encapsulation and release efficiency than traditional encapsulation systems.[4,80,85] Reducing the size of encapsulates to the nanoscale provides prolonged gastrointestinal retention time, which is attributed to bio-adhesive improvements in the mucus covering the intestinal epithelium.[86] Nanoencapsulation in liposomes enhances nisin stability, availability, and distribution, controlling unwanted bacteria in food with extended shelf life.[80] There are several food additives already approved for nanoencapsulation such as carrageenan, chitosan, gelatin, PLA, polyglycolic acid, and alginate. Bioactive packaging materials can make food healthier and also help to control oxidation, to prevent off-flavor formation and undesirable texture of food.[4,85]

There is a need for further work in nano-active-packaging using natural antimicrobials instead of conventional ones to improve the shelf life and safety of perishable foods. Diffusion of active compounds from nanomaterials into food will be the main challenge for the food scientists in the near future.

10.2.4 Nanosensors in Intelligent Food Packaging Systems

Intelligent packaging is a novel interactive packaging system carrying out intelligent functions such as detecting, sensing, recording, tracing, and communicating in order to inform consumers about the quality and safety of the food inside the package and to warn them about possible problems. Intelligent packaging system monitors the condition of packaged food to give information about the quality of the packaged food during storage and transportation. Indicators or sensors in the form of a package label or printed on packaging films can monitor changes in the gas composition, storage temperature, and quality of the product.[87–91]

There are basically two types of intelligent packaging. The first one is based on measuring the condition of the package on the outside such as time–temperature indicators. The second one is dependent on the measuring quality of the product inside the packaging like gas indicators, freshness indicators, and biosensors.[90,92] In the latter case, there is direct contact with the food or with the headspace, and there is always need for a marker indicative of the quality and safety of the packed food.[90]

Time–temperature indicators can provide temperature history of the packaged product during storage and transportation. TTIs are placed outside the pack and can be defined as small measuring devices that show a time- and temperature-dependent irreversible change (e.g., a color change).[90] TTIs may also used as freshness indicator to estimate the remaining shelf life of perishable products.[93,94] These indicators are already commercialized for products that especially require cold chain products. The wide usage of TTIs is suggested to monitor the various steps of the real distribution chain of fresh products, namely, transport, bench, and home storage.[92,94]

An ideal indicator for the quality control of packaged food is to indicate lack of freshness or the spoilage of the product, in addition to temperature abuse or package leaks.[89] A freshness indicator determines the quality of packaged foods by measuring the metabolites formed through the deterioration process in the food and therefore inside the package. Most of the freshness indicators are based on the detection of volatile compounds such as CO_2, diacetyl, amines, ammonia, and hydrogen sulfide produced during the aging of foods.[89,90,95]

Nanosensors can detect certain chemical compounds, pathogens, or their toxins in food products providing real-time status of food freshness eliminating the need for inaccurate expiration dates since nanosensors are able to respond to spoilage products, microbial contamination, and environmental changes (such as temperature, humidity, and oxygen exposure level).[96,97] The recent developments of this technology include oxygen indicators, freshness indicators, and pathogen sensors. Oxygen sensors are very important to assure oxygen free food package especially in the case of vacuum or modified atmosphere packaging.[54] There are few developments based on TiO_2 nanoparticles[60] or SnO_2 nanocrystals[98] as oxygen indicators.

The sensing capability of nanocomposites can be based on conductivity changes due to gas or chemical interactions with nanofiller or conjugated polymer, pH changes, electrochromic or electro-optical property change, catalytic activity, or biological recognition.[99]

Odor sensors (electronic noses) and taste sensors (electronic tongues) are expected to be used in applications of food science and technology as biomimetic systems. These sensors could be used in inspection of food and grading quality and safety of food based on odor and taste, checking mayonnaise for rancidity, fish inspection, automated flavor control, beverage container inspection, fermentation control, monitoring cheese ripening, etc.[100]

Sensors based on conducting nanoparticles embedded into a polymer matrix to detect foodborne pathogens or their toxins are still in development stage. There are few patents but no commercial application yet. DNA-based biochips to detect presence of harmful bacteria in food are also under development.[54] Biosensors using fluorescent dye particles attached to the bacterial antibodies are at research stage. When specific bacteria are present in food, the nano-sized dye particles become visible. Sensors have been developed for *Staphylococcus* enterotoxin B, *E. coli*, *Salmonella* spp, and *Listeria monocytogenes*.[101]

Gas sensors could be used to determine food deterioration by detecting spoilage gases produced in the package headspace using conducting polymer nanocomposites (CPCs) or metal oxides. CPC sensors containing carbon black and polyaniline were developed to detect foodborne pathogens (*Bacillus cereus*, *Vibrio parahemolyticus*, and *Salmonella* spp.) by producing a specific response pattern for each microorganism.[102] The freshness of chicken meat was investigated based on the smell when the output data of metal oxide gas sensors were processed with a neural network.[103] An electronic tongue consisting of an array of nanosensors extremely sensitive to gases released by spoilage microorganisms was developed to produce a color change indicating the food is spoiled.[60,104]

Nanosensors embedded into a packaging film are able to detect food spoilage microorganisms and trigger a color change to inform the consumer about the status of the food quality or the shelf life. Nano Bioswitch/Release-on-Command is an example for intelligent preservative packaging technology, which releases a preservative if food begins to spoil. Nanoscale sensing devices can enable the food or food ingredients to be traced back to the source of origin when these devices are attached to the food or food packaging.[20]

Enzyme immobilization into nanoclays in food packaging is a new approach to control the release of enzyme molecules since nanoscale immobilization would increase the available surface contact area and modify the mass transfer.[105,106] The modified silica nanoparticles with immobilized enzymes could be used for biosensing applications.[60]

10.2.5 Commercial Nanomaterials in Food Packaging

Although there are many advantages of nanocomposites, there are uncertainties in the use of nanotechnology either as food or food contact materials considering the safety/toxicology, risk assessment, and regulation issues. Despite all these uncertainties, there are already many products in the market.[107]

PET bottles produced by nGimat coating technology, which contain 30–60 nm thick layers of silicium-based nanoparticles, increased shelf life from 10 to 30 weeks for carbonated beverages. Another application is a silicon oxide coating layer by plasma deposition of less than 100 nm inside PET bottles. It is reported by the company (SIG Chromoplasts P) that the PET increased the shelf life for 12 oz carbonated soft drink bottles to more than 25 weeks.[54] The company Nanocor (Chicago, United States) is producing nanocomposites containing clay for use in plastic beer bottles with 6 month shelf life.[108] The clay nanoparticles embedded in the plastic bottles reduce gas permeability minimizing the loss of carbon dioxide from the beer and ingress of oxygen into the bottle.[1] Miller Brewing Co. in the United States and Hite Brewery Co. in South Korea are reported to be using this technology in their beer bottles.[21]

There are few commercial nanoproducts in the area of active nanomaterials based on the use of nanosilver. Several companies such as Sharper Image® and BlueMoonGoods in the United States, Quan Zhou Hu Zeng Nano Technology in China, and A-DO Global in South Korea used nanosilver in plastic food packages to protect food from microbial spoilage. Examples include "FresherLonger® Miracle Food Storage Containers" and "FresherLonger® Plastic Storage Bags" produced by Sharper Image, "Nano Silver Food Containers" by A-DO, and "Nano Silver Baby Milk Bottle" by

Baby Dream® Co. Ltd. FresherLonger™ storage containers with silver nanoparticles in a PP was designed to inhibit microbial growth. Silver zeolites are also commercially used as active nanoparticles approved by the Food and Drug Administration (FDA) for food contact use.[21,54] Agion Technologies have approval from the European Food Safety Agency (EFSA) to use silver zeolites for food packaging.

Another application of active nanomaterials is to use active oxygen scavengers with passive nanocomposite clays to provide improved barrier properties with active nanotechnology.[54] Ageis® OX (Honeywell) is an oxygen scavenging nylon resin produced as high oxygen barrier layer for PET containers such as beer bottles. This material is also good barrier for carbon dioxide and comparable to glass in terms of cost and performance. The shelf life is suggested as 6 months for beer filled in this material. Ageis HFX has also an oxygen scavenging nylon with high gas barrier properties, which is produced as juice, tea, and condiment bottles. Ageis CSDE produced for carbonated soft drinks and water is a non-scavenging resin with high carbon dioxide barrier properties.[62] These nanocomposite bottles have the advantage of less weight, reducing the transportation cost.[108]

Durethan® (Bayer AG) is a new hybrid plastic comprising polyamide and layered silicate barriers. Clay produced by Nanocor is incorporated into the plastic to increase barrier properties and enhance gloss and stiffness. This material can be used in packaging applications where conventional Polyamide (PA) is too permeable and EVOH coatings is too expensive.[21] This nanoclay composite material can be used as interior coating for paperboard cartons to keep juice fresh.[54]

Biomark developed by AgroMicron is a nanobioluminescence detection spray containing a luminescent protein that has been engineered to bind to the surface microorganisms such as *Salmonella* and *E. coli*. When binding occurs, it emits a visible glow, which changes according to the amount of bacterial contamination.[108] Nanosensors developed a nanoporous silicon-based biosensor to detect *Salmonella* and *E. coli*. A prototype nanobiosensor was tested to detect multiple pathogens such as *Bacillus cereus* and *E. coli* more accurately.[101]

10.3 Interactions between Nanomaterials and Food

10.3.1 Possible Interactions of Nanoparticles with Food Components

Nanoscale materials have the ability to interact with proteins and other biopolymers. The catalytic activity of nanoparticles especially in oxidative reactions may lead to the formation of reactive oxygen species and the oxidation stress. Nanoparticles with electron acceptors could react with hydroxyl groups, which are mainly found in polysaccharides. Inorganic nanoparticles with adsorbed cations may interact with aromatic rings such as benzene and heterocyclic aromatic rings, which are mainly found in proteins and nucleic acids. These aromatic rings can also be found in minor food components such as in antioxidants, pigments, and flavors. The enzyme activity will decrease when interaction occurs with a nanoparticle. Sulfur-containing groups like thiol groups, which are mainly found in specific proteins, are known to interact with gold and silver. Nanoparticles can also cause formation of different polymorphs and recrystallization of fat, which may lead to changes in food consistency and sensory properties. The interaction of nanoparticles with water and minerals can occur. Interaction of nanoparticles with water may bring significant changes in food due to water adsorption capacity of some nanoparticles such as SiO_2. The interactions with vitamins may modify their bioavailability. The sensory quality of food can be affected due to interaction of nanoparticles with pigments and flavors in food.[109]

A study reported that there was no apparent effect of a novel chitosan-Ag nanocomposite on the intracellular protein in GFP of the bacteria, *E. coli*,[81] although previous studies claimed that Ag nanoparticles could interact with the sulfur-containing intracellular proteins in bacteria.[57,110]

Although the effects of nanoparticles depend on number and size of nanoparticles present in the food matrix, nanoscale materials are indicated to have ability to react with almost all food components, especially with proteins. Thus, the interaction of nanoparticles, migrated from nano-packaging materials into food, with food components needs further investigation to make sure that quality and safety of the food products are maintained.

10.3.2 Migration from Nanomaterials into Food

Nanocomposites have been shown to provide barriers to gas diffusion; however, there are only few studies on the migration of additives from nanocomposites into food. It is important to determine migration of chemicals from food packaging material into the food since exposure to toxic compounds may have potential adverse effects on human health. That is why the safety of nanoproducts has received increasing attention.[91,107]

The overall migration results of potato starch/clay nanocomposite contact with vegetables (lettuce and spinach) at dried films satisfied overall migration limit of 60 mk/kg required for plastics. Although the migration of minerals such as Fe and Mg from the biodegradable nanomaterial to packaged vegetables is insignificant, Si which is the main component of nanoclay showed a 3–5 fold higher migration level.[33]

The potential rate of migration and equilibrium distribution of engineered nanoparticles (ENP) from food packaging materials to food were predicted based on mathematical modeling, taking into account of the physicochemical properties of both ENP and the packaging materials.[111] The results showed that any detectable migration of ENPs from packaging to food can occur if ENPs are very small with a radius on the order of 1 nm from polymer matrices with low dynamic viscosity or low density. This was the case where nanosilver was incorporated into LDPE, high-density PE (HDPE), and PP. In the case of bigger ENPs, which are bound in polymer matrices with a relatively high dynamic viscosity such as PS and PET, the migration was not detectable. The same authors reported that no detectable migration of clay minerals from beer bottles, which had nanoclay composites embedded between PET layers, was observed. They also tested the other food container of PP-nanosilver composite for migration. Low level of migration of silver was predicted, which was lower than the limit of quantification of the method. This theoretical study indicated that migration of nanoparticles in nanocomposites strongly depends on the particle size and the density of the polymer matrix. The smaller size of nanoparticles and smaller density of polymers will cause higher migration.[111]

It is important to characterize migrates from nanocomposites containing clay as fillers considering food-safety point of view.[112] A migration of clay nanoparticles from bionanocomposite, polylactide (PLA) with 5% montmorillonite clay (Cloisite 30B), into food simulant (95% ethanol) was tested using asymmetrical flow field-flow fractionation (AF4) with multi-angle light-scattering detection (MALS) and inductively coupled plasma mass spectrometry (ICP-MS). The migration of the nanoparticles in the range of 50–800 nm in radius was detected by AF4-MALS analyses; however, the full AF4-MALS-ICP-MS system showed that none of the characteristic clay minerals was detectable, indicating absence of clay nanoparticles in the migrate.[113]

The silver ion releasing capability of PLA composites with commercial silver zeolites was determined by Fernandez et al.[66]. In this study, the silver ion releasing capacity of the composites of PLA with silver zeolites in tryptone soy broth (TSB) and different food simulants (distilled water, 3% acetic acid, and 10% and 95% ethanol) was quantified by graphite furnace atomic absorption spectroscopy (GFAAS). Two types of films produced by solution-casting and melt-mixing/compression molding were compared. The silver releasing capacity of films formed by melt mixing was lower than the cast films. However, migrated ions were in the range of the legal limit of 0.05 mg Ag$^+$/kg food stated by EFSA.[114]

A recent study on migration was done by Pereira de Abreu et al.[115]. A migration of different additives from polyamide and nanocomposite polyamide films was determined by High performance liquid chromatography-Ultraviolet (HPLC-UV) fluorescence system. The migration of caprolactam, 5-choloro-2(2,4-dichlorophenoxy) phenol (triclosan), and trans, trans-1,4-diphenyl-1,3-butadiene (DPBD) from polyamide and polyamide-nanoclays to different types of food simulants (3% acetic acid, water, isooctane, and 10% ethanol) was tested. The migration values were at the level of detection limits for caprolactam (0.5 mg/L), triclosan (0.02 mg/L), and DPBD (0.01 mg/L). The study showed that migration was lower in nanocomposite polyamide film than in polyamide film probably due to barrier effect of nanoclay in the polyamide matrix.[115]

Any potential risk due to nanotechnology-derived food contact materials is dependent on the migration behavior of nanomaterials in food packaging. The migration potential of small nanoparticles in the lower nm range is expected to occur from packaging to food since these particles are not bound to the polymer matrix with low dynamic viscosity.[116] Different methods were applied and proved to be effective to detect and characterize nanomaterials in food. However, the methods to quantify nanoparticles are still rare.[112] More studies are needed to assure safety of food contact nanomaterials.

10.4 Consumer Perception of Nanotechnology in Food Packaging Applications

Public perception studies of nanotechnology in the United States and Europe indicated that public knowledge regarding nanotechnology is very limited.[117,118] The public seems to be less optimistic about nanotechnology in Europe than in the United States.[119]

Siegrist et al.[120] performed a survey to determine laypeople's (375 individual) and experts' (46 individual) risk perceptions associated with 20 different nanotechnology applications including food packaging and biosensors. Experts were in the domain of nanotechnology who read scientific publications about nanotechnology. The results of the survey showed that the laypeople perceived higher levels of risk than experts who had more trust in governmental agencies that protect people's health from nanotechnology risks than the public.[120]

A survey performed with 153 participants evaluated perception of nanomaterials such as antibacterial food packaging material used to increase the shelf life of meat and nano-engineered food materials such as nanocoating to protect tomatoes from oxygen and humidity, a bread containing nanocapsules of omega-3-fatty acids, and a juice with beta-carotene encapsulated in starch. The study showed that the participants have hesitation to buy nanotechnology foods or food with nanotechnology packaging. However, nanotechnology in the food packaging is perceived as more beneficial than nanotechnology inside a food.[121]

Siegrist et al.[122] examined perceived risks and benefits of 19 different applications of nanotechnology in food and food packaging by 337 participants. According to the study, 41% of the variance of perceived risks was associated with nanotechnology inside the food and 26% of the variance of perceived risks was associated with nanotechnology in food packaging applications. Age and gender did not significantly influence the risk perception of nanotechnology applications. The study revealed that nanotechnology in food packaging was perceived as less problematic than nanotechnology in foods in the public view.[122]

These studies indicated that consumers could accept innovations in nanotechnology related to packaging than those related to foods. Consumers need to be educated about the requirement of new technologies, innovations, or materials either in food or food packaging area.

10.5 Conclusions

Scientific studies proved that nanomaterials with improved barrier and mechanical properties could be more effective in food packaging to maintain quality and increase the shelf life of food. However, it is important to explore safety aspects associated with the use of nanocompounds in food packaging in terms of health and environmental risks. Health effects of these particles are especially important where packaging materials have direct contact with the food. The possible risks of nanomaterials to human health and environment were already indicated in the literature since the toxicity of these nanomaterials can differ from that of similar bulk materials due to the greater surface-area-to-volume ratio of nanoscale materials.[48,54,86,123–126] Metal oxide nanoparticles were blamed to alter cellular function due to their size and surface area; on the contrary, these materials are believed not to induce stress in biological systems due to oxidized state of the metal.[126] FDA is currently evaluating the potential toxic/carcinogenic effects caused by exposure to nanoparticles associated with food containers, food preparation surfaces, and food wraps.[80] Perspectives on FDA's regulation on nanotechnology considering emerging challenges and potential solutions are documented by Sandoval.[127]

There is need for more studies in the area of interactions of nanomaterials with food, especially migration and risks for human health and environment, before these materials are approved for food contact use. In summary, there are potential benefits of nanotechnology in food packaging but safety/toxicology issues, environmental impacts, economics, and consumer acceptance will determine the success of this technology in food packaging applications in the future.

Acknowledgment

The author would like to give her deep appreciation to Dr. Clara Silvestre, editor of the book, for her invitation to write this book chapter and her comments and corrections to improve the chapter.

References

1. S. Neethirajan and D.S. Jayas, Nanotechnology for the food and bioprocessing industries, *Food and Bioprocess Technology*, 4 (2011) 39.
2. J. Weiss, P. Takhistov, and J. McClements, Functional materials in food nanotechnology, *Journal of Food Science*, 71 (9) (2006) 107.
3. M. Jamshidian, E.A. Tehrany, M. Imran, M. Jacquot, and S. Desobry, Poly-lactic acid: Production, applications, nanocomposites, and release studies, *Comprehensive Reviews in Food Science and Food Safety*, 9 (2010) 552.
4. M. Garcia, T. Forbe, and E. Gonzalez, Potential applications of nanotechnology in the agro-food sector. *Ciencia e Technologia de Alimentos*, 30(3) (2010) 573.
5. Z. Ayhan, Potential applications of nanotechnology in food packaging. In 1st international congress on food technology. November 3–6, 2010. Antalya, Turkey.
6. O. Esturk and Z. Ayhan, Nanocomposites and food packaging. In *VI. International Packaging Congress and Exhibition*. Istanbul, Turkey, Vol. 1, pp. 215–224, September 16–18, 2010.
7. L.E. Nielsen, Models for the permeability of filled polymer systems. *Journal of Macromolecular Science, Part A: Pure and Applied Chemistry*, 1 (1967) 929.
8. G.W. Beall, New conceptual model for interpreting nanocomposite behavior. In Pinnavaia T.J. and Beall G.W., eds. *Polymer–Clay Nanocomposites*. New York: John Wiley & Sons, Inc., pp. 267–279, 2000.
9. R.K. Bharadwaj, Modelling the barrier properties polymer-layered silicate nanocomposites. *Macromolecules*, 34 (2001) 9189.
10. R.K. Bharadwaj, A.R. Mehrabi, C. Hamilton, M.F. Murga, A. Chavira, and A.K. Thompson, Structure-property relationship in cross-linked polyester-clay nanocomposites. *Polymer*, 43 (2002) 3699.
11. P. Mangiacapra, G. Gorrasi, A. Sorrentino, and V. Vittoria, Biodegradable nanocomposites obtained by ball milling of pectin and montmorillonites. *Carbohydrate Polymers*, 64(4) (2006) 516.
12. P. Jawahar and M. Balasubramanian, Preparation and properties of polyester based nanocomposite gel coat system. *Journal of Nanomaterials*, 4 ID 21656 (2006).
13. C. Lotti, C.S. Isaac, M.C. Branciforti, R.M.V. Alves, S. Liberman, and R.E.S. Bretas, Rheological, mechanical and transport properties of blown films of high-density polyethylene nanocomposites. *European Polymer Journal*, 44 (2008) 1346.
14. H.C. Koh, J.S. Park, M.A. Jeong, H.Y. Hwang, Y.T. Hong, and S.Y. Ha, Preparation and gas permeation properties of biodegradable polymer/layeredsilicate nanocomposite membranes. *Desalination*, 233 (2008) 201.
15. D. Adame and G.W. Beall, Direct measurement of the constrained polymer region in polyamide/clay nanocomposites and the implications for gas diffusion. *Applied Clay Science*, 42 (2009) 545.
16. A. Morgan, D.G. Priolo, and J.C. Grunlan, Transparent clay-polymer nano brick wall assemblies with tailorable oxygen barrier, *Applied Materials and Interfaces*, 2 (2010) 312.
17. J.M. Lagaron, M.D. Sanchez-Garcia, and E. Gimenez, Novel PET nanocomposites of interest in food packaging applications and comparative barrier performance with biopolyester nanocomposites. *Journal of Plastic Film Sheeting*, 23 (2007) 133.

18. D. Cava, E. Giménez, R. Gavara, and J.M. Lagaron, Comparative performance and barrier properties of biodegradable thermoplastics and nanobiocomposites versus PET for food packaging applications. *Journal of Plastic Film Sheeting*, 22 (2006) 265.

19. M.A. Priolo, D. Gamboa, and J.C. Grunlan, Transparent clay-polymer nano brick wall assemblies with tailorable oxygen barrier. *Applied Materials and Interfaces*, 2 (2010) 312.

20. D.A. Pereira de Abreu, P.P. Losada, I. Angulo, and J.M. Cruz, Development of new polyolefin films with nanoclays for application in food packaging. *European Polymer Journal*, 43 (2007) 2229.

21. Q. Chaudhry, M. Scotter, J. Blacburn, B. Ross, A. Boxall, L. Castle, R. Aitken, and R. Watkins, Applications and implications of nanotechnologies for the food sector. *Food Additives and Contaminants*, 25(3) (2008) 241.

22. Y.Y. Huang and Q.H. Hu, Effect of a new fashion nano-packaging on preservation of quality of gren tea. *Food Science*, 27(4) (2006) 244.

23. M. Avella, G. Bruno, M.E. Errico, G. Gentile, N. Piciocchi, A. Sorrentino, and M.G. Volpe, Innovative packaging for minimally processed fruits. *Packaging Technology and Science*, 20 (2007) 325.

24. J.H. Johnston, J.E. Grindrod, M. Dodds, and K. Schimitschek, Composite nano-structured calcium silicate phase materials for thermal buffering in food packaging. *Current Applied Physics*, 8 (2008) 508.

25. F.M. Yang, H.M. Li, Z.H. Xin, L.Y. Zhao, Y.H. Zheng, and Q.H. Hu, Effect of nano-packing on prseervation quality of fresh strawberry (Fragaria ananassa Duch. Cv Fengxiang) during storage at 4°C. *Journal of Food Science*, 75(3) (2010) 236.

26. X. Li, W. Li, Y. Jiang, Y. Ding, J. Yun, Y. Tang, and P. Zhang, Effect of nano-ZnO active packaging on the quality of fresh-cut 'Fuji' apple. *International Journal of Food Science and Technology*, 46 (2011) 947.

27. A. Sorrentino, G. Gorrasi, and V. Vittoria, Potential perspectives of bio-nanocomposites for food packaging applications. *Trends in Food Science and Technology*, 18 (2007) 84.

28. A. Sorrentino, M. Tortora, and V. Vittoria, Diffusion behaviour in polymer-clay nanocomposites. *Journal of Polymer Science, Part B: Polymer Physics*, 44 (2006) 265.

29. J.L. Pandey, K.R. Reddy, A.P. Kumar, and R.P. Singh, An overview on the degradability of polymer nanocomposites. *Polymer Degradation and Stability*, 88 (2005) 234.

30. B. Chen and J.R.G. Evans, Thermoplastic starch-clay nanocomposites and their characteristics. *Carbohydrate Polymers*, 61 (2005) 455.

31. S.A. McGlashan and P.J. Halley, Preparation and characterization of bipdegradable starch-based nanocomposites materials. *Polymer International*, 52 (2003) 1767.

32. H.-M. Park, X. Li, C.-Z. Jin, C.-Y. Park, W.-J. C and H.-C. H, Preparation and properties of biodegradable thermoplastic starch/clay hybrids, *Macromolecular Materials and Engineering*, 287 (2002) 553.

33. M. Avella, J.J. De Vlieger, M.E. Errico, S. Fischer, P. Vacca, and M.G. Volpe, Biodegradable starch/clay nanocomposite films for food packaging applications. *Food Chemistry*, 93 (2005) 467.

34. S. Torres-Giner, E. Gimenez, and J.M. Lagaron, Characterization of the morphology and thermal properties of Zein Prolamine nanostructures obtained by electrospinning. *Food Hydrocolloids*, 22 (2008) 601.

35. J.M. Lagaron, E. Gimenez, M.A. Sanchez-Garcia, M.J. Ocio, and A. Fendler, Second generation nanocomposites: A must in passive and active packaging and biopackaging applications. In *The 15th IAPRI World Conference in Packaging*, October, Tokyo, Japan, 2006.

36. C. Kriegel, K.M. Kit, D.J. McClements, and J. Weiss, Influence of surfactant type and concentration on electrospining of chitosan-poly(Ethylene oxide) blend nanofibers. *Food Biophysics*, 50(1) (2009) 189.

37. N. Sozer and J.L. Kokini, Nanotechnology and its applications in the food sector. *Trends in Biotechnology*, 27(2) (2009) 82.

38. L. Cabedo, J.L. Feijoo, M.P. Villanueva, J.M. Lagaron, and E. Gimenez, Optimization of biodegradable nanocomposites based on aPLA/PCL blends for food packaging applications. *Macromolecular Symposia*, 233 (2006) 191.

39. Y. Kim, R. Jung, H.S. Kim, and H.J. Jin, Transparent nanocomposites prepared by incorporating microbial nanofibrils into poly(-lactic acid). *Current Applied Physics*, 9 (2009) S69.

40. H. Wei, Y. YanJun, L. NingTao, and W. LiBing, Application and safety assesment for nano-composite materials in food packaging. *Chinese Science Bulletin*, 56(12) (2011) 1216.

41. J.W. Rhim, S.I. Hong, H.M. Park, and P.K.W. Ng, Preparation and characterization of chitosan-based nanocomposite films with antimicrobial activity. *Journal of Agricultural and Food Chemistry*, 54(16) (2006) 5814.

42. M.R. de Moura, R.J. Avena-Bustillos, T.H. McHugh, J.M. Krochta, and L.H.C. Mattoso, Properties of novel hydroxypropyl methycellulose films containing chitosan nanoparticles. *Journal of Food Science*, 73(7) (2008) 31.

43. G.A. Burdock, Safety assessment of hydroxypropyl methylcellulose as a food ingredient. *Food and Chemical Toxicology*, 45(12) (2007) 2341.

44. M. Turowski, B. Deshmukh, and R.A. Harfmann, A method for determination of soluble dietary fiber in methycellulose and hydroxypropyl methylcellulose food gums. *Journal of Food Composition and Analysis*, 20(5) (2007) 420.

45. H. Moller, S. Grelier, P. Pardon, and V. Coma, Antimicrobial and physicochemical properties of chitosan-HPMC based films. *Journal of Agricultural Food Chemistry*, 52(2) (2004) 6585.

46. V. Coma, A. Martial-Gros, S. Garreau, A. Copinet, F. Salin, and A. Deschamps, Edible anti-microbial films based on chitosan matrix. *Journal of Food Science*, 67(3) (2002) 1162.

47. H.K. No, S.P. Meyers, W. Prinyawiwatkul, and Z. Xu, Applications of chitosan for improvement of quality and shelf life of foods: A review. *Journal of Food Science*, 72(5) (2007) 87.

48. R. De Lima, L. Feitosa, A. De E.S. Pereira, M.R. De Moura, F.A. Aouada, L.H.C. Mattoso, and L.F. Fraceto, Evaluation of the genotoxicity of chitosan nanoparticles for use in food packaging films. *Journal of Food Science*, 75(6) (2010) 89.

49. J.P. Zheng, P. Li, Y.L. Ma, and K.D. Yao, Gelatine/montmorillonite hybrid nanocomposite. Preparation and properties, *Journal of Applied Polymer Science*, 86 (2002) 1189.

50. M. Darder, M. Colilla, and E. Ruiz-Hitzky, Biopolymer-clay nanocomposites based on chitosan intercalated in montmorillonite. *Chemistry of Materials*, 15(2003) 3774.

51. H.M.C. Azeredo, L.H.C. Mattoso, D. Wood, T.G. Williams, R.J. Avena-Bustillos, and T.H. McHugh, Nanocomposite edible films from mango puree reinforced with cellulose nanofibers. *Journal of Food Science*, 74(5) (2009) 31.

52. L. Zisheng, X. Xiaoling, X. Tingqiao, and X. Jing, Effect of chitosan coating with nano-CaCO₃ on quality of fresh cut yam. *Transactions of the Chinese Society for Agricultural Machinery*, 40(4) (2009) 125.

53. L. Zi-sheng and Z. Li, Effect of chitosan/SiOx complex on quality and physiology of fresh-cut bamboo shoot. *Scienta Agricultura Sinica*, 43(22) (2010) 4694.

54. C. Silvestre, D. Duraccio, and S. Cimmino, Food packaging based on polymer nanomaterials. *Progress in Polymer Science*, 36 (12) (2011) 1766–1782.

55. S.Y. Liau, D.C. Read, W.J. Pugh, J.R. Furr, and A.D. Russell, Interaction of silver nitrate with readily identifiable groups: Relationship to the antibacterial action of silver ions. *Letters in Applied Microbiology*, 25 (1997) 279.

56. R. Kumar and H. Münstedt, Silver ion release from antimicrobial polyamide/silver composites. *Biomaterials*, 26 (2005) 2081.

57. J.R. Morones, J.L. Elechiguerra, A. Camacho, K. Holt, J.B. Kouri, and J.T. Ramirez, The bactericidal effect of silver nanoparticles. *Nanotechnology*, 16 (2005) 2346.

58. I. Sondi and B. Salopek-Sondi, Silver nanoparticles as antimicrobial agent: A case study on E. coli as a model for Gram-negative bacteria. *Journal of Colloid Interface Science*, 275 (2004) 177.

59. Q. Li, S. Mahendra, D.Y. Lyon, L. Brunet, M.V. Liga, D. Li, and P.J.J. Alvarez, Antimicrobial nanomaterials for water disinfection and microbial control: Potential applications and implications. *Water Research*, 42 (2008) 4591.

60. H.M.C. Azeredo, Nanocomposites for food packaging applications. *Food Research International*, 49 (2009) 1240.

61. A. Rodriguez, C. Nerin, and R. Battle, New cinnamon-based active paper packaging against *Rhizopusstolonifer* food spoilage. *Journal of Agricultural and Food Chemistry*, 56(15) (2008) 6364.

62. Anonymous, Nanotechnology in packaging: A revolution in waiting. *Food Engineering and Ingredients*, 33(3) (2008) 6.

63. A. Emamifar, M. Kadivar, M. Shahedi, and S. Soleimanian-Zad, Evaluation of nanocomposite containing Ag and ZnO on shelf life of orange juice. *Innovative Food Science and Emerging Technology*, 11(4) (2010) 742.

64. A.L. Incoronata, A. Conte, G.G. Buonocore, and M.A. Del Nobile, Agar hydrogel with silver nanoparticles to prolong the shelf life of Fior di Latte cheese. *Journal of Dairy Science*, 94(4) (2011) 1697.

65. A. Fernandez, E. Soriano, G. Lopez-Carballo, P. Picouet, E. Lloret, R. Gavara, and P. Hernandez-Munoz, Preservation of aseptic conditions in absorbent pads by using silver nanotechnology. *Food Research International*, 42 (2009) 1105.

66. A. Fernandez, P. Picouet, and E. Lloret, Reduction of the spoilage-related microflora in absorbent pads by silver nanotechnology during modified atmopshere packaging of meat. *Journal of Food Protection*, 73(12) (2010) 2263.

67. A. Fernandez, P. Picouet, and E. Lloret, Cellulose-silver nanoparticle hybrid materials to control spoilage-related microflora in absorbent pads located in trays of fresh-cut melon. *International Journal of Food Microbiology*, 142 (2010) 222.

68. A. Emamifar, M. Kadivar, M. Shahedi, and S. Soleimanian-Zad, Effects of nanocomposite packaging containing Ag and ZnO on inactivation of *Lactobacillus plantarum* in orange juice. *Food Control*, 22 (2011) 408.

69. M.A. Del Nobile, M. Cannarsi, C. Altieri, M. Sinigaglia, P. Favia, G. Iacoviello, and R. D'Agostino, Effect of Ag-containing nano-composite active packaging system on survival of *Alicyclobacillus acidoterrestris*. *Journal of Food Science*, 69(8) (2004) 379.

70. A.L. Incoronato, G.G. Buonocore, A. Conte, M. Lavorgna, and M.A. Del Nobile, Active systems based on silver-montmorillonite nanoparticles embedded into bio-based polymer matrices for packaging applications. *Journal of Food Protection*, 73(12) (2010) 2256.

71. X.L. Li, M. Zhang, X. Duan, and A.S. Mujumdar, Effect of nano-silver coating on microbial control of microwave-freeze combined dried sea cucumber. *International Agrophysics*, 25(2) (2011) 181.

72. J. An, M. Zhang, S. Wang, and J. Tang, Physical, chemical and microbiological changes in stored green asparagus as affected by coating of silver nanoparticles-PVP. *LWT*, 41 (2008) 1100.

73. H. Jiang, S. Manolache, A.C.L. Wong, and F.S. Denes, Plasma-enhanced deposition of silver nanoparticles onto polymer and metal surfaces for the generation of antimicrobial characteristics. *Journal of Applied Polymer and Science*, 93 (2004) 1411.

74. Y. Kampman, E. De Clerck, S. Kohn, D.K. Patchala, R. Langerock, and J. Kreyenschmidt, Study on the antimicrobial effect of silver-containing iner liners in refrigerators. *Journal of Applied Microbiology*, 104 (2008) 1808.

75. Y. Matsumura, K. Yoshikata, S. Kunisaki, and T. Tsuchido, Mode of bactericidial action of silver zeolite and its comparison with that of silver nitrate. *Applied Environmental Microbiology*, 61 (2003) 4278.

76. C. Chawengkijwanich and Y. Hayata, Development of TiO$_2$ powder-coated food packaging film and its ability to inactivate *Esherichia coli* in vitro and in actual tests. *International Journal of Food Microbiology*, 123 (2008) 288.

77. H. Li, F. Li, L. Wang, J. Sheng, Z. Xin, L. Zhao, H. Xiao, Y. Zheng, and Q. Effect of nano-packing on preservation quality of Chinese jujube (*Ziziphus jujuba Mill. Var. Inermis (Bunge) Rehd*). *Food Chemistry*, 114 (2009) 547.

78. L. Qi, Z. Xu, X. Jiang, C. Hu, and X. Zou, Preparation and antibactarial activity of chitosan nanoparticles. *Carbohydrate Research*, 339 (2004) 2693.

79. E.I. Rabea, M.E. Badawy, C.V. Stevens, G. Smagghe, and W. Steurbaut, Chitosan as antimicrobial agent: Applications and mode of action. *Biomacromolecules*, 4(6) (2003) 1457.

80. M. Imran, A.-M. Revol-Junelles, A. Martyn, E.A. Tehrany, M. Jacquot, M. Linder, and S. Desorby, Active food packaging evolution: Transformation from micro-to nanotechnology. *Critical Reviews in Food Science and Nutrition*, 50 (2010) 799.

81. P. Sanpui, A. Murugadoss, P.V. Durga Prasad, S.S. Ghosh, and A. Chattopadhyay, The antibacterial properties of a novel chitosan-Ag-nanoparticle composite. *International Journal of Food Microbiology*, 124 (2008) 142.

82. J. Rivett and D.V. Speer, Oxygen scavenging film with good interplay adhesion. US Patent 75141512 (2009).

83. L. Xiao-e, A.N.M. Green, S.A. Haque, A. Mills, and J.R. Durrant, Light-driven oxygen scavenging by titania/polymer nanocomposite films. *Journal of Photochemistry and Photobiology A: Chemistry*, 162 (2004) 253.

84. A. Mills, G. Doyle, A.M. Peiro, and J. Durrant, Demonstration of a novel, flexible, photocatalic oxygen-scavenging polymer film. *Journal of Photochemistry and Photobiology A: Chemistry*, 177 (2006) 328.

85. A. Lopez-Rubio, R. Gavara, and J.M. Lagaron, Bioactive packaging: Turning foods into healthier foods through biomaterials. *Trends in Food Science and Technology*, 17 (2006) 567.

86. H. Bouwmeester, S. Dekkers, M. Noordam, W.I. Hagens, A.S. Bulder, C. De Heer, S.E.C.G. Ten Voorde, S.W.P. Wijnhoven, and H.J.P. Marvin, Review of health safety aspects of nanotechnologies in food production. *Regulatory Toxicology and Pharmacology*, 53 (2009) 52.

87. R. Ahvenainen and E. Hurme, Active and smart packaging for meeting consumer demands for quality and safety. *Food Additives and Contaminants*, 14 (1997) 753.

88. M. Smolander, E. Hurme, and R. Ahvenainen, Leak indicators for modified atmosphere packages. *Trends in Food Science and Technology*, 4 (1997) 101.

89. N. De Kruijf, M. Van Beest, R. Rijk, T. Sipilainen-Malm, P.P. Losada, and B. De Meulenaer, Active and intelligent packaging: Applications and regulatory aspects. *Food Additives and Contaminants*, 19 (2002) 144.

90. A.R. De Jong, H. Boumans, T. Slaghek, J. Van Veen, R. Rijk, and M. Van Zandvoort, Active and intelligent packaging for food: Is it the future? *Food Additives and Contaminants*, 22(10) (2005) 975.

91. B. Farhang, Nanotechnology and applications in food safety. In Barbosa-Canovas G., Mortimer A., Lineback D., Spiess W., Buckle K., and Colonna P., eds. *IUFoST World Congress Book: Global Issues in Food Science and Technology*, Chapter 22. Elsevier Inc., New York, US, pp. 401–410, 2009.

92. S.F. Schilthuizen, Communication with your packaging: Possibilities for intelligent functions and identification methods in packaging. *Packaging Technology and Science*, 12 (1999) 225.

93. K.L. Yam, P.T. Takhistov, and J. Miltz, Intelligent packaging: Concepts and applications. *Journal of Food Science*, 70(1) (2005) 1.

94. M. Riva, L. Piergiovanni, and A. Schiraldi, Performances of time-temperature indicators in the study of temperature exposure of packaged fresh foods. *Packaging Technology and Science*, 14 (2001) 1.

95. Z. Ayhan, Application of modified atmosphere packaging with new concepts for respiring foods. *Acta Horticulturae*, 876 (2010) 137.

96. R. Ahvenainen, *Novel Food Packaging Techniques*. Bosa Roca, FL: CRC Press Inc., 2003.

97. F. Liao, C. Chen, and V. Subramanian, Organic TFTs as gas sensors forelectronic nose applications. *Sensors and Actuators B: Chemical*, 107(2) (2005) 849.

98. A. Mills and D. Hazafy, Nanocrystalline SnO_2-based, UVB-activated, colourimetric oxygen indicator. *Sensors and Actuators B: Chemical*, 136 (2009) 344.

99. D.R. Paul and L.M. Robeson, Polymer nanotechnology: Nanocomposites. *Polymer*, 49 (2008) 3187.

100. M. Ghasemi-Varnamkhasti, S.S. Mohtasebi, and M. Siadat, Biomimetric-based odor and taste sensing systems to food quality and safety characterization: An overview on basic principles and recent achievements. *Journal of Food Engineering*, 100 (2010) 377.

101. Y. Liu, S. Chakrabartty, and E. Alocilja, Fundamental building blocks for molecular biowire based forward error-correcting biosensors. *Nanotechnology*, 18 (2007) 1.

102. K. Arshak, C. Adley, E. Moore, C. Cunniffe, M. Campion, and J. Harris, Characterisation of polymer nanocomposite sensors for quantification of bacterial cultures. *Sensors and Actuators B*, 126 (2007) 226.

103. A. Galdikas, A. Mironas, V. Senuliene, A. Setkus, and D. Zelenin, Response time based output of metal oxide gas sensors applied to evaluation of meat freshness with neural signal analysis. *Sensors and Actuators B*, 69 (2000) 258.

104. C. Ruengruglikit, H. Kim, R.D. Miller, and Q. Huang, Fabrication of nanoporous oligonucleotide microarrays for pathogen detection and identification. *Polymer Preprints*, 45 (2004) 526.

105. A. Fernandez, D. Cava, M.J. Ocio, and J.M. Lagaron, Perspectives for biocatalyst in food packaging. *Trends in Food Science and Technology*, 19(4) (2008) 198.

106. J.W. Rhim and P.K.W. Ng, Natural biopolymers-based nanocomposite films for packaging applications. *Critical Reviews in Fodo Science and Nutrition*, 47(4) (2007) 411.

107. K.T. Lee, Quality and safety aspects of meat products as affected by various physical manipulations of packaging materials. *Meat Science*, 86 (2010) 138.

108. A.L. Brody, B. Bugusu, J.H. Han, C.K. Sand, and T.H. McHugh, Innovative food packaging solutions. *Journal of Food Science*, 73(8) (2008) 107.

109. P. Simon and E. Joner, Conceivable interactions of biopersistent nanoparticles with food matrix and living systems following from their physicochemical properties. *Journal of Food and Nutrition Research*, 47(2) (2008) 51.

110. Q.L. Feng, J. Wu, G.Q. Chen, F.Z. Cui, T.N. Kim, and J.O. Kim, A mechanistic study of the antibacterial effect of silver ions on *Esherichia coli* and *Staphylococcus aureus*. *Journal of Biomedical Materials Research*, 52 (2000) 662.

111. P. Simon, Q. Chaudhry, and D. Bakos, Migration of engineered nanoparticles from polymer packaging to food-a physicochemical view. *Journal of Food and Nutrition Research*, 47(3) (2008) 105.

112. C. Blasco and Y. Pico, Determining nanomaterials in food. *Trends in Analytical Chemistry*, 30(1) (2011) 84.

113. B. Schmidt, J.H. Petersen, C.B. Koch, D. Plackett, N.R. Johansen, V. Katiyar, and E.H. Larsen, Combining asymmetrical flow field-flow fractionation with light-scattering and inductively coupled plasma mass spectrometric detection for characterization of nanoclay used in biopolymer nanocomposites. *Food Additives and Contaminants Part A—Chemistry Analysis Control Exposure & Risk Assessment*, 26(12) (2009) 1619.

114. A. Fernandez, E. Soriano, P. Hernandez-Munoz, and R. Gavara, Migration of antimicrobial silver from composites of polylactide with silver zeolites. *Journal of Food Science*, 75(3) (2010) 186.

115. D.A. Pereira de Abreu, J.M. Cruz, I. Angulo, and P.P. Losada, Mass transport studies of different additives in polyamide and exfoliated nanocomposite polyamide films for food industry. *Packaging Technology and Science*, 23 (2010) 59.

116. Q. Chaudry and L. Castle, Food applications of nanotechnologies: An overview of opportunities and challanges for developing countries. *Trends in Food Science and Technology*, DOI:10.1016/j.tifs.2011.01.001 (2011).

117. M. Cobb and J. Macoubrie, Public perception about nanotechnology: Risks, benefits and trust. *Journal of Nanoparticle Research*, 6 (2004) 395.

118. C.-J. Lee, D.A. Scheufele, and B.V. Lewenstein, Public attitudes toward emerging technologies. *Science Communication*, 27 (2005) 240.

119. G. Gaskell, T. Ten Eyck, J. Jackson, and G. Veltri, Imaging nanotechnology: Cultural support for technological innovation in Europe and the United States. *Public Understanding of Science*, 14 (2005) 81.

120. M. Siegrist, C. Keller, H. Kastenholz, S. Frey, and A. Wiek, Laypeople2S and experts' perecption of nanotechnology hazards. *Risk Analysis*, 27(1) (2007) 59.

121. M. Siegrist, M.-E. Cousin, H. Kastenholz, and A. Wiek, Public acceptance of nanotechnology foods and food packaging: The influence of affect and trust. *Appetite*, 49 (2007) 459.

122. M. Siegrist, N. Stampfli, H. Kastenholz, and C. Keller, Perceived risks and benefits of different nanotechnology foods and nanotechnology food packaging. *Appetite*, 51 (2008) 283.
123. C.F. Chau, S.H. Wu, and G.C. Yen, The development of regulations for food nanotechnology. *Trends in Food Science and Technology*, 18(5) (2007) 269.
124. D.B. Warheit, C.M. Sayes, K.L. Reed, and K.A. Swain, Health effects related to nanoparticle exposures: Environmental, health and safety considerations for assessing hazards and risks. *Pharmacology and Therapeutics*, 120(1) (2008) 35.
125. J. Fabrega, J.C. Renshaw, and J.R. Lead, Interactions of silver nanoparticles with Pseudomonas putida biofilms. *Environmental Science and Technology*, 43(23) (2009) 9004.
126. C. Sayes and I. Ivanov, Comparative study of predictive computational models for nanoparticle-induced cytotoxicity. *Risk Analysis*, 30(11) (2010) 1723.
127. B.M. Sandoval, Perspectives on FDA's regulation of nanotechnology: Emerging challenges and potential solutions. *Comprehensive Reviews in Food Science and Food Safety*, 8 (2009) 375.

11

Photodegradation of Poly(Lactic Acid)/
Organo-Modified Clay Nanocomposites
under Natural Weathering Exposure

Mustapha Kaci, Aida Benhamida, Lynda Zaidi,
Naima Touati, and Chérifa Remili

CONTENTS

11.1 Introduction

Nowadays, aliphatic polyesters such as poly(lactic acid) (PLA) encounter various applications due to their biodegradable and/or biocompatible character [1]. PLA is widely used in medical applications such as wound closure, surgical implants, resorbable sutures, tissue culture, and controlled release system [2]. However, PLA has many other desirable properties that make it a very interesting material for commodity applications. Indeed, improvements in the synthesis of lactic acid have been largely responsible for the reduction in cost of PLA that in turn has led to a much wider variety of available applications such as packaging materials for food and beverages, plastic bags, thin film coatings, or rigid thermoform [3]. Furthermore, PLA is intensively studied as an alternative solution to partially solve the ecological problem of plastic waste accumulation, with a special focus on packaging [1,2,4]. In spite of these advantages, the application of PLA is somewhat limited by its limited range of physical properties such as thermal stability and gas barrier properties and its brittle nature [1,5]. Therefore, PLA-based materials still need to be improved to fulfill, for instance, the requirements for food packaging applications.

In recent years, several PLA-based technologies have emerged with an emphasis on achieving chemical, mechanical, and biological properties equivalent or superior to conventional polymers [6]. Among these methods, the polymer nanotechnology can provide new food packaging materials [7,8]. In this respect, the field of polymer nanocomposites, primarily based on layered silicates, such as montmorillonite (nanoclay), has drawn increasing attention from industry [2]. The literature [9–16] reported that the dispersion of the high aspect ratio nanoclay in the polymer matrix has been shown to provide remarkable improvements on the mechanical, fire retardant, rheological, gas barrier, and optical properties, especially at low clay loading levels (as low as 1 wt.%) in comparison with more conventional microcomposites (30 wt.% of fillers).

PLA-layered silicate nanocomposites are processed on large scale production lines through the melt intercalation technique involving injection molding, film extrusion, blow molding, thermoforming, fiber spinning, and film forming [17]. Melt intercalation has become the standard for the preparation of such nanomaterials due mainly to the absence of a solvent that makes this technique an environmentally sound and economically favorable one for industries from a waste perspective [11]. The preparation of PLA-based nanocomposites through melt intercalation has been previously described [18,19]. PLA-organoclay nanocomposites have been successfully prepared by a melt intercalation method using Cloisite 30B as organoclay, and the material morphology obtained indicates that the dispersion of mineral platelets within the PLA matrix is relatively homogeneous, as revealed by wide angle X-ray scattering (WAXS), transmission electron microscopy (TEM), and rheology [19] (see Figure 11.1).

On the other hand, the study of degradation of polymer nanocomposites is an extremely important area from the scientific and industrial point of view. Chemical degradation of polymers is an irreversible change and is a very important phenomenon, which affects the performance of all plastic materials in daily life and leads finally to the loss of functionality [20]. Therefore, the material usefulness depends on its durability in a particular environment in which materials are used or their interaction with environmental

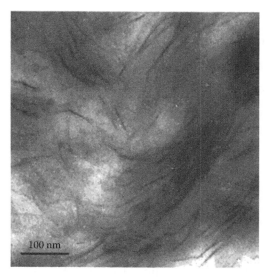

FIGURE 11.1
TEM micrograph of PLA nanocomposite filled with Cloisite 30B (1 wt.%).

factors [21]. The study of durability/degradability of polymer-nanoparticu-
late systems under environmental conditions will give an insight to their
applications as well as limitations.

Photooxidative degradation is the process of decomposition of the material by the action of light, which is considered as one of the primary sources
of damage of polymeric substrates in ambient conditions [22]. Although a
significant scientific activity has been carried out on the photooxidation of
nanocomposites with classical polymer matrices such as polypropylene,
polyethylene, and polycarbonate [23], the literature is rather scarce with
regard to nanobiocomposites. So far, the major part of the research work
dealing with nanobiocomposite materials have been focused mainly on
the preparation methods as well as the structure/property relationships,
especially the nanodispersion effect of the nanofiller on the functional
properties. Other studies regarding the susceptibility of PLA nanocomposites for both hydrolytic and thermal degradation are reported in the
literature [24].

In this chapter, in addition to a literature review regarding PLA nanocomposites and their applications in food packaging, a series of experimental data on the natural weathering effects on PLA/organo-modified clay
nanocomposites is presented. Modifications in the chemical structure and
physical properties occurring in the nanocomposite samples as a result of
photodegradation are discussed on the basis of both the neat polymer and
the nanoclay loading ratios. Further, photodegradation mechanisms taking
place in PLA nanocomposites are also debated.

11.2 Theoretical Background

11.2.1 Literature Review on PLA and PLA/Clay Nanocomposites

11.2.1.1 PLA Synthesis and Properties

The present industrial production of lactic acid is based on microbial carbohydrate fermentation because it is chemically and economically more feasible compared with the chemical route and enables the production of optically pure lactic acid [25]. Due to its availability on the market and its low price, PLA has one of the highest potentials among biopolyesters, particularly for packaging and medical applications [26].

The literature reported that the optical purity of the reagent lactic acid is the key factor during PLA production because small amounts of enantiomeric impurities drastically change the polymer properties like crystallinity or biodegradation rate.

Lactic acid, the monomeric building block of PLA, exists as two optical isomers, L- and D-lactic acid shown in Figure 11.2. The L-lactic acid rotates the plane of polarized light clockwise and D-lactic acid rotates it counter clockwise [6]. The monomers used for ring-opening polymerization of lactides are synthesized from glycolic acid, DL-lactic acid, L-lactic acid, or D-lactic acid. Among them, only L-lactic acid is optically active and produced by fermentation using *Lactobacilli*. The raw materials for this fermentation are corn, potato, sugar cane, sugar beat, etc. The widely used catalyst for ring-opening polymerization of PLA is stannous octoate and the regulator of chain length is laurel alcohol [4].

According to the literature [17], the synthesis of PLA into high molecular weight can follow two different routes of polymerization, as shown in Figure 11.3. Lactic acid is condensation polymerized to yield to a low molecular weight polymer that is glassy and brittle and, subsequently, unuseful for any application. The molecular weight of this condensation polymer is low due to the viscous polymer melt, the presence of water and impurities, the low concentration of reactive end groups, and the "back-biting" equilibrium reaction that forms the six-member lactide ring.

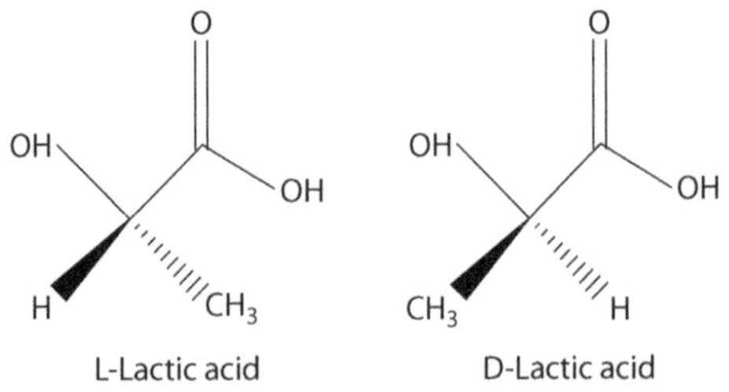

FIGURE 11.2
Lactic acid optical monomers.

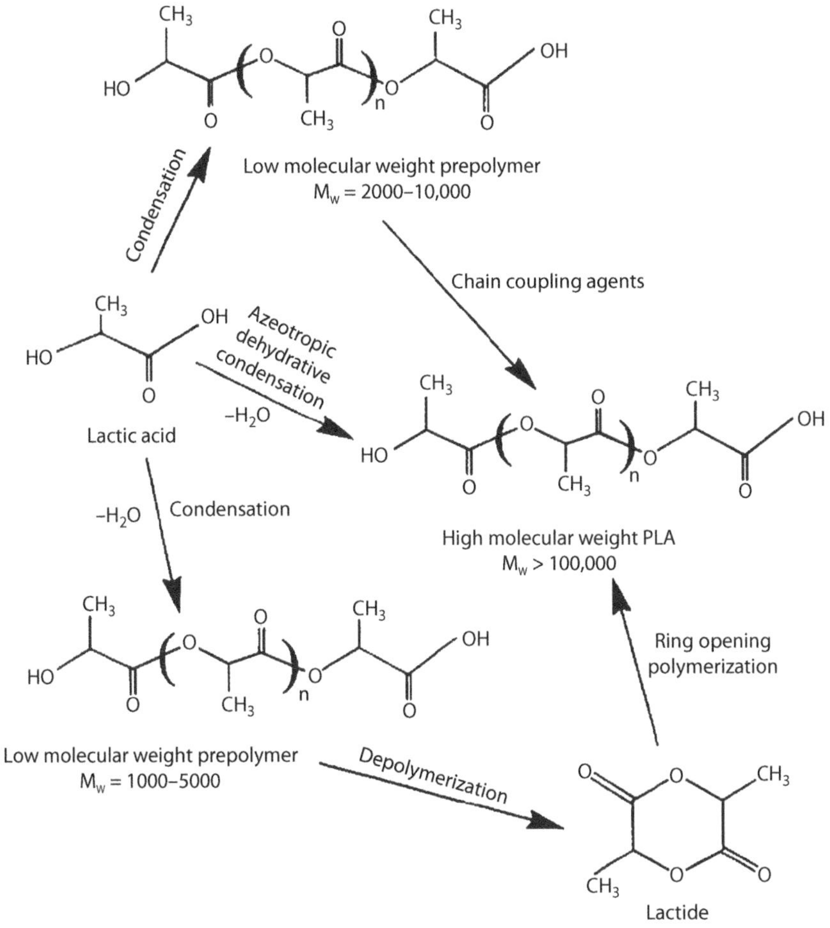

FIGURE 11.3
Synthesis methods for high molecular weight PLA.

The other route of producing PLA is to collect, purify, and ring-open polymerize (ROP) lactide to yield $M_w > 100,000$ PLA. The lactide method was the only method of producing pure, high molecular weight PLA until the commercial process developed by Mitsui Toatsu Chemicals wherein lactic acid and catalysts are azeotropically dehydrated in a refluxing, high boiling, and aprotic solvent under reduced pressures to obtain PLA with $M_w > 300,000$ [17].

The azeotropic condensation polymerization is a method to obtain high molecular weight PLA without the use of chain extenders or adjuvants. The general procedure is well reported in literature [27]. However, this procedure results in residual catalyst, which may cause many problems during further processing (degradation, irreproducible hydrolysis rates, etc.). The catalyst can be deactivated by the addition of phosphoric or pyrophosphoric acid. This provides a polymer with improved weathering resistance and heat and storage stability. Another possibility to remove the residual catalyst is by precipitation followed by filtration through the addition of strong acids such as sulfuric acid. The level of catalysts can be reduced to 10 ppm or less.

Polylactide is well known for its good processability, biocompatibility, and biodegradability mainly by simple hydrolysis. PLA can be quite different in chemical and physical properties because of the presence of a pendent methyl group on the alpha carbon atom. This structure causes chirality at a carbon of lactic acid and thus L, D, and DL isomers are possible. Poly(L-lactic acid), poly(D-lactic acid), and poly(DL-lactic acid) are synthesized from L(−), D(+), and DL-lactic acid monomers, respectively. A wide range of degradation rates and physical and mechanical properties can be achieved by varying its molecular weights and composition in its copolymers.

PLLA has a melting point of 170°C–183°C and a glass-transition temperature of 55°C–65°C [28,29], while PDLLA has a (T_g) of 59°C [30]. Density of PLLA ranges from 1.25 to 1.29 g/cm^3, while for PDLLA, it is 1.27 g/cm^3 [31]. PLLA is crystalline whereas PDLLA is a completely amorphous biodegradable polymer. Because of the crystallinity, poly(L-lactide) of same molecular weight has better mechanical properties than poly(DL-lactide). PLLA has more ordered and compact structure and hence it has better mechanical properties and longer service time. However, the annealed PLLA has better mechanical properties than un-annealed PLLA [32] because of higher degree of crystallinity resulted by annealing. The degree of crystallinity depends on many factors, such as molecular weight, thermal and processing history, and the temperature and time of annealing treatments. According to the literature [33,34], the calculated values for the heat of fusion of 100% crystalline PLLA are generally comprised between 135 and 203 j/g.

The solubility of lactic acid–based polymer is highly dependent on the molar mass, degree of crystallinity, and other co-monomers present in the polymer. Good solvents for poly(L-lactide) are chlorinated or fluorinated organic solvents, 1–4 dioxane, and furan. For poly(DL lactide), in addition to the previously mentioned ones, there are many other organic solvents like acetone, pyridine, ethyl lactate, tetrahydrofuran, xylene, ethyl acetate, dimethylsulfoxide, N,N-dimethylformamide, and methyl ethyl ketone. The typical nonsolvents for lactic acid–based polymers are water; alcohols like methanol, ethanol, and propylene glycol; and unsubstituted hydrocarbons like hexane, heptane, etc. [35].

The mechanical properties of PLA can be varied to a large extent ranging from soft, elastic plastic to stiff and high strength plastic. With the increase of molecular weight, the mechanical properties also increase. Indeed, Loo et al. [36] reported that an increase of molecular weight of PLLA from 23 k to 67 k yields to an increase in the values of flexural strength from 64 to 106 MPa, but the tensile strength remains the same at 59 MPa. In the case of poly(DL-lactide), when the molecular weight is increased from 47.5 k to 114 k, the values of tensile and flexural strengths increase slightly from 49 to 53 MPa and 84 to 88 MPa, respectively. The various properties of PLA are listed in Tables 11.1 and 11.2 [37,38].

TABLE 11.1

Mechanical Properties of PLA (Naturework™ Cargill Dow)

	PLA Grads	
Properties	**2002D**	**2100D**
Specific gravity	1.24 (D792)	1.30 (D792)
Melt index, g/10 min, (190°C/2.16 K)	4–8 (D1238)	5–15 (D1238)
Clarity	Transparent	Opaque
Tensile strength at break (MPa)	53 (D882)	56 (D638)
Tensile yield strength (MPa)	60 (D882)	62 (D638)
Tensile modulus (GPa)	3.5 (D882)	3.5 (D638)
Tensile elongation (%)	6.0 (D882)	3.0 (D638)
Notched izod impact, (J/m)	12.81 (D256)	19.8 (D638)
Shrinkage	Similar to PET	—

Applications	Dairy containers
	Food service ware
	Transparent food containers
	Blister packaging
	Cold drink cups
		Plates and bowls for hot
		  food service ware
		Shallow draw
		  microwavable trays

Source: Doi, Y. and Steinbuchel, A., Biopolymers, in Kawashima, N., Ogawa, S., Obuchi, S., and Matsuo, M., eds., *Polylactic Acid "LACEA"*, Vol. 4, Wiley-VCH Inc., Weinheim, Germany, p. 251, 2002. With permission.

TABLE 11.2

Mechanical Properties of PLA (LACEA Mitsui Chemicals)

		PLA		Commodity Plastics		
Properties		**Standard**	**IRG[a]**	**GPPS**	**PET**	**PBT**
Tensile strength	(MPa)	68	44	45	57	56
Elongation at break	(%)	4	3	3	300	—
Flexural strength	(MPa)	98	76	76	88	—
Flexural modulus	(MPa)	3700	4700	3000	2700	2340
Izod impact	(J/m)	29	43	21	59	53
Vicat softening point	(°C)	58	114	98	79	170
Density	(kg/m³)	1.26	1.48	1.05	1.4	—

Source: Gupta, A.P. and Kumar, V., *Eur. Polym. J.*, 43, 4053, 2007. With permission.
GPPS, General purpose polystyrene (PS).
[a] Impact-resistant grads.

11.2.1.2 Nanoclay

The phyllosilicates represent a wide family in which clays with different structures, textures, or morphologies can be found. For instance, the montmorillonite and the synthetic laponite clays are anisotropic particles with a thickness of 1 nm but a width of hundreds and tens of nanometers, respectively [39].

The phyllosilicates mainly present three organization levels depending on the observation scale: (i) the layer, (ii) the primary particle, and (iii) the aggregate (see Figure 11.4).

  i. The layer is equivalent to a disc or a platelet having a width varying from 10 nm to 1 mm and a thickness of 1 nm. These layers, and especially the widest, are flexible and deformable.
 ii. The primary particle is composed of 5–10 stacked platelets. The cohesion of the structure is assured by Van der Waals and electrostatic attraction forces between the cations and the platelets. The stacking of these particles is perpendicular to the z direction and is disordered in the plane (x, y). The structure thickness is around 10 nm.
iii. The aggregate is the association of primary particles orientated in all the directions. The size of the aggregates varies from 0.1 to 10 μm.

11.2.1.2.1 Phyllosilicate Structure

The phyllosilicate crystal structure is based on the pyrophyllite structure $Si_4Al_2O_{10}(OH)_2$ and can be described as a crystalline 2:1 layered clay mineral with a central alumina octahedral sheet sandwiched between two silica tetrahedral sheets corresponding to seven atomic layers superposed as illustrated in Figure 11.5 [40]. This structure becomes (Si_8) $(Al_{4-y}$ $Mg_y)O_{20}$ $(OH)_4$, M_y^+ for the montmorillonite or (Si_8) $(Al_{6-y}$ $Li_y)O_{20}$ $(OH)_4$, M_y^+ for the hectorite. These differentiations are mainly due to the isomorphic substitutions that take place inside the aluminum oxide layer [41]. These substitutions induce a negative charge inside the clay platelet, which is naturally counter balanced by inorganic cations ($Li+$, $Na+$, Ca_2+, $K+$, Mg_2+, etc.) located into the interlayer spacing. The global charge varies depending on the phyllosilicates.

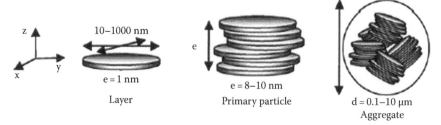

FIGURE 11.4
Phyllosilicate multiscale structure.

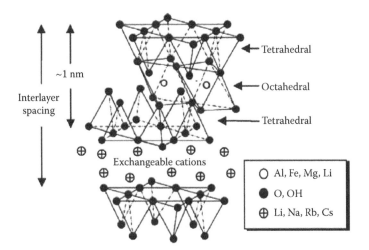

FIGURE 11.5
Structure of 2:1 phyllosilicates.

TABLE 11.3

Classification of 2:1 Phyllosilicates

Charge per Unit Cell	Di-Octahedral Phyllosilicate	Tri-Octahedral Phyllosilicate
Smectites		
0.4–1.2	Montmorillonite (Si_8) $(Al_{4-y}Mg_y)O_{20}(OH)_4, M_x^+$	Hectorite (Si_8) $(Al_{6-y}Li_y)$ $O_{20}(OH)_4, M_y^+$
	Beidellite $(Si_{8-x}Al_x)$ $Al_4O_{20}(OH)_4, M_x^+$	Saponite $(Si_{8-x}Al_x)$ (Mg_6) $O_{20}(OH)_4, M_x^+$
1.2–1.8	Illites $(Si_{8-x}Al_x)$ $(Al_{4-y}M_y^{2+}O_{20}(OH)_4, K_{x+y}^+$	Vermiculite $(Si_{8-x}Al_x)$ $(Mg_{6-y}M_y^{3+}O_{20}(OH)_4, K_{x+y}^+$
Micas		
2	Muscovite (Si_6Al_2) (Al_4) $O_{20}(OH)_2, K_2^+$	Phlogopites (Si_6Al_2) (Mg_6) $O_{20}(OH)_2, K_2^+$
4	Margarite (Si_6Al_4) (Al_4) $O_{20}(OH)_2, Ca_2^+$	Clintonite (Si_6Al_4) (Mg_6) $O_{20}(OH)_2, Ca_2^+$

For the smectite and the mica families, this charge varies from 0.4 to 1.2 and from 2 to 4 per unit cell, respectively (see Table 11.3). The charge amount is characterized by the cationic exchange capacity (CEC) and corresponds to the amount of monovalent cations necessary to compensate the platelets' negative charge, which is usually given in milliequivalent per 100 g (meq/100 g). For instance, the montmorillonite CEC varies from 70 to 120 meq/100 g depending on their extraction site [42].

The distance observed between two platelets of the primary particle, named interlayer spacing or d-spacing (d_{001}), depends on the silicate type. This value does not entirely depend on the layer crystal structure but also on the type of the counter-cation and on the hydration state of the silicate.

For instance, $d_{001} = 0.96$ nm for anhydrous montmorillonite with sodium as counterion, but $d_{001} = 1.2$ nm in usual conditions. This increase is linked to the adsorption of one layer of water molecules between the clay platelets [10].

11.2.1.2.2 Phyllosilicate Swelling Properties

The phyllosilicate multi-scale structure has different porosity levels, which drive its swelling properties. The water absorption occurs thanks to the intercalated cation hydration, which lowers the attractive forces between the clay layers [43], and also thanks to the water capillarity phenomena, which take place in the interparticle and interaggregate porosities [44,45]. For a given pressure, this swelling is characterized by a d_{001} increase until an equilibrium distance [46]. In general, the smaller the cation and the lower its charge, the higher the clay swelling. In the case of montmorillonite, the swelling decreases depending on the cation chemical type according to the following trend: Li+ > Na+ > Ca_2+ > Fe_2+ > K+ [47–49]. The potassium cation is a specific case because its size is equal to the dimension of the platelet surface cavity. Thus, the potassium is trapped into these cavities, leading to a lowering of its hydration ability.

11.2.1.2.3 Phyllosilicate Organomodification

To enhance the intercalation/exfoliation process in a polymer matrix, a chemical modification of the clay surface, with the aim to match the polymer polarity, is often carried out [10,50]. The cationic exchange is the most common technique, but other original techniques as the organosilane grafting [51,52], the use of ionomers [53,54], or block copolymer adsorption [55] are also used.

The cationic exchange consists in the inorganic cation substitution by organic ones. These cations are often alkylammonium surfactants having at least one long alkyl chain. Phosphonium salts are also interesting clay modifiers, thanks to their higher thermal stability, but they are not often used [56]. The ionic substitution is performed in water because of the clay swelling, which facilitates the organic cations' insertion between the platelets. Then, the solution is filtered, washed with distilled water (to remove the salt formed during the surfactant adsorption and the surfactant excess), and lyophilized to obtain the organo-modified clay. In addition to the modification of the clay surface polarity, organomodification increases the d_{001}, which will also further facilitate the polymer chains intercalation [57]. Various commercially available organo-modified montmorillonites (OMMTs), which mainly differ in the nature of their counter-cation and their CEC, are produced with this technique such as Cloisite 15A, 20A, and 30B, or Nanofil1 804.

11.2.1.3 Preparation Methods of Polymer/Clay Nanocomposites

The nanofiller incorporation into the polymer matrix can be carried out with three main techniques [10]: (i) the in situ polymerization, (ii) the solvent intercalation, or (iii) the melt intercalation process. Depending on the process conditions and on the polymer/nanofiller affinity, different morphologies can be obtained. These morphologies can be divided in three distinct main categories: (i) microcomposites, (ii) intercalated nanocomposites, or (iii) exfoliated nanocomposites [10,58]. For microcomposites, the polymer chains have not penetrated into the interlayer spacing and the clay particles are aggregated. In this case, the designation as nanocomposite is abusive. In the intercalated structures, the polymer chains have diffused between the platelets leading to a d_{001} increase. In the exfoliated state, the clay layers are individually delaminated and homogeneously dispersed into the polymer matrix. Intermediate dispersion states are often observed, such as intercalated-exfoliated structures. This classification does not take into account the dispersion multi-scale structure, such as percolation phenomenon, preferential orientation of the clay layers, etc. [50].

11.2.1.3.1 In Situ Polymerization Process

In this method, layered silicates are swollen into a monomer solution. Then, the monomer polymerization is initiated and propagated. The macromolecule's molecular weight increases, leading to a d_{001} increase and sometimes to an almost fully exfoliated morphology for some studied systems [50].

11.2.1.3.2 Solvent Intercalation Process

This elaboration process is based on a solvent system in which the polymer is soluble and the silicate layers are swellable. The polymer is first dissolved in an appropriate solvent. In parallel, the clay (organo-modified or not) is swollen and dispersed into the same solvent or another one to obtain a miscible solution. Both systems are pooled together leading to a polymer chain intercalation. Then, the solvent is evaporated to obtain nanocomposite materials. Nevertheless, for non-water-soluble polymers, this process involves the use of a large amount of organic solvents, which is environmentally unfriendly and cost prohibitive. Moreover, a small amount of solvent may remain in the final product at the polymer/clay interface creating lower interfacial interaction between the polymer and the clay surfaces [59]. Thus, this technique is mainly used in academic studies. Since some polysaccharides, such as chitosan or pectin, cannot be melt processed due to high thermal or thermo-mechanical degradations, the solvent process has been extensively used to produce polysaccharide/clay hybrid materials.

11.2.1.3.3 Melt Intercalation Process

Both the polymer and the clay are introduced simultaneously into a melt mixing device (extruder, internal mixer, etc.). According to Dennis et al. [60], in addition to the polymer/nanofiller affinity, two main process parameters favor the nanodispersion of the nanoclay. These parameters, which are the driving force of the intercalation–exfoliation process into the matrix, are (i) the residence time and (ii) the shearing. Shearing is necessary to induce platelet delamination from the clay tactoids. The extended residence

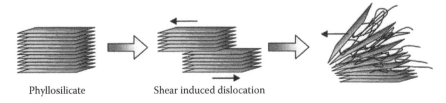

Phyllosilicate   Shear induced dislocation

FIGURE 11.6
Mechanism leading to clay exfoliation under shearing.

time is needed to allow the polymer chains' diffusion into the interlayer gallery and then to obtain an exfoliated morphology as illustrated in Figure 11.6. Therefore, it is necessary to balance the process parameters to minimize the chain degradation and to obtain a well-exfoliated morphology.

11.2.1.4 PLA/Clay Nanocomposites in Food Packaging

Most materials currently used for food packaging are nondegradable, generating environmental problems. Several biopolymers such as PLA have been exploited to develop materials for eco-friendly food packaging. However, the use of PLA, for instance, has been limited because of their usually poor mechanical and barrier properties, which may be improved by adding reinforcing nanofillers. Indeed, the use of fillers with at least one nanoscale dimension (nanoparticles) produces nanocomposites.

Nanoparticles have proportionally larger surface area than their microscale counterparts, which favors the filler–matrix interactions and the performance of the resulting material. Besides nanoreinforcements, nanoparticles can have other functions when added to a polymer, such as antimicrobial activity, enzyme immobilization, biosensing, etc. [61].

The combination of PLA and montmorillonite-layered silicate may result in a nanocomposite with good barrier properties that is suitable for film packaging material. The modulus of PLA would be increased by the addition of montmorillonite. However, this could also decrease the toughness of the materials. There are various technical approaches to achieve a balance of good strength and toughness for PLA nanocomposites. The addition of polyethylene glycol could act as a good plasticizer in a PLA/clay system [62]. In this respect, a comprehensive review is provided by Ray and Okamoto [10] for the preparation, characterization, materials properties, crystallization behavior, melt rheology, and foam processing of pure polylactide (PLA) and PLA/layered silicate nanocomposites. The authors reported that this new class of nanocomposite materials frequently exhibits remarkable improvements in its properties when compared with those of virgin PLA. The improved properties may include a high storage modulus both in the solid and melt states, increased tensile and flexural properties, decreased gas permeability, increased heat distortion temperature, and increased rate of biodegradability of pure PLA.

11.2.1.5 Recent Research Works on PLA Nanocomposites

Research works carried out in the last decade on PLA nanocomposites for food packaging applications are reported. Among these, photodegradation of PLA has been studied in the presence of TiO_2 nanoparticles as photocatalysts, which are known to decompose various organic chemicals like aldehyde, toluene, and polymers such as polyethylene (PE), polypropylene (PP), poly(vinyl chloride) (PVC) and polystyrene (PS). In this connection, TiO_2 nanoparticles were prepared and the surface of TiO_2 was modified using propionic acid and n-hexylamine with the modified TiO_2 uniformly dispersed into PLA matrices without aggregation [63]. The authors studied the photodegradation of PLA-TiO_2 nanocomposites under UV light and concluded that the photodegradability of nanocomposites can be efficiently promoted.

Melt intercalation is a method where the blending of polymer and silicate layers is followed by molding to form a polymer-layered silicate nanocomposite. In general, for intercalation, polymers and layered hosts are annealed above the softening point of the polymer. Chow and Lok [64] used this method for studying the effect of maleic anhydride-grafted ethylene propylene rubber (EPM-g-MA) on the thermal properties of PLA/organo-montmorillonite (OMMT) nanocomposites. It was found that the addition of the organoclay and EPM-g-MA did not influence much both the glass transition temperature (T_g) and the melting temperature (T_m) of PLA nanocomposites. The degree of crystallinity of PLA increased slightly in the presence of OMMT; it had been supposed that the organoclay could act as a nucleating agent to increase the crystallinity of PLA. In contrast, the addition of EPM-g-MA may restrict the crystallization process and crystal formation of PLA, which subsequently reduces the degree of crystallinity of PLA nanocomposites. Finally, they claimed that the thermal stability of PLA/organoclay was greatly enhanced by the addition of EPM-g-MA.

Kim et al. [65] studied the effect of bacterial cellulose on the transparency of PLA/bacterial nanocomposites. Bacterial cellulose has shown good potential as a reinforcement or in preparing optically transparent materials due to its structure, which consists of ribbon-shaped fibrils with diameters ranging from 10 to 50 nm. They found that light transmission of the PLA/bacterial cellulose nanocomposite was quite high due to the size effect of the nanofibrillar bacterial cellulose. Additionally, the tensile strength and Young's modulus of the PLA/bacterial cellulose nanocomposite were increased by 203% and 146%, respectively, compared with those of PLA.

Carbon nanotubes (CNTs) have been the subject of much attention because of their outstanding performance including excellent mechanical, electrical, and thermal properties. The most promising area of nanocomposite research involves the reinforcement of polymers using CNTs as reinforcing filler [66]. In this respect, Li et al. [67] introduced functionalized multiwalled carbon nanotubes (f-MWCNTs) into PLA to investigate the effect of such filler on the crystallization behavior of PLA. It was found that the addition of f-MWCNTs accelerates the crystallization of PLA significantly and induces formation of homogeneous and very small spherulites. The results of polarized optical microscopy showed that the average spherulite diameter is about 200 μm, but for nanocomposites it was very difficult to differentiate the spherulites one by one.

11.2.2 Natural Weathering and Aging of Polymeric Materials

Natural weathering (or outdoor weathering) and exposure in artificial accelerated weathering equipment have been used extensively to assess the durability of polymers. An increasing number of weathering stations all over the world are available for lifetime determinations of polymeric materials in various climates [68].

Under service conditions, polymeric materials are subjected to one or more stresses that accelerate the process of aging. As a matter of fact, the deleterious effect of natural weathering on polymeric materials has been ascribed to a complex set of reactions in which both the absorption of ultraviolet light and the presence of oxygen are participating events. As a result, the process has been termed oxidative photodegradation or photooxidation [69].

Oxidative photodegradation depends upon several factors, probably the most important of which is the sunlight that is responsible for the initial or primary photoprocess in polymer degradation. The exposure of polymers to natural weathering gives rise to a variety of chemical and physical changes, which slowly accumulate and lead to visible physical effects such as discoloration, surface cracking, or deterioration of mechanical properties. These visible results of oxidative photodegradation naturally cause some concern to the plastic consumer [70].

The sun's spectrum at the earth's surface extends down to approximately 290 μm. Ultraviolet (UV) at 300 μm has 400 kJ/mol of energy, which is enough to break carbon–carbon bonds [71]. The degree of degradation depends on the material susceptibility and on environmental factors, such as temperature, sunlight, oxidation, moisture, and microbiologic attack.

Natural weathering of plastics can be carried out according to ASTM standard practice D 1435 [72]. The samples are mounted on ASTM 45° racks, facing south. The material used to construct the racks is untreated wood. Thecontrol and the withdrawn samples are retained, covered in inert wrapping, at standard conditions of 23°C ± 1°C and 50% ± 2% humidity before analysis [73]. The literature [68] reported also some weathering tests performed on polymeric materials, in closed borosilicate and quartz glass vessels and in open glass vessels, in each case at an exposure angle of 45° directed to the south. It is argued that the use of quartz glass, although requiring more delicate handling, is to make sure that all wavelengths occurring in the terrestrial sunlight would be included. Furthermore, all the closed-vessel experiments contain a synthetic mixture of nitrogen and oxygen, simulating the composition of air (artificial air).

Although there are a very few reports on natural weathering degradation of PLA-based materials, on the other hand, in their study, Pluta et al. [74] evaluated the stability of PLA/calcium sulfate composites stored for 1 year under atmospheric conditions in relation to the filler content (20 and 40 wt.%) and the unfilled PLA. The results indicated that the molecular weight of the aged PLA and PLA composite materials did not decrease up on aging, at least after 6 months, typical period for a potential application such as packaging.

Other test methods have been developed to simulate natural weather at an accelerated rate, such as ASTM D 2565, G53, and D822 and Society of Automotive Engineers (SAE 1985) so that long-term weathering effects can be estimated in a shorter time [70]. However, such artificial tests do not necessarily give the exact response that a plastic material will exhibit in actual use. The literature [75] reported that the correlation of accelerated test results to actual end-use weathering is somewhat tenuous and gave the general rule: the less accelerated the test, the better correlation to actual part weathering.

Among the laboratory testing equipment, a Sairem-Sepap 12.24 apparatus (Materiel Physico-Chimique, France) is one of the most used to study polymer aging at 60°C under accelerated weathering conditions [75,76]. The apparatus is equipped with four mercury medium-pressure lamps (400 W) filtered by a borosilicate glass bulb ($\lambda > 400$ nm) simulating outdoor exposure. To establish the sensitivity of a given polymer to specific spectral bands, the activation spectra can be determined using a series of cutoff filters with sharp cutoffs at different wavelengths.

For instance, Copinet et al. [77] reported some experimental results on the simultaneous action of temperature, relative humidity, and UV light on the degradation of PLA films using a QUV panel (Sodexim S.A Muizon, France) as the accelerated weathering equipment. In this equipment, there are two spectrolines of 4 long-wavelength UV (315 nm) lamps (Model UVB-313 Sodexim S.A Muizon, France). The aging was carried out in a QUV device with UVB-315 lamps at different combinations of relative humidity (30%, 50%, or 100%) and temperature (30°C, 45°C, and 60°C) combinations. The results obtained showed that the degradation rate of PLA was enhanced by increasing temperature and relative humidity, factors responsible for a faster reduction of the average-weight molecular weight, the T_g and the (%) elongation at break. Moreover, UV treatment accelerated these phenomena.

Solarski et al. [78] investigated the aging of two grades of PLA filaments with different stereochemical compositions and the influence of the incorporation of clay into the PLA matrix. The aging conditions of PLA and PLA nanocomposite filaments have been performed in a climatic room (Excal 2221-HA from Climats) at 50°C and a relative humidity of 95%. The study showed that to modulate the lifetime of PLA filaments, the low D-content should be favored for long-time applications, whereas for short-time applications, PLA with higher D-content and filled with organoclay can be used.

More recently, Islam et al. [79] studied the influence of accelerated aging on the physicomechanical properties of untreated and alkali-treated industrial

hemp fiber reinforced PLA composites. The aging technique consists of plac-
ing the samples in an accelerated weathering tester (Model QUV/spray with
solar eye irradiance control) in accordance with ASTM G 154 00a: Standard
Practice for Operating Fluorescence Light Apparatus for UV Exposure of
Non-Metallic Materials. A fluorescent bulb (UVA) with 0.68 W/m^2 irradiance
(at 340 nm) was used with cycles of 1 h UV irradiation, followed by 1 min of
spray with de-ionized water, and a subsequent 2 h condensation while maintaining a temperature of 50°C. The results indicated the overall deterioration
of mechanical properties for the whole composites upon exposure to accelerated aging environment, however, much more pronounced for the untreated
ones. Further, untreated composites were found to be more resistant after
250 h of accelerated aging, while the treated composites showed more resistance after 500–1000 h of exposure.

11.3 Photooxidation of PLA and PLA/Clay Nanocomposites

11.3.1 Photooxidation Mechanisms

The photooxidation mechanism of PLA and PLA nanocomposites as well as
the influence of radiation intensity on the photodegradation rate has been
discussed in various publications [20,63,80–84,88]. The nature of the main
oxidation products is now well established although the mechanisms by
which oxidation occurs are somewhat controversial. These previous studies report two primary competitive processes to explain the photooxidative
degradation of PLA involving a photolysis reaction leading to breakage of
the backbone C–O bond and a photooxidation of the polymer leading to
the formation of hydroperoxides and their subsequent decomposition to
carboxylic acid and diketone end groups. Further, the literature [20] refers
also to another photodegradation mechanism that proceeds via Norrish II.
As a matter of fact, Norrish II mechanism was suggested recently by Tsuji
et al. [80,81] and Nakayama et al. [63] to explain the photodegradation of
PLA. Accordingly, Norrish II reactions proceed at ester group and ethylidene
group adjacent to the ester oxygen, as shown in Figure 11.7. This mechanism causes chain cleavage and the formation of C=C double bonds and the
carboxylic acid OH stretching. These structural changes have been reported
earlier by Ikada [82] and seem to be consistent with the infrared (IR) data
reported by Kaci et al. [20].

On the other hand, Janorkar et al. [83] suggested the occurrence of two
mechanisms for the degradation of PLA under UV irradiation as described
in Figure 11.8. In the mechanism (a), there is a photolysis reaction leading
to breakage of the backbone C–O bond, while the mechanism (b) involves
photooxidation of PLA, which leads to the formation of hydroperoxide derivative and its subsequent degradation to compounds containing a carboxylic

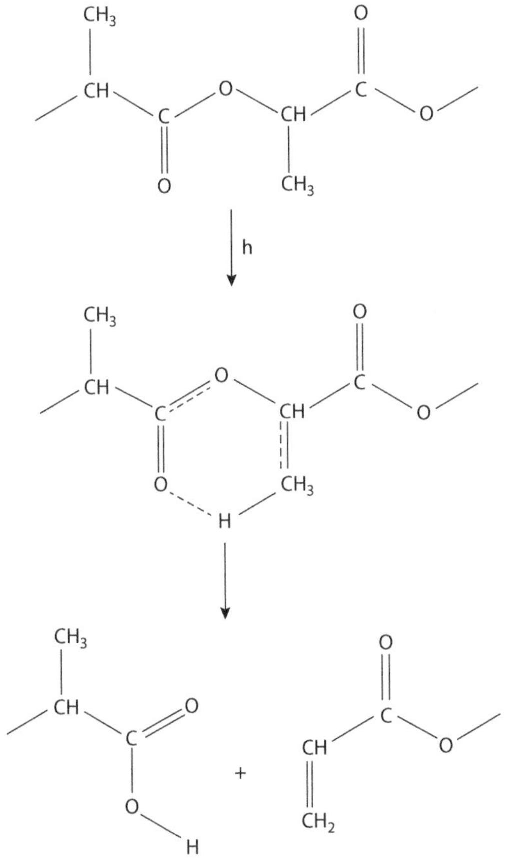

FIGURE 11.7
Norrish II mechanism of PLA.

acid and diketone end groups. Furthermore, the photolysis of the diketone may lead to the homolytic cleavage of the C–O bond between the two carbonyl groups, resulting in two carbonyl radicals. This radical pair can undergo cage escape to form several photodecomposed products.

More recently, Gardette et al. [84] reported some experimental data on the photodegradation of PLA. The authors indicated that under conditions of photooxidation with UV irradiation at wavelengths above 300 nm, the PLA undergoes a process of degradation. As a result, several photoproducts are formed, and anhydride functions have been detected. Photooxidation proceeds mainly via chain scission mechanism. Figure 11.9 shows the oxidation mechanism accounting for the main routes of degradation. This mechanism involves a classical hydrogen abstraction on the polymeric backbone at the tertiary carbon in the α-position of the ester function leading to the formation of macroradicals. It is postulated that initiation of the photochemical reaction results from the presence of chromophoric defects in the polymer

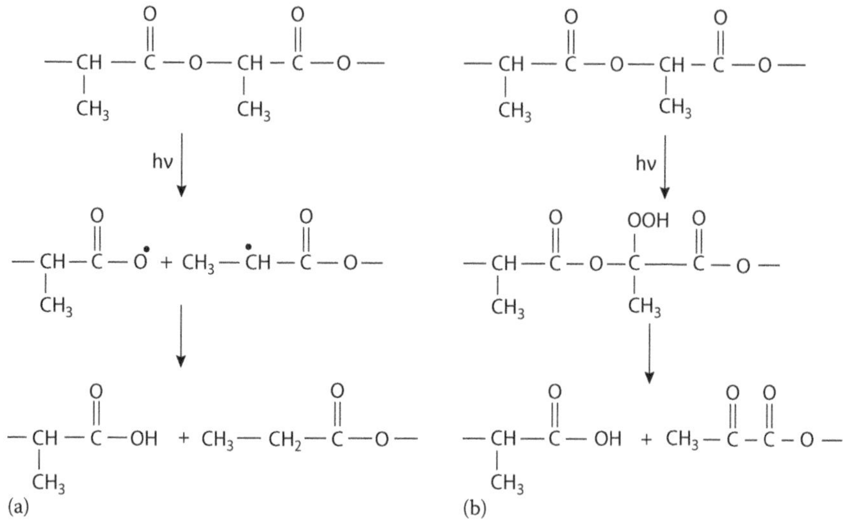

FIGURE 11.8
Plausible photodegradation mechanisms occurring in PLA. Mechanism (a): photolysis reaction inducing a cleavage of the backbone C–O bond. Mechanism (b): photooxidation of PLA leading to the formation of hydroperoxide derivatives.

at very low concentrations. This mechanism is in contradiction with the one reported previously in the literature, which is associated with a Norrish II-type photocleavage [16,63,80–82]. Gardette and his coworkers argued that the involvement of the Norrish II reactions has already been reported in the case of polyester, but this concerned aromatic polyesters such as poly(ethylene terephtalate) (PET) or poly(butylene terephtalate) (PBT) [86]. These polymers can directly absorb the light in the range of 300–330 nm, which is likely to induce Norrish cleavage. However, in the publication of Ikada [82], the PLA samples have been subjected to a light source emitting in the UV domain from 230 nm and placed in quartz cells during irradiation. These conditions are not at all relevant to outdoor exposure. Indeed, in such experimental conditions, the direct photoscission of the aliphatic ester group is likely plausible and can take place, since the ester groups of the polymer can absorb the incident light, but this is not the case under exposure to natural conditions or at wavelengths above 300 nm, which are not absorbed by PLA.

In Figure 11.9, Gardette et al. [84] argued that the radicals that are formed as a result of chain scission react with oxygen, leading to a peroxy radical that gives hydroperoxide by abstraction of a labile hydrogen atom, thus propagating the chain oxidation reactions. Once formed, the decomposition of hydroperoxides can lead to the formation of alkoxy and hydroxyl radicals. The alkoxy radical is the key intermediate in the reaction and can decompose through ß-scission. Regarding the structure of the alkoxy radical, three different ß-scissions may occur. Two of them generate chain scissions, that is, 1 and 3, which have been observed by size exclusion chromatography analysis.

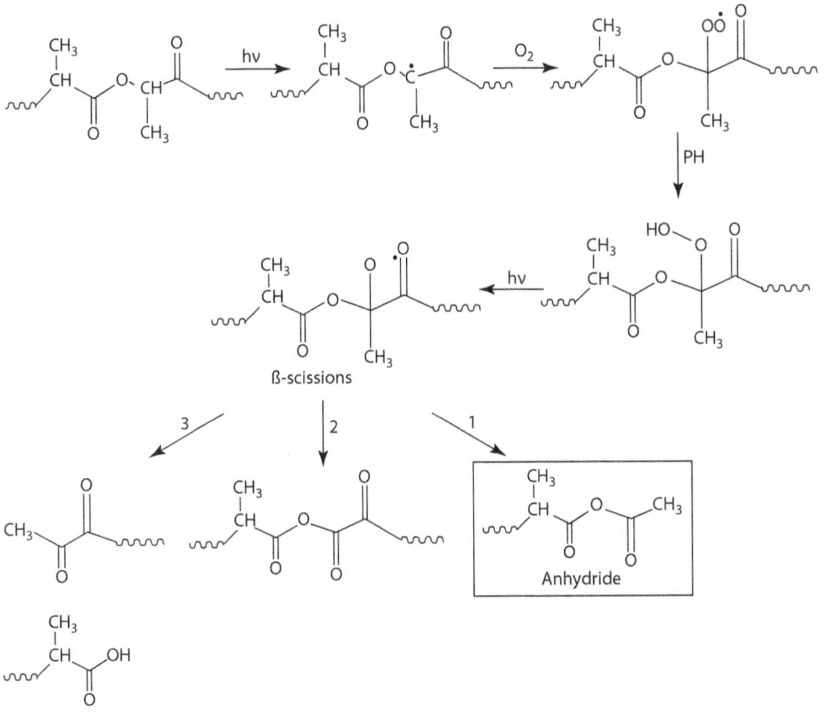

FIGURE 11.9
Main route of PLA photooxidation.

However, only reaction (1) produces anhydrides. This reaction can thus be considered to be the major mechanism operating during the photooxidation of PLA under irradiation at wavelengths above 300 nm. Further, the authors confirmed the formation of anhydrides by subjecting the photooxidized film samples to NH_3-saturated atmosphere [84]. Infrared analysis of irradiated films shows the disappearance of the absorption band at λ_{max} = 1845 cm^{-1} corresponding to anhydride and the formation of an absorption band at 1624 cm^{-1} and a shoulder at about 1670 cm^{-1}, which can be ascribed to the formation of carboxylic salt and primary amide group on the basis of the well-known reaction between amines and anhydrides reported by Bocchini et al. [87]. The reaction mechanism is described in Figure 11.10.

Similar conclusions are deduced for the exposed PLA nanocomposite samples. According to Kaci et al. [20], the comparison of the different Fourier transform infrared (FT-IR) spectra of the nanocomposite samples with the

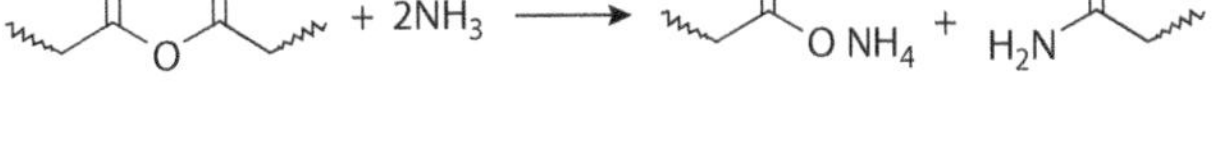

FIGURE 11.10
Reaction of anhydride with ammonia.

neat PLA before exposure and after 60 days under natural weathering conditions revealed no shift of the absorption band position in the presence of the nanoclay (Cloisite 30B). However, higher band intensity was observed for the nanocomposite samples compared with the matrix. Furthermore, it was noted also that the absorption band intensity increases with increase in clay loadings. In conclusion, the introduction of organophilic clay into PLA matrix promotes the degradation rates of the nanocomposite materials; this effect was found to be much pronounced at higher loading levels, that is, 3 and 5wt.%.

The photodegradation of PLA nanocomposites in accelerated UV conditions was also studied by Bocchini et al. [87]. Different oxidation rates have been measured on PLA-based nanocomposites, depending on the type of nanofiller. For all the nanocomposite samples, the FT-IR spectra exhibit the presence of an absorption band at λ_{max} = 1845 cm^{-1} and a broad band localized in the 4000–3000 cm^{-1} region. The addition of nanofillers without organic modifiers does not induce any significant qualitative changes in the infrared spectra evolution compared to the neat PLA. On the other hand, in PLA/C20A nanocomposite, the CH$_2$ stretching bands at λ_{max} = 2922 and λ_{max} = 2853 cm^{-1} assigned to the organic modifier completely disappear in the first hours of oxidation (Figure 11.11), evidencing that the organic modifier of C20A is easily photodegraded. After 170h of UV irradiation, PLA/C20A films were cracked, whereas no significant damages were observed for reference PLA and other PLA composites. The rapid decomposition of the organic modifier accounts for a photodegradation activity of montmorillonite surface. Therefore, it is obvious to predict a higher oxidation rate for the OMMT containing nanocomposites.

The rate of formation of anhydride is approximately constant with time and strictly depends on the type of nanofiller, the ranking in oxidation rates of the different samples being PLA/C20A > PLA/CNa > PLA, as shown in Figure 11.12.

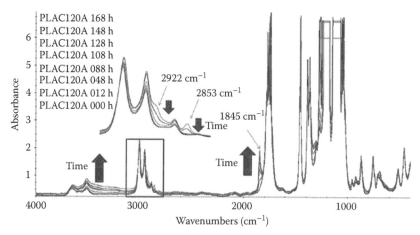

FIGURE 11.11
FT-IR spectra of PLA/C20A as a function of UV exposure time.

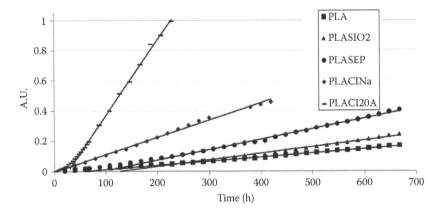

FIGURE 11.12
Evolution of anhydride absorption band intensity at $\lambda_{max} = 1845\,cm^{-1}$ as a function of UV exposure time.

These differences can be explained taking into account the differences between the fillers such as chemical composition, UV-Vis absorbance, and dispersion. Moreover, Bocchini et al. [83] argued that no evidence of low photostability of alkyl chains in onium salts are reported in literature, this effect is ascribed to a catalytic effect of montmorillonite, possibly linked to the presence of transition metals in the clay structure. This effect of iron impurities is, thus, likely to apply in PLA decomposition by increasing the decomposition rate of hydroperoxide, thus accelerating the overall radical process. Catalytic effects appear to play the most important role in PLA nanocomposites containing montmorillonite, which is supposed to have a higher amount of transition metal impurities or at least is more accessible to PLA macromolecular chains, owing to the higher polymer contact surface area. The presence of transition metal impurities in the natural mineral, especially iron, and the catalytic effect of metallic compounds are previously reported in the literature and are well known [88,89]. Metal ions can cause an acceleration of the oxidation of polymers by various processes including the decomposition of hydroperoxides (Figure 11.13). This effect of iron impurities is, thus, likely to apply in PLA decomposition by increasing the decomposition rate of hydroperoxide, thus accelerating the overall radical process.

The importance of catalytic site accessibility is evidenced by the difference between PLA/CNa and PLA/C20A. Indeed, the better dispersion in PLA/C20A results in a higher interface surface; thus, the overall photodegradation efficiency of transition metal impurities is increased. An additional

$$R\text{-}OOH + M^{n+} \longrightarrow R\text{-}O\cdot + M^{n+1+} + OH^-$$
$$R\text{-}OOH + M^{n+1+} \rightleftharpoons R\text{-}O\cdot + M^{n+} + OH^+$$

FIGURE 11.13
Photocatalytic decomposition of hydroperoxides by transition metal ions.

contribution to the higher oxidation rate for PLA/C20A compared to PLA/CNa is in the initiation activity of catalyzed decomposition of organic modifier. It constitutes a supplementary source of radicals that are likely to initiate the oxidation of PLA, leading to an increase of the overall rate [87].

The photodegradation of PLA/CaSO$_4$ composites was also investigated by Gardette et al. [84] under accelerated UV conditions. The study showed that the composite samples when subjected to UV-light undergo photooxidation resulting in the formation of carbonyl products, identified as anhydride functions observed in the neat PLA. The formation of these photoproducts implies a mechanism of chain scission of PLA, accompanied by a decrease of polyester molecular weight. From FT-IR analysis, the presence of a dispersed phase (filler) does not modify the photooxidation mechanism of PLA, since the same products of oxidation were observed. However, the presence of CaSO$_4$ and the amount of the filler, as well as the particle size, have an influence on the rates of photooxidation; the higher the level of filler in the composites, the faster is the degradation.

The authors reported that a peculiar behavior was observed for the PLA/CaSO$_4$ composites based on a variation of absorbance at 1845 cm^{-1} as a linear function of irradiation time (Figure 11.14a and b). The curves of the photooxidation rates exhibit no induction period, and the photooxidation products are accumulated from the beginning of irradiation. Furthermore, it was also observed that the photooxidation of PLA starts immediately upon irradiation regardless of what the filler content is. It can be seen that the increase of absorption band intensity of PLA and PLA/filler composites is linear with irradiation time. However, the most important result concerns the influence of the filler on the rates of photooxidation of PLA. As shown in Figure 11.14a and b, the slope of the kinetic curves increases with the filler content. This means that the rate of degradation of PLA increases with the amount of the filler, which suggests that CaSO$_4$ behaves as a prooxidant. This effect could be explained by the presence of chromophoric impurities in the composites that can have an inductive effect.

11.3.2 Evaluation of Property Changes

11.3.2.1 Molecular Weight

The chemical structure of macromolecules in a degraded polymer sample may be substantially different from that of the virgin material because chemical degradation causes several changes in the molecular structure such as chain scission, cross-linking, or the introduction of other chemical species [90]. Several authors [20,84,90] have reported that the photodegradation mechanism of PLA and PLA nanocomposites induces changes in the molecular weight distribution (MWD) and the mass average molecular weight of the irradiated materials. As a matter of fact, the MWD curves have been found to shift toward the low molecular weight

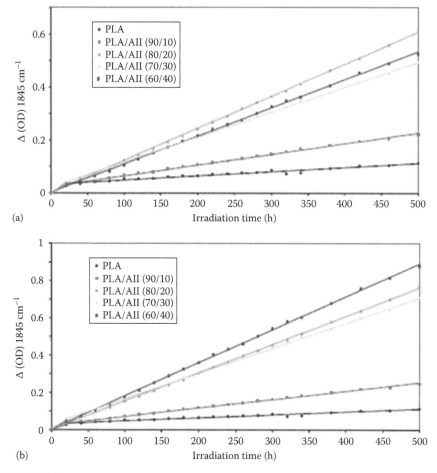

FIGURE 11.14
(See color insert.) Evolution of anhydride absorption band intensity at $\lambda_{max} = 1845\,cm^{-1}$ as a function of UV exposure time for PLA and PLA composites with different filler ratios. (a) Without correction of filler amount. (b) With correction of filler amount.

scale indicating the formation of low molecular weight compounds in the polymer during the photooxidation.

In this respect, Gardette et al. [84] indicated that the photooxidation of PLA leads to the formation of low molecular weight oligomers as a result of chain scission mechanism. Further, the authors reported that the distribution of photoproducts within the oxidized samples is homogeneous and without gradients, mainly attesting to a uniform degradation of the bulk PLA-based products. In the case of PLA/CaSO$_4$ composites, similar trend is observed as that observed for the neat PLA, which is characterized by a decrease of the average-weight molecular weight and an increase of the polydispersity index with UV exposure time. This is attributed to occurrence of chain scissions in the PLA matrix during pho-tooxidation. This result clearly shows that the mineral filler has no influ-ence on the type of chemical reactions involved in the photodegradation mechanism of PLA; however, only an accelerating effect on the photooxi-dation rate is noted.

Kaci et al. [20] have studied the effects of natural weathering exposure on the molecular weight of PLA and PLA/C30B nanocomposites at vari-ous clay loadings, that is, 1, 3, and 5 wt.%. The experimental data in the form of histograms are shown in Figures 11.15 and 11.16, which correspond to the average-weight molecular weight and the polydispersity index, respectively. From Figure 11.15, it can be observed that both PLA and PLA nanocomposite samples exhibit a significant reduction in $\bar{M}_w$ with exposure time, being much faster for the PLA nanocomposites. Moreover, the reduc-tion in molecular weight seems to increase with increase in clay loadings. On the other hand, the reduction of $\bar{M}_w$ associated with an enhanced poly-dispersity of the nanocomposite samples indicates that chain scission plays an important role among the degradation mechanisms. This is in agreement with the literature data. Also, under natural weathering exposure, both PLA and PLA/C30B nanocomposites undergo photooxidative degradation by free radical process, which can lead to a breakdown of the polymer back-bone and to the formation of lower molecular weight species as well (see schemes illustrated in Figures 11.8 and 11.9). Consequently, these structural modifications are responsible for the deterioration of the physicomechani-cal properties of the samples.

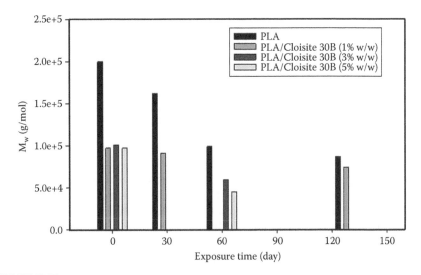

FIGURE 11.15

Average-weight molecular weight ($\bar{M}_w$) of PLA and PLA/C30B nanocomposite as a function of exposure time in natural weathering.

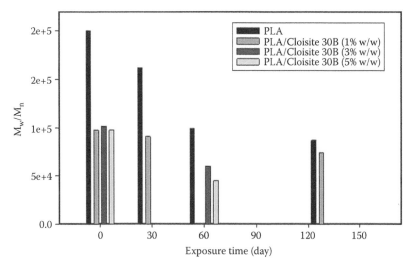

FIGURE 11.16

Polydispersity index ($\bar{M}_w/\bar{M}_n$) of PLA and PLA/C30B nanocomposites as a function of exposure time in natural weathering.

11.3.2.2 Crystallinity

Crystallinity is one of the most important morphological characteristics in the studies of polymeric materials. It is dependent on the composition of the polymers and crystallization conditions and affects many of the macroscopic properties, especially mechanical properties [91]. In literature, an increase of the crystallinity index as a function of aging time (80 days in a climatic room at 50°C and RH = 95%) for PLA and PLA/Bentone 104 nanocomposite filaments has been reported [78]. For the samples investigated herein, the results presented in Figure 11.17 indicate that before exposure to natural weathering, the crystalline index (X_c) of PLA/30B nanocomposites is lower than that of the neat polymer and this property decreases gradually with increasing clay content. This behavior may be explained as a result of a polymer chain disorder, which is favored by the presence of higher amounts of nanoclay [92]. Accordingly, this clearly indicates that the nanofiller does not act as a nucleating agent as it would be expected as impurity and seem to have no influence on the crystallization of PLA in nanocomposites. Bouza et al. [93] have attributed the reduction of the PLA crystallization rate to the improvement of the nanofiller dispersion into the polymer matrix.

Under natural weathering, it is observed, however, that there is a fast increase of X_c in the initial 30 days for all the irradiated samples before it remains fairly constant up to 60 days. In the initial exposure period of 30 days, the increase of X_c for the neat PLA and the PLA/Cloisite 30B at 1 wt.% is around 14%, whereas for 3 and 5 wt.% filled PLA nanocomposites, X_c is more than 19%. According to the literature [78,92], the increase of X_c is

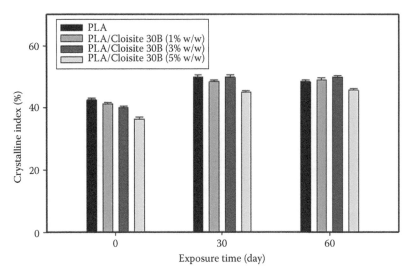

FIGURE 11.17
Crystalline index of PLA and PLA/C30B nanocomposites as a function of exposure time in natural weathering.

explained by the fact that amorphous regions are easily accessible to degradation during photooxidation process resulting in chain scissions. This will decrease the density of entanglements in the amorphous phase allowing the shorter molecules to crystallize due to higher mobility leading to an increase of the degree of crystallinity. This phenomenon is well known as a secondary crystallization (or chemicrystallization) [91].

11.3.2.3 Thermal Properties

To observe the effect of natural weathering on the glass transition temperature (T_g) of the various PLA/Cloisite 30B nanocomposites with respect to the neat PLA, T_g was measured during the second heating scan and the data are presented in Figure 11.18. The results indicate a decrease of T_g with exposure time for all the irradiated samples, however, being much pronounced for the 3 and 5 wt.% PLA nanocomposites. The decrease of T_g is explained by the decrease of the number average molecular weight ($\bar{M}_n$) of the exposed samples [92]. On the other hand, it appears that the natural weather has no significant effect on the melting temperature (T_m) for both the neat PLA and the PLA nanocomposites. This may be attributed to the fact that no change has occurred in the lamellar thickness [94].

11.3.2.4 Mechanical Properties

The deterioration of mechanical response is the most evident consequence due to polymer photodegradation. Accordingly, it was reported

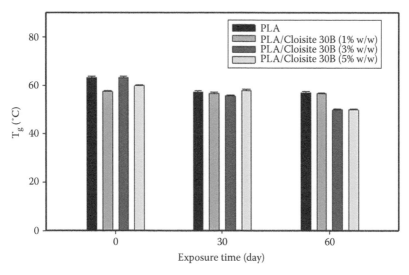

FIGURE 11.18
Glass transition temperature (T_g) of PLA and PLA/C30B nanocomposites as a function of exposure time in natural weathering.

that most of the tensile properties of PLA films were affected by UV irradiation. As a result, significant changes in elastic modulus, yield stress, tensile strength, and strain at break are highlighted, and both Young's modulus and yield stress decrease exponentially with UV irradiation time [90].

On the other hand, indentation testing is an increasingly popular area of study because of the potential to apply the technique to polymer films and polymer composites where monolithic mechanical testing is not always feasible [95].

In this respect, the effect of natural weather on the modulus and hardness determined by nanoindentation has been reported by Kaci et al. [20]. The variation of modulus and hardness under natural weathering exposure for both neat PLA and various PLA/Cloisite 30B nanocomposites are shown in Figures 11.19 and 11.20, respectively. The results indicate that both modulus and hardness slightly increase with exposure time for both the neat PLA and the PLA/C30B nanocomposite samples. Further, the increase is also dependent on the nanoclay content ratios. As a matter of fact, the addition of 1 wt.% of clay yields to an increase of the modulus by 5.4% and the hardness by 4.4% with respect to the neat PLA. For 5 wt.% of clay, the modulus is improved by 12.5%, while the hardness is enhanced by 15.9%. According to Dakhal et al. [96], the increase in nanohardness can be correlated with the dispersion of clay and its nanostructure; the higher the d-spacing values, better is the dispersion of the nanoclay.

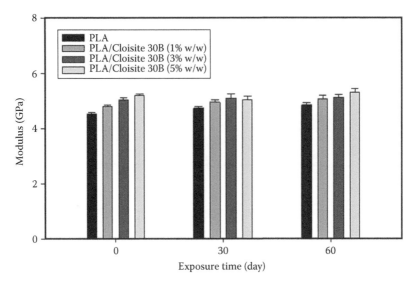

FIGURE 11.19
Modulus (by nanoindentation) of PLA and PLA/C30B nanocomposites as a function of exposure time in natural weathering.

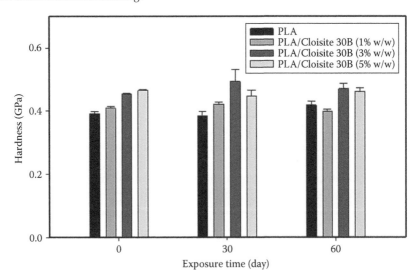

FIGURE 11.20
Hardness (by nanoindentation) of PLA and PLA/C30B nanocomposites as a function of exposure time in natural weathering.

Under natural weathering, the increase in modulus of PLA is estimated to 6%, up to 60 days of exposure, whereas it is almost 2% for the 5wt.% PLA nanocomposite samples. Similar trend is observed for hardness. The reason of enhancement of both modulus and hardness may be related to the increase in the crystallinity index during UV irradiation rendering the aged materials more rigid.

11.4 Conclusions

The photooxidation of PLA nanocomposites showed that UV light irradiation leads to the formation of carbonyl groups, anhydrides, and hydroperoxides. The formation of these photoproducts may be the result of mechanism of chain scission of PLA, accompanied by a decrease of molecular weight as shown by SEC measurements. The results of photooxidation of the PLA/Cloisite 30B nanocomposites showed that the presence of organophilic montmorillonite does not modify the photooxidation mechanism of PLA, since the same products of oxidation were observed. However, the presence of Cloisite 30B and the amount of the filler had an influence on the rates of photooxidation; this effect is much more pronounced for the samples containing 3 and 5 wt.%. The differential scanning calorimetry (DSC) results indicated an increase in crystallinity with time of exposition under natural weathering conditions, whereas T_g decreases for the whole samples due to the reduction of the molecular weight.

References

1. M.A. Paul, M. Alexandre, P. Degée, C. Henrist, A. Rulmont, and P. Dubois, *Polymer*, 44 (2003) 443.
2. P. Krishnamachari, J. Zhang, J. Lou, J. Yan, and L. Uitenham, *Int. J. Polym. Anal. Charact.*, 14 (2009) 336.
3. J.M. Becker, R.J. Pounder, and A.P. Dove, *Macromol. Rapid Commun.*, 31 (2010) 1923.
4. Y. Ikada and H. Tsuji, *Macromol. Rapid Commun.*, 21 (2000) 117.
5. K. Hamad, M. Kaseem, and F. Deri, *J. Mater. Sci.*, 46 (2011) 3013.
6. R.M. Rasal, A.V. Janorkar, and D.E. Hirt, *Prog. Polym. Sci.*, 35 (2010) 338.
7. C. Silvestre, D. Duraccio, and S. Cimmino, *Prog. Polym. Sci.*, 36 (2011) 1766.
8. N.B. Hatzigrigoriou and C.D. Papaspyrides, *J. Appl. Polym. Sci.*, 122 (2011) 3720.
9. M. Alexandre and P. Dubois, *Mater. Sci. Eng.*, 28 (2000) 1.
10. S.S. Ray and M. Okamoto, *Prog. Polym. Sci.*, 28 (2003) 1539.
11. S.S. Ray and M. Bousmina, *Prog. Mater. Sci.*, 50 (2005) 962.
12. A. Leszczynska, J. Njuguna, K. Pielichowski, and J.R. Banerjee, *Thermochim. Acta*, 454 (2007) 1.
13. N.Tz. Dintcheva, S. El Malaika, and F.P. La Mantia, *Polym. Degrad. Stab.*, 94 (2009) 1571.
14. D. Wu, Li. Wu, La. Wu, and M. Zhang, *Polym. Degrad. Stab.*, 91 (2006) 3149.
15. A.R. McLauchlin and N.L. Thomas, *Polym. Degrad. Stab.*, 94 (2009) 868.
16. S. Bourbigot, F. Samyn, T. Turf, and S. Duquesne, *Polym. Degrad. Stab.*, 95 (2010) 320.
17. D. Garlotta, *J. Polym. Environ.*, 9 (2001) 63.
18. S.S. Ray, K. Yamada, M. Okamoto, and K. Ueda, *Nano Lett.*, 2 (2002) 1093.
19. L. Zaidi, S. Bruzaud, A. Bourmaud, P. Médéric, M. Kaci, and Y. Grohens, *J. Appl. Polym. Sci.*, 116 (2010) 1357.

20. L. Zaidi, M. Kaci, S. Bruzaud, A. Bourmaud, and Y. Grohens, *Polym. Degrad. Stab.*, 95 (2010) 1751.
21. J.K. Pandey, K.R. Reddy, A.P. Kumar, and R.P. Singh, *Polym. Degrad. Stab.*, 88 (2005) 234.
22. B. Singh and N. Sharma, *Polym. Degrad. Stab.*, 93 (2008) 561.
23. N. Touati, M. Kaci, S. Bruzaud, and Y. Grohens, *Polym. Degrad. Stab.*, 96 (2011) 1064.
24. H. Tsuji and T. Tsuruno, *Polym. Degrad. Stab.*, 95 (2010) 477.
25. S. Inkinen, M. Hakkarainen, A.C. Albertsson, and A. Södergard, *Biomacromolecules*, 12 (2011) 523.
26. P. Bordes, E. Pollet, and L. Avérous, *Prog. Polym. Sci.*, 34 (2009) 125.
27. M. Ajioka, K. Enomoto, K. Suzuki, and A. Yamaguchi, *J. Environ. Polym. Degrad.*, 3 (1995) 225.
28. M. Hakkarainen, S. Karlsson, and A.C. Albertsson, *Polymer*, 41 (2000) 2331.
29. H. Tsuji and K. Ikarashi, *Biomaterials*, 25 (2004) 5449.
30. S.Y. Kim, I.G. Shin, and Y.M. Lee, *J. Control. Release*, 56 (1998) 197.
31. A.J. Domb, J. Kost, and D.M. Wiseman, Handbook of biodegradable polymers. In: Perrin D.A. and English J.P., eds. *Polyglycolode and Polylactide*. Narwood Academic Publishers, Amsterdam, the Netherlands (1997), p. 3.
32. G. Perego, G.D. Cella, and C. Basitoli, *J. Appl. Polym. Sci.*, 59 (1996) 37.
33. G.L. Loomis, J.R. Murdoch, and K.H. Gradner, *Polym. Prepr.*, 2 (1990) 55.
34. T. Miyata and T. Masuko, *Polymer*, 39 (1998) 5515.
35. M.H. Hartmann, *Biopolymers from Renewable Resources*. In D.L. Kaplan, ed. Springer-Verlag, Berlin, Germany (1998), pp. 367–411.
36. J.S.C. Loo, C.P. Ooi, and F.Y.C. Boey, *Biomaterials*, 26 (2005) 1359.
37. Y. Doi and A. Steinbuchel, Biopolymers. In: Kawashima N., Ogawa S., Obuchi S., and Matsuo M., eds. *Polylactic Acid "LACEA"*, vol. 4. Wiley-VCH Inc., Weinheim, Germany (2002), p. 251.
38. A.P. Gupta and V. Kumar, *Eur. Polym. J.*, 43 (2007) 4053.
39. F. Chivrac, E. Pollet, and L. Avérous, *Mater. Sci. Eng.*, 67 (2009) 1.
40. A.P. Kumar, D. Depan, N.S. Tomer, and R.P. Singh, *Prog. Polym. Sci.*, 34 (2009) 479.
41. F. Thomas, L.J. Michot, D. Vantelon, E. Montarges, B. Prelot, M. Cruchaudet, and J.F. Delon, *Colloid Surf. A Physicochem. Eng. Aspects*, 159 (1999) 351.
42. G. Sposito and D. Grasso, *Surfact. Sci. Ser.*, 85 (1999) 107.
43. J. Méring, *Trans. Faraday Soc.*, 42B (1946) 205.
44. P.F. Luckham and S. Rossi, *Adv. Colloid. Interface Sci.*, 82 (1999) 43.
45. J.M. Cases, I. Brend, G. Besson, M. François, J.P. Uriot, F. Thomas, and J.E. Poirier, *Langmuir*, 8 (1992) 2730.
46. D.H. Powell, K. Tongkhao, S.J. Kennedy, and P.G. Slade, *Physica B*, 241/243 (1998) 387.
47. D.H. Powell, H.E. Fischer, and N.T. Skipper, *J. Phys. Chem. B*, 102 (1998) 10899.
48. R. Tettenhorst, *Am. Miner.*, 47 (1962) 769.
49. J.C. Dai and J.T. Huang, *Appl. Clay Sci.*, 15 (1999) 51.
50. R.A. Vaia and E.P. Giannelis, *Macromolecules*, 30 (1997) 8000.
51. G. Lagaly, *Appl. Clay Sci.*, 15 (1999) 1.
52. Z. Shen, G.P. Simon, and Y.B. Cheng, *Polymer*, 43 (2002) 4251.
53. H.R. Fischer, L.H. Gielgens, and T.P.M. Koster, *Acta Polym.*, 50 (1999) 122.
54. C.A. Wilkie, J. Zhu, and F. Uhl, *Polym. Prepr.*, 42 (2001) 392.
55. G. Lagaly, *Solid State Ionics*, 22 (1986) 43.

56. Y.H. Jin, H.J. Park, S.S. Im, S.Y. Kwak, and S. Kwak, *Macromol. Rapid Commun.*, 23 (2002) 135.
57. H.R. Dennis, D.L. Hunter, D. Chang, S. Kim, J.L. White, J.W. Cho, and D.R. Paul, *Polymer*, 42 (2001) 9513.
58. J.W. Cho and D.R. Paul, *Polymer*, 42 (2001) 1083.
59. A. Guilbot and C. Mercier, Starch. In: G.O. Aspinall, ed. *Molecular Biology*, vol. 3. Academic Press Inc., New York (1985), pp. 209–282.
60. G. Della Valle, A. Buleon, P.J. Carreau, P.A. Lavoie, and B. Vergnes, *J. Rheol.*, 42 (1998) 507.
61. A. Lopez-Rubio, E. Almenar, P. Hernandez-Munoz, J.M. Lagaron, R. Catala, and R. Gavara, *Food Rev. Intern.*, 20 (2004) 357.
62. M. Shibata, Y. Someya, M. Orihara, and M. Miyoshi, *J. Appl. Polym. Sci.*, 99 (2006) 2594.
63. N. Nakayama and T. Hayashi, *Polym. Degrad. Stab.*, 92 (2007) 1255.
64. W.S. Chow and S.K. Lok, *J. Therm. Anal. Calorim.*, 95 (2009) 627.
65. Y. Kim, R. Jung, H.S. Kim, and H.J. Jin, *Curr. Appl. Phys.*, 9 (2009) 69.
66. J.Y. Kim, H.S. Park, and S.H. Kim, *J. Appl. Polym. Sci.*, 103 (2007) 1450.
67. Y. Li, Y. Wang, L. Liu, L. Han, F. Xiang, and Z. Zhou, *J. Polym. Sci. Part A Polym. Chem.*, 47 (2009) 326.
68. J. Sampers, *Polym. Degrad. Stab.*, 76 (2002) 455.
69. A.M. Trozzolo, Stabilization against oxidative photodegradation. In: *Polymer Stabilization*. Wiley-Interscience, New York (1971, © 1972).
70. H. Al-Madfa, Z. Mohamed, and M.E. Kassem, *Polym. Degrad. Stab.*, 62 (1998) 105.
71. S.H. Hamid, F.S. Qureshi, M.B. Amin, and A.G. Maadhah, *Polym. Plast. Technol. Eng.*, 28 (1989) 475.
72. ASTM Standard D-1435, Outdoor weathering of plastics (1979), 8.01, pp. 639–644.
73. F.S. Qureshi, M.B. Amin, A.G. Maadhah, and S.H. Hamid, *Polym. Plast. Technol. Eng.*, 28 (1989) 649.
74. M. Pluta, M. Murariu, M. Alexandre, A. Galeski, and P. Dubois, *Polym. Degrad. Stab.*, 93 (2008) 925.
75. R.E. Lavengood and F.M. Silver, *Eng. Plastics*, Vol. 2. ASTM International, West Conshohocken, PA (1988), p. 63.
76. M. Scoponi, F. Pradella, H. Kaczmarek, R. Amadelli, and V. Carassiti, *Polymer*, 37 (1996) 903.
77. A. Copinet, C. Bertrand, S. Govindin, V. Coma, and Y. Couturier, *Chemosphere*, 55 (2004) 763.
78. S. Solarski, M. Ferreira, and E. Devaux, *Polym. Degrad. Stab.*, 93 (2008) 707.
79. M.S. Islam, K.L. Pickering, and N.J. Foreman, *Polym. Degrad. Stab.*, 95 (2010) 59.
80. H. Tsuji, Y. Echizen, and Y. Nishimura, *Polym. Degrad. Stab.*, 91 (2006) 1128.
81. H. Tsuji, Y. Echizen, and Y. Nishimura, *J. Polym. Environ.*, 14 (2006) 239.
82. E. Ikada, *J. Photopolym. Sci. Technol.*, 10 (1997) 265.
83. A.V. Janorkar, A.T. Metters, and D.E. Hirt, *J. Appl. Polym. Sci.*, 106, (2007), 1042.
84. M. Gardette, S. Thérias, J.L. Gardette, M. Murariu, and P. Dubois, *Polym. Degrad. Stab.*, 96, (2011), 616.
85. A. Rivaton, *Polym. Degrad. Stab.*, 41 (1993) 297.
86. T. Grosstete, A. Rivaton, J.L. Gardette, C.E. Hoyle, M. Ziemer, and D.R. Fagerburg, *Polymer*, 41 (2000) 3541.

87. S. Bocchini, K. Fukushima, A. Di Blasio, A. Fina, A. Frache, and F. Geobaldo, *Biomacromolecules*, 11 (2010) 2919.
88. M.F. Brigatti, L. Medici, and L. Poppi, *Appl. Clay Sci.*, 11 (1996) 43.
89. B. Ranby and J.F. Rabek, *Photo-Stabilisation of Polymers*, John Wiley & Sons, Ltd., London, U.K. (1975), p. 118.
90. S. Belbachir, F. Zairi, G. Ayoub, U. Maschke, M. Nait-Abdelaziz, J.M. Gloaguen, M. Benguediab, and J.M. Lefebvre, *J. Mech. Phys. Sol.*, 58 (2010) 241.
91. M. Kaci, T. Sadoun, and S. Cimmino, *Int. J. Polym. Anal. Charact.*, 6 (2001) 455.
92. N. Vasanthan and O. Ly, *Polym. Degrad. Stab.*, 94 (2009) 1364.
93. R. Bouza, S.G. Pardo, L. Barral, and M.J. Abad, *Polym. Comp.*, 30 (2009) 880.
94. T.M. Wu and C.Y. Wu, *Polym. Degrad. Stab.*, 91 (2006) 2198.
95. G.M. Spinks, H.R. Brown, and Z. Liu, *Polym. Test.*, 25 (2006) 868.
96. H.N. Dhakal, Z.Y. Zhang, and M.O.W. Richardson, *Polym. Test.*, 25 (2006) 846.

12

Recycling of Nanocomposites

Marek A. Kozlowski and Joanna Macyszyn

CONTENTS

12.1 Introduction

Directive 2008/98/EC of the European Parliament and of the Council of 19 November 2008 on waste and repealing certain directives establishes a legal framework for the treatment of waste within the European community [1]. It aims at protecting the environment and human health through the prevention of the harmful effects of waste generation and waste management. Earlier community has accepted within the Sixth Program for Environment (2002–2012) seven Thematic Strategies, one of which (adopted in 2005) has been devoted to prevention and recycling of waste. Its goal express the idea of sustainable development that has been formulated in the watchword of *Europe—society applying recycling, avoiding waste generation, and using waste as secondary raw material.*

That strategy has also been applied to plastic waste. Polymers belong to materials whose consumption is steadily increasing in a global dimension. Although the economical crisis has markedly influenced most industries, the branch of polymer manufacturing and processing of plastics suffered much less than the others. One of the reasons is an uneven development of plastics in different regions and high demand for polymers in the fast developing countries, where the consumption of plastics is substantially lower than that in the leading economies, who suffered much heavier crisis. The economical reasons with higher emphasis on the reasonable consumption of resources and more ecological approach to business with higher respect to

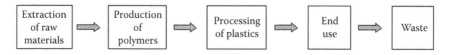

FIGURE 12.1
Life cycle of polymeric materials.

the environment have focused more attention to recycling. Multiple use of materials in their life cycle (Figure 12.1) saves resources and money, as well as reduces amount of waste dumped in landfills.

Recycling is the most favorite option of waste management, which in case of polymers may be performed by means of

- Mechanical recycling (reprocessing)
- Chemical recycling (recovery of raw materials)
- Organic recycling (composting)

Multiple reprocessing is a dominating way of plastic waste recycling, since its technology is well established, is economically viable, and offers a wide range of modifications of the end-use properties of articles made of recyclates. In Europe, the total production of plastics in 2010 accounted for 57 million tons, of which 24.7 million tons became post-consumer waste. Out of these, 14.3 million tons was recovered and 10.4 million tons was landfilled [2]. Part of the waste was recovered by end users (reuse), substantial part was reprocessed (mechanical recycling), and only small amounts were recovered to raw materials (chemical recycling) according to Figure 12.2. Mechanical recycling of plastics reached 6 million tons due to the activity of citizens, packaging collection organizations, and recycling companies.

An important factor for the market acceptance of products made of recyclates is their quality. Properties of polymers depend on several factors, first of all on their chemical structure, molecular weight, polymer chain architecture, stereoisomery, and crystallinity. Polymer macromolecules form long chains, whose length and conformation markedly influence the final material characteristics. However, neither the structure nor crystallinity of polymers after processing or after the end use phase is the same as

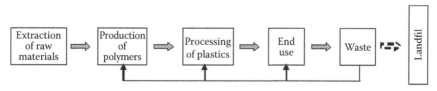

FIGURE 12.2
Preferable options of life cycle.

that formed at the polymerization stage. Current structure depends on the total thermomechanical history of the polymer that it underwent after the production stage.

Majority of polymers form chains of different length with –C–C– backbone. Considering that the energy needed to break a single C–C bond is ca. 347 kJ/mol, several forms of energy that affect a polymer during its life time (temperature, mechanical stress, UV) may be high enough to cause C–C bond scission in the polymer chain. That results usually in the decrease of the molecular weight (M_w) and in higher crystallinity of a polymer, since it is related to the length and mobility of macromolecules. Less frequently also branching and cross-linking take place due to available free radicals, depending on the polymer chemical structure.

Length of polymer chains is related to their properties. Viscosity of polymers (η) depends proportionally to the molecular weight according to Mark–Houwink equation:

$$\eta = K\,M^\alpha \tag{12.1}$$

where K and α are constants depending on the nature of the polymer and solvent.

However, mechanical properties of high molecular weight polymers (P_∞) are significantly higher than that of oligomers according to Equation 12.1:

$$P = P_\infty - C/M_w \tag{12.2}$$

where
  P is the property (strength, Young's modulus, elongation)
  C is the constant depending on a polymer and selected property

That dependence is similar to the change in Young's modulus with temperature. Until a critical level of the molecular weight (M_{wcr}) selected mechanical properties keep high, almost constant values, however below M_{wcr} they drop dramatically.

Change in the molecular weight of polymers after recycling should accordingly reflect in their viscosity and mechanical properties. It has been shown [3,4] that selected properties of polymers are sensitive to mechanical recycling in the extent dependent on the nature of the polymer and conditions of recycling (temperature, stress level). Melt viscosity, elongation at break, crystallinity, and Young's modulus have been shown as the most changing parameters of polymers after reprocessing. Nevertheless, the end use properties may remain within the region of interest to many applications, depending on the polymer nature and processing history.

Polymer composites reinforced with fibers exhibit deterioration of mechanical properties after recycling due to mechanical stresses occurring during grinding, mixing, injection molding or extrusion, that cause a degradation of the polymer matrix, the fibers breaking and reduce the matrix-fiber interfacial adhesion [5–7]. In this respect, nanocomposites are more interesting, since they are reported as nonsensitive to recycling due to the nanometer size of fillers, which do not break during reprocessing.

Nanocomposites constitute a two-phase material, in which nanoparticles of different size and shape are dispersed within the polymer matrix. Recently, nanofillers are becoming popular, because they can be used in much smaller quantities compared to conventional composites, allowing to reduce the weight of products. Incorporation of nanofillers into a polymer matrix in a very low quantity (less than 5 wt.%) brings about an improvement in properties comparable to that achieved with 10 times higher loadings of traditional fillers [8–10].

Introduction of different types of nanoparticles to waste polymers was applied to up-grade the properties [11–16]. For example, PET nanocomposite with layered silicates as nanofillers exhibited high stiffness flexibility [17].

Mechanical recycling of nanocomposites sometimes even improves selected properties of materials. Karahaliou and Tarantili [18] have shown that nanoparticles behave as a reinforcement for terpolymer ABS (acrylonitrile-butadiene-styrene) and stabilize the composite's properties during mechanical recycling with a twin-screw extruder. In the second cycle of extrusion, an improvement in the dispersion of montmorillonite (MMT) nanofiller in the matrix was observed. Even after five recycling cycles, no significant changes were observed. In the second cycle of extrusion, an improvement in the dispersion of montmorillonite (MMT) nanofiller in the matrix was observed thermal gravimetric analysis (TGA) study showed no degradation of the material after several extrusion processes; as well as rheological, mechanical, and thermal properties and color of the nanocomposites remained almost stable.

Nanocomposites consisting of polyamide 6 matrix and nanoclay have been reprocessed for five times by injection molding. The molecular weight was reduced, as revealed by viscosity measurements; however, no change in the Young's modulus was observed [19].

Jurado et al. [20] confirmed that the number of processing cycles has a significant influence on exfoliation in the polyamide 6 nanocomposite filled with 6 wt.% of nanoclay. Up to 30% of the recycled material could be added to polyamide (PA) nanocomposite, not affecting its mechanical properties.

The effect of the number of cycles and the type of extrusion of polyamide 6 nanocomposites filled with layered silicate (3 and 6 wt.%) was studied by Russo et al. [21]. They have evidenced a better dispersion of MMT in a matrix using the twin-screw extruder. Mechanical tests have shown an increase in the elastic modulus with increasing filler content. It was concluded that the silicate reinforced composites may constitute an interesting option for the stiffness enhancement of polymers during recycling. Performance of recycled nanocomposites depends on the degree of degradation of polymer matrix, the loading of silicate filler, a number of cycles, and the kind of equipment used for reprocessing.

Touati et al. [22] have performed multiple processing of polypropylene nanocomposite filled with 5 wt.% of organically modified MMT, with and without addition of a compatibilizer (PP-g-MA). Nanoparticles have shown an improved intercalation, especially after the fourth cycle. Addition of the compatibilizer allowed to develop the combined intercalated/exfoliated structure already after the second cycle. However, a better dispersion of nanoclay in the nanocomposite does not necessarily improve its thermal and mechanical properties due to the negative effect of polymer degradation.

Massardier and others [23] have studied PP/ethylene propylene rubber (EPR) polymer blends filled with $CaCO_3$ nanoparticles with compatibilizers, used to manufacture automobile fenders. As a result of recycling with the twin-screw extruder, an increase in the crystallinity and Young's modulus was observed.

Thermoplastic elastomers (TPO) composed mainly of polypropylene and elastomer (EPDM and EPR) are often used in the automotive industry. Recycling of TPO/MMT nanocomposite by multiple extrusion has been studied by Thompson and Yeung [24]. They observed the delamination of MMT platelets in the nanocomposites, as well as the oxidative degradation of a matrix after recycling. The degradation began after the second cycle of extrusion process. The presence of MMT filler in nanocomposites increased their resistance to degradation, whereas the compatibilizer (maleic anhydride–grafted PP) facilitated chain breakage. Despite the degradation during recycling, the mechanical and rheological properties of nanocomposites remained unchanged. What is worth to be noticed is, all recycled samples revealed improved properties, compared to the pristine polymer matrix.

12.2 Recycling of PLA Nanocomposites

Biodegradable polymers frequently have properties comparable to commodity but their applications are intended mainly for the single use food packaging, containers, cups, plates, and cutlery because of their unique sensitivity to undergo microbially induced degradation [4,18–21]. Thus, the preferable recycling of biodegradable polymers should be composting (organic recycling) [25–29]. However, during processing some amount of waste is remaining, which could be mechanically recycled similarly to the majority of commodity plastics at the processing site.

The extent of changes in the characteristics of biodegradable polymers and nanocomposites due to mechanical recycling has been presented in the example of polylactic acid (PLA 3051D, NatureWorks, USA) and composites containing 5 wt.% of the platelet clay MMT (Nanomer I.30, organically modified MMT, Nanocor, USA). Nanocomposites were prepared by melt mixing, using a master batch technology, followed by a dilution with the matrix polymer to a desired filler loading with the twin-screw extruder.

PLA is a biodegradable material produced from natural, renewable raw materials. It is characterized as a high gloss and transparency material, with high barrier properties for gases, good mechanical properties, and good processability. PLA is used in many industrial sectors, mainly in the packaging industry and in medicine. It has been used for producing implants, screws, or sutures for a long time. Due to its relatively high price, it is beneficial to recycle the polymer. PLA 3051D is a crystalline biodegradable thermoplastic polymer with a relatively high melting point (150°C–180°C) and with a glass transition point around 54°C.

Recycling was performed with the twin-screw extruder CTW100 (Haake, Germany) at 160°C. Extrusion was repeated for 10 times, before each cycle the material was thoroughly dried.

Structural changes in polylactide caused by recycling have been studied with differential scanning calorimetry (DSC) using an instrument DSC Q20 (TA Instruments, United States). The materials were first heated above the melting temperature (T_m) (to 200°C), then cooled and scanned again by a second heating. The first heating scan provides basic characteristics of the polymer material. The cooling scan follows the heat effects as the molecules lose mobility. From a dependency of the heat flow on temperature, the glass temperature (second-order transition) and cold crystallization temperature (first-order transition) were evaluated. The enthalpy of melting was used to measure crystallinity, since the crystalline polymers due to their limited molecular motion exhibit lower specific heat capacity.

Melt viscosity of neat PLA and that after 1st, 5th, and 10th cycle has been presented in Figure 12.3. One can observe a decrease in the melt viscosity of PLA after consecutive extrusion cycles, which suggests a decrease in the molecular weight. Only after the 10th cycle the melt viscosity slightly increased in comparison to the 5th cycle, which could be due to branching of the macromolecules as the chain scission proceeds.

The results of DSC measurements for recycled PLA have been presented in Figure 12.4. One can observe that the glass transition temperature (T_g) does not change with consecutive extrusion cycles in contrary to the cold crystallization temperature (T_{cc}), which shifted to a lower temperature (Table 12.1). Crystallization degree increased with recycling, which evidences a higher order of macromolecules that is possible with their shorter length. That finding confirms a conclusion on the chain scission from the melt viscosity measurements.

The oxygen permeability measurements revealed enhanced barrier properties with recycling (Figure 12.5), which correlates with higher crystallinity as has been presented earlier. Polymer crystals formed within an amorphous material constitute impermeable regions that restrict oxygen penetration.

Melt viscosity curves of polylactide and PLA nanocomposites with MMT have been presented in Figure 12.6. One can observe a fundamental

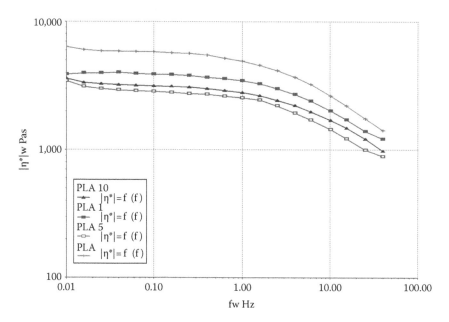

FIGURE 12.3
Melt viscosity of polylactide.

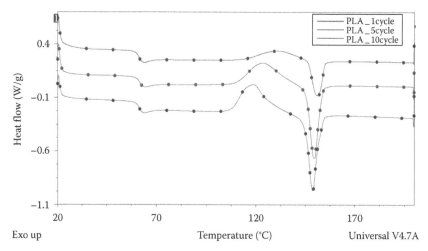

FIGURE 12.4
DSC thermograms of recycled PLA (1-5-10 cycle).

change in the flow characteristics. For neat polylactide, a Newtonian plateau was observed at low deformation rates followed by a pseudoplastic
decrease in viscosity. PLA/MMT exhibited pseudoplasticity in the entire
shear deformation range with the yield stress expected at very low deformation rate. Such relationship is typical for polymer composites due to

TABLE 12.1

PLA Characteristics from DSC

Cycle No	T_g, °C	T_{cc}, °C	T_m, °C	Crystall., %
PLA_1	61.6	130.8	151.4	15.0
PLA_5	62.4	123.6	149.5	28.1
PLA_10	62.2	118.5	148.7	29.1

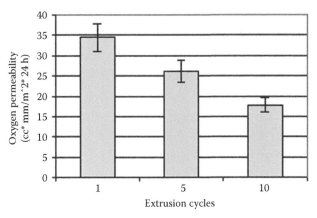

FIGURE 12.5
Oxygen permeability for recycled PLA.

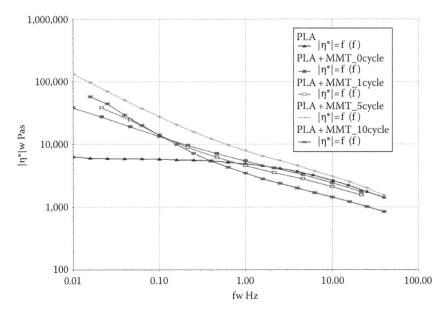

FIGURE 12.6
Melt viscosity of PLA/MMT nanocomposite.

a solid filler presence and interfacial interactions between the polymer matrix and a filler. Such behavior is clearly evident just with 5 wt.% of MMT, which evidences its regular distribution within PLA matrix. The increase of the composite melt viscosity with consecutive extrusion cycles suggests an increased intercalation of the polymer chains between clay platelets and higher total interfacial adhesion due to higher surface of clay available to polymer molecules. Only after 10th extrusion a remarkable decrease in the melt viscosity was found, which suggests an advanced mechanochemical degradation.

Higher elasticity of the composite melt in comparison to the pristine polymer has been more clearly presented in Figure 12.7. The change in G′ modulus observed at low deformation rates between PLA and PLA/MMT composites is of three order of magnitude and increased with recycling. That confirms the supposition on a high amount of solid particles (clay) being splitted out of the clay stacks as a function of a thermomechanical history of the material.

The results of DSC measurements for recycled PLA have been presented in Figure 12.8. One can observe that the cold crystallization temperature was shifted to lower temperature with increasing thermomechanical history of the materials. Crystallization degree increased with the extrusion cycles, which supports a previous remark on a chain scission and higher order of macromolecules in the materials (Table 12.2).

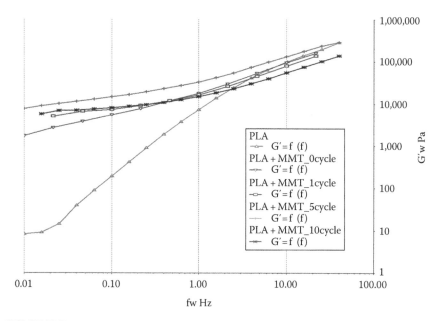

FIGURE 12.7
Storage modulus G′ for PLA and recycled PLA/MMT nanocomposites.

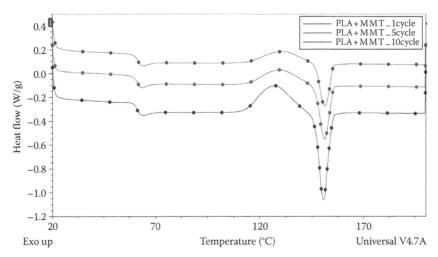

FIGURE 12.8
DSC thermograms of recycled PLA/MMT nanocomposites (1-5-10 cycle).

TABLE 12.2

PLA/MMT Composite Characteristics from DSC

Cycle No	$T_{g'}$ °C	$T_{cc'}$ °C	$T_{m'}$ °C	Crystall., %
PLA + MMT_1	61.7	130.6	151.3	15.0
PLA + MMT_5	61.3	129.7	151.0	18.6
PLA + MMT_10	61.1	126.8	150.6	30.7

The oxygen permeability dependence of PLA/MMT nanocomposites on the extrusion cycles (Figure 12.9) exhibited much more distinct drop after the fifth cycle than after the first one, which cannot be explained only by a higher crystallinity degree due to polylactide chain scission. The possible reason could be a progressive penetration of PLA chains between silicate layers which is the easier the shorter the macromolecules are.

In order to characterize the morphology of PLA/MMT nanocomposites, wide angle x-ray scattering (WAXS) analysis was performed. XRD data for the filler, e.g., MMT modified with $C_{18}H_{37}NH_2$, have revealed that the mean interlayer spacing of (001) plane amounts to 18 Å (Figure 12.10).

The data for polylactide (Figure 12.11) have shown that multiple extrusion does not change a structure significantly. Just an increasing distance between PLA chains could be speculated.

However, XRD patterns for PLA/MMT nanocomposites have shown additional peaks of intensity (Figure 12.12). Besides the peaks for PLA and MMT, very high intensity peaks have been found at low angles, which prove that intercalated nanocomposites have been tested. Interlayer spacing of the (d001) plane for the nanocomposites recycled for 1-5-10 times was of 33.8 Å—35 Å—34.8 Å, respectively, whereas the intensity of the peaks was 27-29-38.

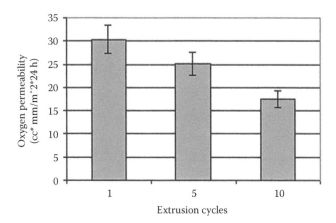

FIGURE 12.9
Oxygen permeability for recycled PLA/MMT nanocomposite.

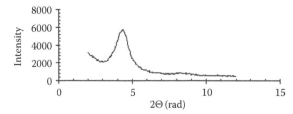

FIGURE 12.10
WAXS data for MMT filler.

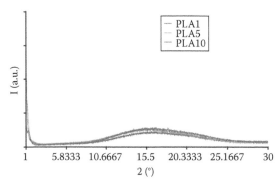

FIGURE 12.11
WAXS data for PLA matrix recycled 1-5-10 times.

The data presented so far suggest that with consecutive extrusions the intercalation proceeds as a result of higher thermomechanical load. However, the process is very complicated, because in parallel with higher stresses affecting MMT aggregates in the molten polymer matrix, imposed by a high melt viscosity (caused by a filler), degradation of the matrix and enhanced

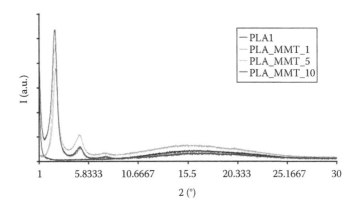

FIGURE 12.12
WAXS data for PLA/MMT nanocomposite recycled 1-5-10 times.

crystallinity also occurs. Therefore, a combined process of intercalation and delamination within the nanocomposite takes place. The processing properties and performance characteristics of such complicated material depend on the entire processing history being related to its structure and heterogeneity. For example, the melt flow rate index (MFR), widely used as a measure of processability of plastics, has been presented in Figure 12.13. One can observe that MFR increases with consecutive extrusions both for neat PLA and for PLA/MMT composites. Melt flow rates for a composite are always slightly higher, but only after the 10th cycle the difference is large, which suggest an advanced degradation of polylactide after the total residence time that is long.

Surface performance of the materials has been presented in Figure 12.14. Gloss of the polymer surface measured at 60° have shown that polylactide losses a gloss with the recycling cycles, which should be expected due to the surface microcracks related to the polymer degradation. However, nanocomposites exhibited higher gloss with consecutive extrusions, which could be

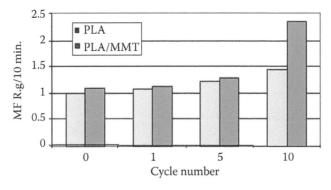

FIGURE 12.13
Melt flow rate of recycled PLA and PLA/MMT nanocomposite.

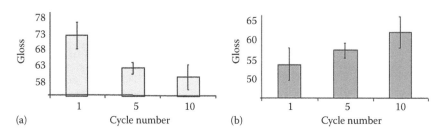

FIGURE 12.14
Gloss of recycled PLA (a) and PLA/MMT nanocomposites (b).

explained by a progressing intercalation and exfoliation. Increasing number of the silicate platelets may reflect light and thus enhance a gloss of the material.

The results presented earlier have revealed that recycling has an impact on morphology and performance of polymer nanocomposites. The processing history developing with consecutive extrusions brings about a degradation of the polymer matrix and influences a level of intercalation and spatial distribution of the nanofiller in a matrix. However, it is of interest how far the composite morphology is depending on a processing history. That issue has been studied in a research and its results have been presented in the next paragraph.

12.3 Influence of a Processing History on the Morphology and Properties of Nanocomposites

Recycling of PLA and polylactide filled with 5 wt.% of MMT or silica nanospheres (SNS) with a diameter of 30 nm (IChP, Warsaw, Poland) was investigated. The composites were prepared by melt mixing of the components with a single-screw extruder Rheomex 252 (Haake, Germany). The screw was equipped with the mixing elements at the metering zone, providing intensive melt mixing.

The initial composites were extruded for 10 times with the same equipment. The materials were thoroughly dried before each extrusion cycle. Melt viscosity of PLA/MMT composite recycled for 1-3-5-7-9 cycles has been presented in Figure 12.15.

Melt viscosity of PLA/MMT composites at low deformation rates increased as a function of the extrusion cycles, similarly to a dependence of the polymer composite viscosity on the filler content. Because in our study the filler content was constant, it suggests an increase of interfacial interactions between the polymer matrix and MMT. That is possible due to the progressive intercalation and possible exfoliation process that increases a total filler surface available to interactions with polymer macromolecules. In the entire range of deformation rates, the melt flow exhibited a pseudoplastic character with the

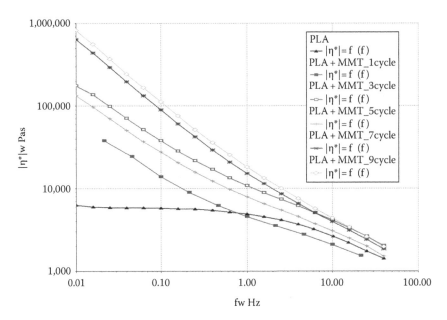

FIGURE 12.15
Melt viscosity of PLA and recycled PLA/MMT composites.

yield stress increasing with recycling cycles. Substantial increase in the melt viscosity of PLA/MMT composites has been observed for the seventh and ninth extrusion cycle, which might be related to a difference in morphology of the composite with an increased thermomechanical history.

Similar observation was noted for the dependency of the storage modulus on the deformation rate in oscillatory measurements (Figure 12.16). All PLA/MMT composites exhibited at low deformation rates a high increase in G′ values in comparison to that of the matrix polymer. That finding has confirmed an increased resistance to deformation of the materials due to the interactions at the polymer–filler interface. However, again for the composites recycled for seven and nine times the enhancement of the storage modulus is one order of magnitude higher that that measured after fifth cycle. That finding suggests a possible difference in the material morphology occurring after seventh extrusion.

The melt viscosity of PLA/SNS composites recycled for 1-3-5-9 times has been presented in Figure 12.17. After the initial viscosity increased due to addition of the filler (first cycle), further results have revealed that the viscosity curves of the PLA/SNS composites recycled for three and five times were almost identical to that of neat polylactide. That may be explained by the polymer degradation, which should be expected after consecutive extrusions.

Results of the viscosity measurements of a composite recycled for seven times were very strange, showing an unusual scatter even after being

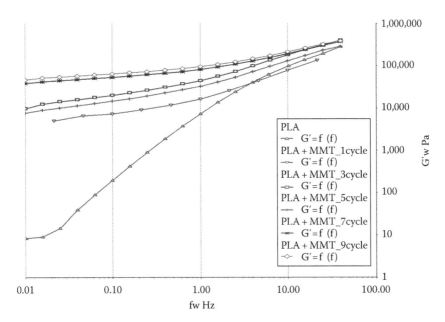

FIGURE 12.16
Storage modulus G′ for PLA and recycled PLA/MMT nanocomposites.

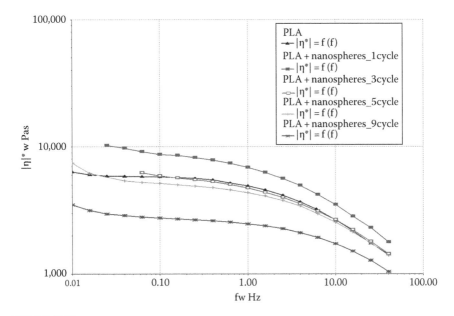

FIGURE 12.17
Melt viscosity of PLA and recycled PLA/SNS composites.

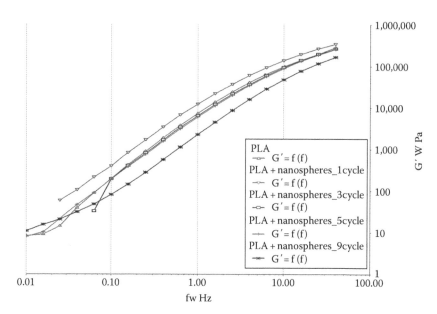

FIGURE 12.18
Storage modulus G′ for PLA and recycled PLA/SNS nanocomposites.

repeated for several times. The viscosity curve after the ninth extrusion was again regular with a shape similar to that of polylactide. In contrary to PLA/MMT composites, the melt viscosity of PLA/SNS composites was inferior to that of PLA, which suggests a lack of the polymer–filler interactions and an advanced degradation of the matrix.

The elastic response to deformations for PLA/SNS melts was again different than that for PLA/MMT composites. It was similar to that of polylactide for all recycling steps (Figure 12.18), suggesting a lack of interactions of PLA with SNS filler particles. Degradation of the matrix polymer seems to be dominating during consecutive extrusions of the composite.

Thermal properties of polylactide as evaluated with DSC did not change significantly during recycling, with an exception of the degree of crystallinity. As mentioned earlier, it should be a result of the polymer chain scission in the extruder (Figure 12.19). A slight shift to lower temperatures of the cold crystallization temperature and melting temperature of PLA as a function of recycling steps has confirmed that conclusion.

The degradation of PLA should be reflected in an increased crystallinity, as it was described earlier. In fact, the dependence of the crystallinity increased asymptotically with the extrusion cycles, reaching 30% after the 10th cycle (Figure 12.20). It seems that after every extrusion some extent of degradation occurs, which increases the mobility of PLA macromolecules and facilitates the creation of highly ordered crystalline regions in the material.

DSC thermograms of recycled PLA/MMT and PLA/SNS composites have a similar character, exhibiting a shift of the melting temperature and the cold

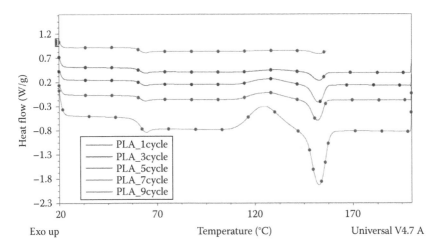

FIGURE 12.19
DSC thermograms of recycled PLA (1-5-7-9 cycle).

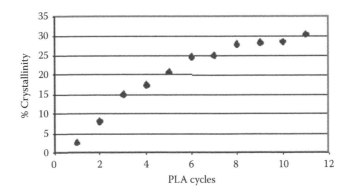

FIGURE 12.20
Crystallinity of recycled PLA.

crystallization temperature to the lower range in the graphs (Figures 12.21 and 12.22). Considering the shape of the melting peak, higher differences should be expected for the crystallinity of the composites after consecutive extrusions.

Crystallinity degree of PLA composites both with the MMT and silica nanospheres increases markedly after first extrusion passes and keeps a constant level (around 25%) up to seventh extrusion (Figures 12.23 and 12.24). Next the crystallinity increases with further passes up to 30% for the composites with nanospheres and up to 35% for PLA composites with MMT. The dependence of the composites crystallinity on recycling is different from that of pristine polylactide; therefore, the filler distribution should be of significance.

Polylactide has good barrier properties to oxygen. Since recycled PLA exhibits higher crystallinity than the virgin polymer, a lower permeability

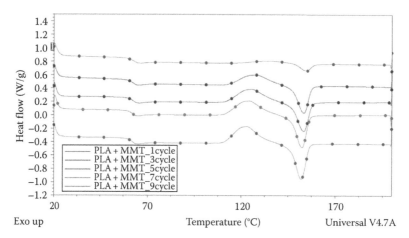

FIGURE 12.21
DSC thermograms of recycled PLA/MMT composite (1-3-5-7-9 cycles).

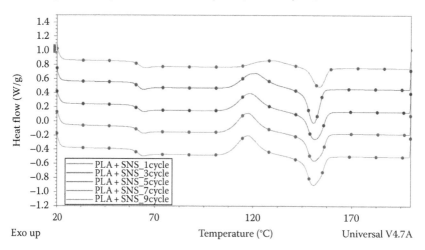

FIGURE 12.22
DSC thermograms of recycled PLA/SNS composite.

of oxygen is expected, since the crystals constitute hindrances for the permeation of the gas molecules through the polymer, which have to follow a more tortuous path. The oxygen permeability of PLA as a function of recycling has been presented in Figure 12.25.

The oxygen permeability decreased with consecutive extrusions from ca. 34 cc*mm/m²*24 h tending asymptotically to a value around 17 cc*mm/m²*24 h. That confirmed the aforementioned expectations and revealed a good correlation with PLA crystallinity presented in Figure 12.20.

The oxygen permeability measurements for PLA/MMT and PLA/SNS composites provided surprising results (Figures 12.26 and 12.27). There was no barrier for oxygen up to the fifth extrusion, after which the

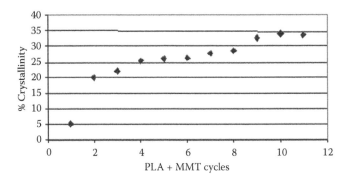

FIGURE 12.23
Crystallinity of recycled PLA/MMT composite.

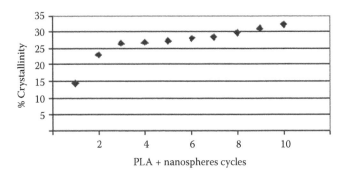

FIGURE 12.24
Crystallinity of recycled PLA/SNS composite.

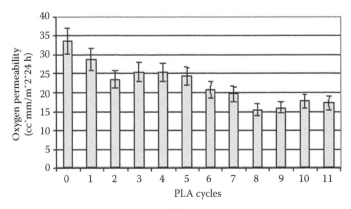

FIGURE 12.25
Oxygen permeability of recycled PLA.

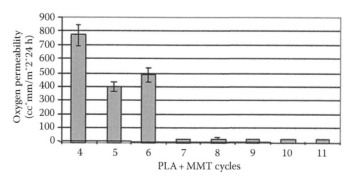

FIGURE 12.26
Oxygen permeability of recycled PLA/MMT composite.

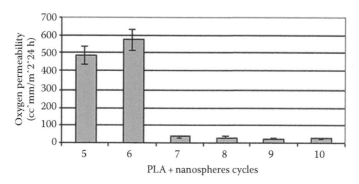

FIGURE 12.27
Oxygen permeability of recycled PLA/SNS composite.

permeability was as high as 500 cc*mm/m²*24 h. Only after the seventh cycle a real barrier to oxygen was created.

Such distinct differences in the materials' performance have to be related to their morphology. Even qualitative observations with a light transmittance have revealed that the composites prepared with a single-screw extruder constitute highly heterogeneous materials. Composite PLA/MMT contains a high number of large aggregates, whose size and number are decreasing with consecutive extrusions (Figure 12.28).

One can conclude that only after seventh extrusion a real nanocomposite should be expected. The materials manufactured till the sixth extrusion should be classified as microcomposites, which properties are different than that of nanomaterials.

Similar observations concern the composite prepared from polylactide and nanospheres (Figure 12.29).

Polylactide is characterized by good mechanical properties. In order to assess the industrial value of PLA composites, their mechanical properties were evaluated as a function of the processing cycles. The extrusion number has no influence on Young's modulus and Charpy impact strength.

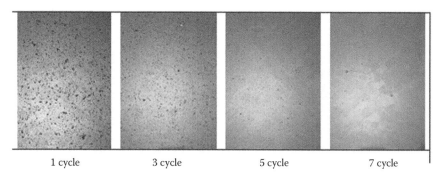

FIGURE 12.28
Distribution of MMT in PLA matrix after recycling.

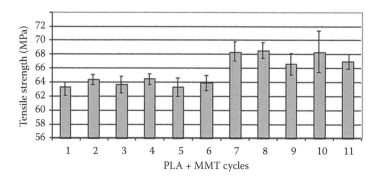

FIGURE 12.29
Distribution of nanospheres in PLA matrix after recycling.

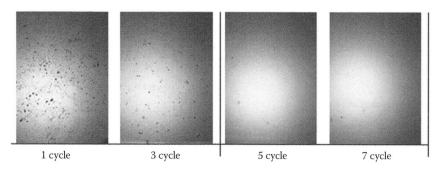

FIGURE 12.30
Tensile strength of recycled PLA/MMT composite.

Those values remain constant because the decrease in the molecular weight of PLA was balanced by an increase in crystallinity. The tensile strength has revealed a correlation with the morphology of materials. The composite PLA/MMT exhibits the tensile strength ca. 63 MPa up to the sixth extrusion. After the seventh cycle, it increases up to 68 MPa (Figure 12.30).

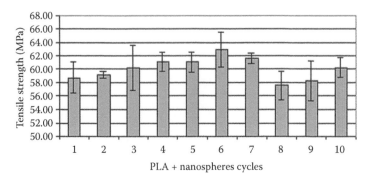

FIGURE 12.31
Tensile strength of recycled PLA/MMT composite.

The tensile strength of PLA/SNS has not revealed a similar correlation with recycling steps. The values remained around 60 MPa with a scatter of 5% (Figure 12.31), thus showing that the composite should be of lower interest to practical applications than the PLA/MMT nanocomposite.

12.4 Conclusions

1. Properties of nanocomposites depend on the polymer matrix nature and the filler geometry and its dispersion and distribution in the matrix.
2. Recycling of polylactide causes a degradation and increase in crystallinity.
3. Multiple extrusions allow for a better distribution of nanoparticles in the matrix.
4. Barrier properties of polylactide nanocomposites depend on a crystallinity of the matrix and on the intercalation/exfoliation extent.
5. Thermomechanical history imposed on the polymer–filler system determines a nature of the composite (micro/nano) and its properties.

Acknowledgments

This research has been performed with a financial support of the Structural Fund, project POIG.01.03.01-00-018/08 (MARGEN) and under activities related to COST Action FA0904. The authors wish to thank S. Frackowiak, A. Iwanczuk, K. Leluk, and M. Siczek for their valuable help in selected measurements and composite processing.

References

1. Directive 2008/98/EC, Official Journal of the European Union, L312/26, 22.11.2008.
2. Plastics—The Facts 2011. An analysis of European plastics production, demand and recovery for 2010. Annual Report, PlasticsEurope, Brussells, 2011.
3. F. La Mantia (Ed.), *Handbook of Plastics Recycling*, Rapra Technology, Shrewsbury, U.K., 2002.
4. M. Kozlowski (Ed.), *Plastics Recycling in Europe*, Oficyna Wyd. P. Wr., Wroclaw, Poland, 2006.
5. S. Pickering, Recycling technologies for thermoset composite materials—Current status, *Composites: Part A*, 37 (2006), 1206–1215.
6. V. Goodship (Ed.), *Management, Recycling and Reuse of Waste Composites*, WP and CRC Press, Cambridge, U.K., 2010.
7. Y. Yang et al., Recycling of composite materials, *Chem. Eng. Process.*, 2011. doi:10.1016/j.cep.2011.09.007.
8. T.J. Pinnavaia and G.W. Beall (Eds.), *Polymer-Clay Composites*, Wiley, Chichester, U.K., 2001.
9. J.H. Koo, *Polymer Nanocomposites. Processing, Characterization and Application*, McGraw-Hill, New York, 2006.
10. S.N. Bhattacharya, M.R. Kamal, and R.K. Gupta (Eds.), *Polymeric Nanocomposites. Theory and Practice*, Hanser, Muenchen, Germany, 2008.
11. Sh. Hamzehlou and A.A. Katbab, Bottle-to-bottle recycling of PET via nano-structure formation by melt intercalation in twin screw compounder: Improved thermal, barrier, and microbiological properties. *J. Appl. Polym. Sci.*, 106 (2007) 1375–1382.
12. M. Kracalik, M. Studenovsky, and J. Mikesowa, Recycled PET-Organoclay Nanocomposites with Enhanced Processing Properties and Thermal Stability. *J. Appl. Polym. Sci.*, 106 (2007) 2092–2100.
13. I. Kelnar, V. Sukhanov, and J. Rotrekl, Toughening of Recycled Poly(ethylene terephthalate) with Clay-Compatibilized Rubber Phase. *J. Appl. Polym. Sci.*, 116 (2010) 3621–3628.
14. Y. Lei, Q. Wu, and C. Clemons, Preparation and Properties of Recycled HDPE/Clay Hybrids. *J. Appl. Polym. Sci.*, 103 (2007) 3056–3063.
15. L. Martin, G. Kortaberria, and A. Vazquez, A Comparative Study of Nanocomposites Based on a Recycled Poly(methyl methacrylate) Matrix Containing Several Nanoclays. *Polym. Comp.*, (2008) 784–790.
16. Y. Yoo and S. Kim, Enhancement of the thermal stability, mechanical properties and morphologies of recycled PVC/clay nanocomposites. *Polym. Bulletin*, 52 (2004) 373–380.
17. M. Kracalik, M. Studenovsky, and J. Mikesowa, Recycled PET Nanocomposites Improved by Silanization of Organoclays. *J. Appl. Polym. Sci.*, 106 (2007) 926–937.
18. E.-K. Karahaliou and P.A. Tarantili, Preparation of poly(acrylonitrile–butadiene–styrene)/montmorillonite nanocomposites and degradation studies during extrusion reprocessing, *J. Appl. Polym. Sci.*, 113 (2009) 2271–2281.
19. I. Goitisolo and J. Eguiazabal, Effects of reprocessing on the structure and properties of polyamide 6 nanocomposites, *Polym. Degrad. Stabil.*, 93 (2008) 1747–1752.

20. M. Jurado, M. Mendizabal, and I. Gaztelumendi, The influence of the recycling on PA nanocomposites. Nanomaterials. 2nd NanoSpain Worshop, Spain 2005.

21. G.M. Russo, V. Nicolais, and L. Di Maio, Rheological and mechanical properties of nylon 6 nanocomposites submitted to reprocessing with single and twin screw extruders, *Polym. Degrad. Stabil.*, 92 (2007) 1925–1933.

22. N. Touati, M. Kaci, S. Bruzaud, and Y. Grohens, The effects of reprocessing cycles on the structure and properties of isotactic polypropylene/cloisite 15A nanocomposites, *Polym. Degrad. Stabil*, 96 (2011) 1–10.

23. N. Mnif, V. Massardier, T. Kallel, and B. Elleuch, Study of the modification of the properties of (PP/EPR) blends with a view to preserving natural resources when elaborating new formulation and recycling polymers, *Polym. Compos.*, 30 (2009) 805–811.

24. M.R. Thompson and K.K. Yeung, Recyclability of a layered silicate thermoplastic olefin elastomer nanocomposite, *Polym. Degrad. Stabil.*, 91 (2006) 2396–2407.

25. M. AvelLa and A. Buzarovska, Eco-challenges of bio-based polymer composites, *Materials*, 2 (2009) 911–925.

26. E. Chiellini and R. Solaro (Eds.), *Biodegradable Polymers and Plastics*, Kluwer, New York, 2003.

27. C. Bastioli, *Handbook of Biodegradable Polymers*, Rapra, Shrewsbury, U.K., 2005.

28. S. Kalia and L. Avérous (Eds.), *Biopolymers: Biomedical and Environmental Applications*, Wiley, Chichester, U.K., 2011.

29. S.S. Ray and M. Bousmina, Biodegradable polymers and their layered silicate nanocomposites: In greening the 21st century materials world, *Prog. Mater. Sci.*, 50 (2005) 962–1079.

13

Polymer Nanocomposite Materials Used for Food Packaging

Erich Kny

CONTENTS

13.1 Introduction

Materials that are used for food packaging have to fulfill more and more, sometimes even controversial, properties. For the material selection of food packages, the full supply chain of food packaging is becoming of increasing importance and has to be taken into account for food packaging developments. The supply chain of food packages starts at raw material selection and origin; continues with the manufacturing of raw materials; processing of materials into semifinished products; the initial customizing into individual packaging products; the final manufacturing processes, e.g., welding, printing, and labeling; the packaging process of the food itself; reuse and recycling issues; and finally ends with decomposition needs and waste management strategies.

A large amount of all polymer materials manufactured (in 2010 the worldwide production is estimated to be 180 million tons or 26 kg average per capita) goes into short-lived materials like packaging (35%). The worldwide consumption of plastics is expected to increase from 180 million tons to 227 million tons in 2020. The major consumption with 36% is in Asia, followed by the United States with 26%, and western Europe with 23%. The majority (53%) of plastic products are still polyolefin (polyethylene [PE] and polypropylene [PP]).[1] Ecological and sustainability issues become therefore more and more of importance and attention (carbon footprint, recycling, reuse, waste management, landfills, pollution of land and specifically the oceans, corporate social responsibility concerns, etc.).

Besides these environmental and sustainability concerns, the cost of food packaging is considerable and is estimated to be in the average of 17% of the food price (NESTLE average[2]), which represents an important cost issue for the food producer with continuing pressure for cost reductions and improvements in packaging efficiency. In addition to that, the consumer expectations are putting additional demand and pressure on properties of food packaging materials. Such consumer expectations partly driven by an increasing greener attitude and environmental sensibility among the general population could be lightweight design, transparency, and visibility of the product, easy to open and resealable features, compostability, freshness of food despite longer shelf life, and environmentally friendly packaging.

New sustainable materials came therefore into play like polymers from renewable resources, which are easy to recycle and are compostable but, however, have to fulfill a whole range of technological properties like strength, transparency, reduced gas diffusion, reduced water vapor penetration, etc.

The use of PNC materials offers technological solutions for some of these demands but is also creating new problems and questions concerning nanotoxicology and possible contamination of food by the nanomaterials and other additive materials used in innovative packaging.

This chapter is intending to give a brief overview about this rapidly evolving and expanding field of the usage of PNC materials for food packaging.

13.2 What Is a Polymer Nanocomposite?

A PNC is a composite or hybrid material. A PNC contains nanoparticles (NPs) that have at least one dimension in the nanosized range (smaller 100 nm) distributed within a polymer matrix. NPs can be spheres, platelets, or rod shaped. This can be clay platelets or in the case of rod-shaped NPs, carbon nanotubes, or in the case of spherical NP, silica NP. To develop and process PNC into successful products, a multidisciplinary expertise and approach is needed.

13.3 Properties of PNC

Even small additions (in the range of a few percents) of NP can alter the properties of the polymer matrix considerably. This is due to the large surface area of NPs, which can range up to several 100 m²/g of NPs depending on the size of individual NPs. The large surface area of the NP interacts with the thin immediate adjacent volume of the polymer matrix and restricts the mobility of the polymer chains, thereby changing polymer properties. Due to the large surface area of NPs, this volume of NP polymer interaction can become very large and will affect a comparatively large volume of the bulk polymer. Therefore, even small additions of NPs into a polymer matrix have much larger effects on the composite properties than the same amount of micro-sized material. In order to achieve the same effects with micro-sized additions into a polymer matrix, a much larger amount of micro-sized material would be needed, which in turn makes processing of the composite much more difficult. PNC materials therefore offer the advantage of not having much changed processing characteristics compared to the unfilled matrix polymer but offer on the other hand considerable property changes of the resulting nanocomposite.

The main properties that are to be improved for food packaging applications are their intrinsic material properties (strength, stiffness, ductility), diffusion barrier for gases (oxygen, carbon dioxide, various other gases used for food conservation), and water vapor penetration as well as thermal properties (thermal stability, thermal conductivity). Another area of interest and of intense development efforts consists in changing properties of the packaging material by functionalizing. Such property improvements could be antibacterial, UV stability, UV protection of packaged food, photocatalytic, deodorizing, printability, and optical (gloss, surface appearance, maintaining transparency).

Many properties can be affected depending on the kind of NP. Usually, more than one property will be changed.

13.3.1 Optical Properties of PNC

For food packaging applications, optical properties of the packing material like transparency, opaqueness, gloss, and UV absorption is of importance. In traditionally filled polymer with particles in the micron size range, the incident light is scattered and the transparency of the initially transparent polymer is reduced. By reducing the particle size of the filler with nanodimensions to values lower than 40 nm, the scattering can considerably be reduced making the nanocomposite transparent without significant loss in transparency compared to the unfilled polymer. However, in order to avoid turbidity in the composite material the distribution of the NPs has to be uniform and the aggregation of NPs has to be avoided by a proper processing technique. With moderate amount of NP fillers (5%) in polymer matrices, the transparency can be retained approximately up to 95%.

13.3.2 Mechanical Properties of PNC

Almost all the NPs added to various petrochemical-based polymers or renewable polymer matrices are affecting significantly the mechanical properties (modulus, tensile elongation, tensile strength) even in small NP additions (below 5%) provided the NPs are properly distributed and well dispersed in the polymer matrix. The observed improvements in mechanical properties are typically up to 150% and can reach in some cases even more than 200%.

This depends on the polymer matrix, the NP, and its proper dispersion and distribution and on the amount of nanosized filler, which is usually between 3 wt% and 8 wt%. Increase in mechanical properties has been achieved, e.g., by 144% of break elongation in polyester urethane with modified nanoclay.[5] An increase of Young's modulus of 70% has been reported by adding mineral clay of 30% to glycerol plasticized starch.[64] Such and even higher improvements in mechanical properties can be obtained with much lower addition of nanoclay particles provided they are properly and fully exfoliated within the polymer matrix. An overview of the property increases by only 4% nanoclay addition in Nylon6 is given in [65,66]. The tensile modulus is increased 91%, the tensile strength 55%, and the heat distortion temperature is increased from 65°C to 145°C.

Huang and Yu[67] determined the tensile properties of starch/nanoclay composites and found that the optimum and maximum increases were obtained with 8 wt% of nanoclay addition. Tensile strength increased by 270% and tensile strain by 25% compared to the properties of the unfilled polymer.

13.3.3 Barrier Properties

Because of a continuing and increasing trend to replace glass and metal containers with excellent barrier properties by lightweight packages made out of polymer foils, an increasing demand for effective and efficient gas and water barrier properties of the polymer materials involved is obvious. The trend away from heavy and voluminous packages made out of glass and metal is not only driven by cost considerations but also by the desire to decrease weight of packaging and environmental concerns. This trend for improved barrier properties is further intensified by the recent move toward naturally

occurring sustainable polymer matrices with a good recycleability and compostability. It is of great importance in food packaging to control the barrier
properties for passing of oxygen from the outside into the package and the
passage of water vapor from the inside to the outside. In some cases, packages are filled with special protective gases or gas mixtures and their diffusion through the package has to be controlled as well. For gaseous drinks
containing citrus fruit extracts, the passage of limonene, a cyclic terpene and
a main ingredient of citrus fruits, has to be controlled as well by using, e.g.,
montmorillonite (MMT) (clay) in a PET matrix.[3] Usually naturally occurring
polymers are hydrophilic and have rather poor barrier properties against the
passage of oxygen and water vapor. The mechanism for improving the barrier properties of polymers are based on the fact that by greater exfoliation of
impermeable clay platelets a longer, more tortuous path is produced for gas
and water molecules as they diffuse through the composite.

The oxygen permeability (OTR) in, e.g., high-density polyethylene (HDPE)
can be reduced by 50% and more by adding organically modified montmorillonite (OMMT).[19,20] Similar results of reduction in OTR have been reported
for low-density polyethylene (LDPE).[17] In a composite of pectin matrix with
nanoclay, the OTR has been reduced down to 75%.[14] By employing a layer-by-
layer technique for creating 40 consecutive layers with high nanoclay contents
(up to 84%) in a branched polyethylene imine matrix, an OTR value below the
detection limit of commercial instruments could be achieved yielding typically reported values better than for commercial SiOx. The authors believe
this is due to the achieved brick wall nanostructure of polymeric matrix
"mortar" and highly oriented exfoliated clay platelets. Despite the high clay
loading, the composite is transparent and has a very high hardness.[15] Similar
results have been reported for very low nonmeasurable oxygen permeation
of PNCs produced by a layer-by-layer technique.[9] Those composites consisted of sodium MMT clay in a cationic polyacrylamide matrix that was
grown on polyethylene terephthalate (PET). The films obtained retained also
an optical transparency of greater than 90%. Water vapor permeability has
been reduced to 71% by addition of 3% cellulose nanowhiskers (CNWs) in a
carrageenan matrix.[40]

13.3.4 Antibacterial Properties

Special antibiotic NPs (Ag, Ag-MMT, ZnO, and TiO_2) are used commonly in
PNC materials in order to make them effective against bacteria, fungi, molds,
and other microorganisms. When chitosan is used as a polymer matrix, it
has also natural antibacterial and antifungal properties. Also the quaternary amines used for functionalizing MMT have been reported to exhibit
antibacterial properties when used in PNC materials. Naturally occurring
antibacterial compounds have also been added to food packaging materials
like carvacrol, which is an essential oil in oregano, thyme, bergamot, and
some other plants.[4]

TABLE 13.1

Property Improvements by Clay NP in Nanocomposites of Different Polymer Matrices

Property	Clay NP in Polymer Matrix and Reference
Thermal stability	PU,[5] MC,[4] PETG,[6] SPI[7]
Transparency	WPI,[8] PAM,[9] PA6,[10] CHI,[11] PVOH[12]
Antimicrobial	MC+CARV,[4] WPI,[8] PLA+Ag,[13] CHI[11]
UV absorption	PCT+CHI[14]
Oxygen diffusion (OTR)	PU,[5] CHI+PCT,[14] PEI,[15] PHB,[16] PAM,[9] LDPE,[17] PA6,[10] PET,[3] PP,[24] EVOH+PLA,[18] HDPE,[19] HDPE,[20] PVOH,[12] PETG[6]
Water vapor permeability	PU,[5] RA,[21] SPI,[7] SPI,[22] CHI+PCT,[14] CS,[23] PHB+PCL,[16] PP,[24] Agar[25]

TABLE 13.2

Property Improvements by ZnO NP and TiO$_2$ NP in Nanocomposites of Different Polymer Matrices

Property	ZnO NP in Polymer Matrix/Reference	TiO$_2$ in Polymer Matrix/Reference
Antimicrobial	LDPE[26]	EVOH[27]
UV absorption	PVC[28] PP[30]	WPI[29]
Oxygen diffusion (OTR)	PP[31]	
Photocatalytic activity		STA[32] EVOH[27]

In Tables 13.1 through 13.4, the properties of PNCs are listed that are most commonly achieved by NPs in various polymer matrices developed or adapted for food packaging applications. Mechanical properties are not listed here specifically because they are influenced by all of the NP additions in all polymers to a certain extent.

Au NPs have been used for deodorizing in a polydimethylsiloxane (PDMS) matrix.[46] Mica NPs have been used as UV absorbent and water vapor barrier in a red-algae-derived carrageenan+zein prolamine (RADC+ZP) matrix.[47] Nisin* NP has been used as antimicrobial agent in a poly (butylene adipate-co-terephthalate) (PBAT) matrix.[48] Carbon NPs have also been shown of having antimicrobial properties in an epsilon polylysine (EP) matrix.[49]

---

* Nisin is a polycyclic antibacterial peptide with 34 amino acids and is used as a food preservative. Commercially, it is obtained from the culturing of the bacterium *Lactococcus lactis* on natural substrates, such as milk or dextrose. Nisin is a rare example of a "broad-spectrum" bacteriocid effective against many gram-positive organisms. As a food additive, nisin has the number E234.

TABLE 13.3

Property Improvements by Cellulose and Chitin NP
in Nanocomposites of Different Polymer Matrices

Property	Cellulose NP in Polymer Matrix/Reference	Chitin NP in Polymer Matrix/Reference
Thermal stability	PHB,[33] PLA,[34] PVOH[35]	GPS[36]
Transparency	AMP[37]	
Antimicrobial		CHI[39]
Oxygen diffusion (OTR)	SHL[38]	
Water vapor permeability	CARR,[40] SHL,[38] MGP,[41] PLA[34]	GPS[36]

TABLE 13.4

Property Improvements by Ag NP in Nanocomposites of Different
Polymer Matrices

Property	Ag NP in Polymer Matrix/Reference
Antimicrobial	LDPE,[26] CHI,[42] PCL+MMT+zein,[43] zein,[44] CHI+STA,[45] CHI[13]
Water vapor permeability	CHI+STA[45]

Abbreviations (Used in Tables 13.1 through 13.4)

AMP	Amylopectin
CARR	Carrageenan
CARV	Carvacrol (5-isopropyl-2- methylphenol)
CHI	Chitosan
CS	Corn starch
EP	Epsilon polylysine
EVOH	Ethylvinyl alcohol
GPS	Glycerol plasticized starch
HDPE	High density polyethylene
MC	Methyl cellulose
MGP	Mango puree film
PA6	Polyamide 6
PAM	Polyacrylamide
PCL	Polyepsilon caprolactone
PCT	Pectin
PDMS	Polydimethylsiloxane
PEI	Polyethylenimine
PET	Polyethylene terephthalate
PETG	Polyethylene terepthalate glycol
PHB	Polyhydroxybutyrate
PLA	Polylactic acid
PP	Polypropylene

PU   Polyester-urethane
PVC   Polyvinylchloride
PVOH   Polyvinyl alcohol
RA   Red algae film
RADC   Red algae derived carrageenan
SHL   Shellac
SPI   Soy protein isolate
STA   Starch
WPI   Whey protein isolate
ZP   Zein prolamine

13.4 Preparation of PNC

In order to make a successful nanocomposite, the uniform distribution of the NPs within the polymer matrix is of greatest importance. NPs have to be distributed evenly but also well dispersed and separated from each other. Only when in the ideal case each NP is embedded within the polymer matrix and is well separated from other dispersed NPs not forming any NP agglomerates or aggregates, the maximum property improvement of the nanocomposite material is obtained (Figure 13.1). In order to achieve such an even distribution of NPs, two strategies are customarily been followed.

13.4.1 Ex Situ Synthesis Method

In the ex situ technique, nanoparticles are first produced and are then distributed into a polymer, a polymer solution, or a monomer solution, which is later polymerized.[50]

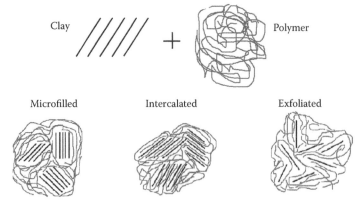

FIGURE 13.1
Schematic representation of intercalated and exfoliated clay.

The ex situ techniques may be also called "top down formation strategy" since preformed NPs delivered either as a solid nanopowder or in a liquid dispersion are then processed with the polymer matrix in order to form a nanocomposite. Obtaining a uniform distribution and dispersion of the NP in the polymer matrix is still the most challenging step despite many improvements in processing techniques and functionalization of NP.

In the ex situ synthesis method, different production routes can be followed.

In the prepolymerization technique, the nanoparticles are mixed with the monomer either with or without a solvent. When a proper distribution and dispersion of the NPs has been reached by a suitable mixing technology, the polymerization reaction is started and a bulk nanocomposite is yielded.

In the post-polymerization technique, the already formed polymer is mixed with the NPs either by melting the polymer or by dissolving the polymer in a solvent in order to aid the mixing. Such techniques are called "melt compounding" and the "solvent method." Melt compounding of the nanofillers into a polymer matrix is performed by processing the mixture through an injection molding machine, extruder, or some other processing machines. The shear forces in such processes are used to distribute the nanofiller uniformly in the polymer matrix.

The prepolymerization method and the melt processing techniques are still the most widely used processes for the preparation of nanocomposite materials.

Both strategies face the problem that the preformed NPs tend to form agglomerates or even aggregates immediately after their formation due to van der Waals forces and their large surface area. In order to overcome this tendency and to support the mixing with the polymer matrix, the NPs are in many cases functionalized by chemically attaching so-called functionalizing groups, making them either hydrophilic or hydrophobic depending on the nature of the polymer matrix. NPs with various functionalized surfaces can already be obtained commercially (e.g., functionalized MMTs). The surface of NP can be altered in many ways starting with charging, chemical bonding of functional groups, physical attachment, etc.

13.4.2 In Situ Synthesis Method

In the in situ technique, NPs are generated inside a polymer matrix or a monomer solution (which is later polymerized) by decomposition of suitable precursors by a chemical, thermal, or photolysis method. The in situ synthesis may also be called "bottom up formation strategy" since the NPs are formed in situ during or before the polymerization process creating a very uniform NP distribution inside the polymer matrix concerning NP size and spatial separation.

For the in situ synthesis method, the following processes can be followed:

The bottom up process works mainly by sol-gel processes and is restricted to metallic NPs (Ag, Au, noble metals, and a range of oxide or sulfide NPs like SiO_2, TiO_2, and others). The sol-gel methods employed are either hydrolytic or nonhydrolytic sol-gel processes.[51]

Reverse microemulsions have also been used in order to synthesize inorganic NP SiO_2 and, e.g., various sulfidic NP have been successfully produced by such techniques.[52] Reverse microemulsions employ aqueous droplets with precursors of the desired NP encapsulated by surfactant molecules and surrounded by an oily phase. A reactant is then added to the oily phase that diffuses to the aqueous droplets initiating the formation of the inorganic NPs. In the case the oily phase consists of the monomer itself, the polymerization process can be performed in a single step synthesis once the desired NPs have been formed.

In order to summarize this short overview about nanocomposite processing and functionalization, the proper dispersion of NP is the most important and still the critical step in nanocomposite production. This requires more efforts for the top down approach and a sequence of well-arranged complicated chemical steps for the bottom up approach.

For a successful "top down" nanocomposite synthesizing approach, the following steps are generally necessary to be carried out:

- Selection of NPs and polymer matrix.
- Surface conditioning or modification of NP in order to make the NP and the polymer matrix system affine to each other.
- Processing of the NP/polymer mixture keeping in mind energy balance and properties of the mixture for further processing (viscosity).
- Master batches are sometimes already commercially available.

"Controlling the proper dispersion of nanoparticles in polymeric matrices is still the most significant impediment to the development of high-performance polymer nanocomposite materials."[53]

13.5 Food Packaging and PNC (Literature Overview)

While the available scientific literature on food packaging is enormous (e.g., in ISI web of science there are listed in July 2011 more than 31,434 papers related to the key word food packaging), the use of PNC for food packaging is a very recent and quite small field of scientific investigation. Only 124 papers can be found in ISI web of science as per July 2011 related to the keywords PNC and food packaging. This small number of publications so far constitutes less than 0.5% of the available literature on food packaging. The number of scientific papers dealing with PNCs used in all different areas of applications amounts at the same time to about 7630. Only after the year 2000, a few publications on this specific topic (PNC and food packaging) start to appear; in 2010, the number of such publications has increased to about 40/year.

Most of the research papers published in the year 2010 (out of a total of 40 publications) originate from the United States (28) followed by Spain (18), China (11), Italy (11), South Korea (8), Brazil (5), Denmark (6), Iran (8), France (9), Canada (5), and India (5). Since cooperation between different laboratories across country borders is in many cases the rule and not the exception, those figures for the involvement of countries sum up to more than the total number of papers published. The number of worldwide patents found on PNC and (food) packaging amounts to about 31 out of a total of 1163 patents on PNCs (according to a search in ESPACENET in September 2011, http://at.espacenet.com). About 10 review papers have been appearing lately on various topics of food packaging and PNC (Table 13.5).

The main NPs used for PNC food packaging that are covered in literature and patents are different types of clay (mainly montmorillonite in natural form (MMT or MMT-Na), montmorillonite organically modified [OMMT], Ag, cellulose nanowhiskers [CNWs], chitin whiskers, ZnO, and TiO_2).

The main polymer matrices considered and investigated are either based on nonrenewable resources (PP, PE, polyamide [PA], PET, polyethylene terephthalate glycol [PETG], polyester-urethane [PU]) or derived from renewable natural sources. Recently, the main scientific interest turns out, however, to improve naturally occurring sustainable polymers (e.g., poly-lactide [PLA], polysaccharides, proteins) with the addition of NPs and to make them useable for the application as food packaging materials.

13.6 Nanoparticles Most Commonly Used in PNC for Food Packaging

Clay minerals have many applications due to their ion exchange properties and their large surface area once fully exfoliated. Clay minerals that are easily obtained at relatively low cost are already used widely in ceramics, as a filler in the paper industry, in the metal industry for refractories and foundry sands, in oil drilling, as a soil additive, as a desiccant and adsorbent and as deodorizer, and for many other uses.[68] Clay is also used increasingly as a nanoadditive in many PNC formulations. Clay is an abundant and relatively cheap resource and also considered being natural and perceived to be less toxic or not at all toxic or harmful to human health.

Clay is a layered silicate and has the possibility to enhance various composite properties significantly compared to the unmodified polymer. The improvement of properties includes barrier properties, flammability resistance, improved stability against thermal and environmental degradations, and

TABLE 13.5

Recent Review Papers Summarizing the State of the Art of PNC Materials for Food Packaging

Year	Author	Topic	Title	Reference
2011	Han Wei, Yu Yan Jun, Li Ning Tao, and Wang Li Bing	Safety of nanocomposites	Application and safety assessment for nanocomposite materials in food packaging	[54]
2011	Ahmed Jasim and Varshney Sunil K.	PLA-based nanocomposites	Polylactides—Chemistry, properties and green packaging technology: A review	[55]
2010	Jamshidian Majid, Tehrany Elmira Arab, Imran Muhammad, Jacquot Muriel, and Desobry Stephane	PLA technology and PLA nanocomposites	Poly-lactic acid: Production, applications, nanocomposites, and release studies	[56]
2010	Siro Istvan and Plackett David	Cellulose-based nanocomposites	Micro fibrillated cellulose and new nanocomposite materials: A review	[57]
2010	Arora Amit, and Padua G. W.	Nanocomposite with clay and carbon fillers	Nanocomposites in food packaging	[58]
2009	de Azeredo Henriette M. C.	Overview of nanoparticles in nanocomposites	Nanocomposites for food packaging applications	[59]
2008	Andersson Caisa	Nanocomposite for paperboard	New ways to enhance the functionality of paperboard by surface treatment: A review	[60]
2008	Zhao Ruixiang, Torley Peter, and Halley Peter J.	Starch/protein based bio-nanocomposites	Emerging biodegradable materials: Starch- and protein-based bio-nanocomposites	[61]
2007	Rhim Jong-Whan	General overview	Potential use of biopolymer-based nanocomposite films in food packaging applications	[62]
2007	Sorrentino Andrea, Gorrasi Giuliana, and Vittoria Vittoria	General overview	Potential perspectives of bio-nanocomposites for food packaging applications	[63]

improvement of biodegradability, reduction of solvent and water uptake, and generally the improvement of mechanical properties like strength, modulus, and hardness (abrasion resistance). The transparency of transparent polymer matrices is not affected significantly by small additions (less than 5 wt%) of layered nanosilicates.

The main reason for this significant improvement of properties is the nanosized dimension of individual clay platelets. The individual clay platelets have a layer thickness of less than 1 nm and very high aspect ratios (100–1000).

Layered silicates have two types of structures: tetrahedral substituted and octahedral substituted and are occurring naturally in multistacked aggregates forming interlayer galleries. The negative charges of the silicates that are caused by isomorphic substitution of atoms by atoms of different charge (e.g., replacement of Al 3+ by Mg 2+ or Fe 2+) are counteracted by alkaline (Na+) or earth alkaline cations, which can move freely between the silicate galleries.

MMT, a member of the smectite family, is a 2:1 clay, meaning that it has two tetrahedral sheets sandwiching a central octahedral sheet.

MMT is the most commonly used layered silicate for the preparation of nanocomposites (Figure 13.2). Other clay types from the smectite family of clays commonly used for the production of PNCs are hectorite and saponite.

In the case of tetrahedral substituted silicate galleries, the negative charge is located at the surface of the silicate layers. This negative charge on the surface

Structure of montmorillonite

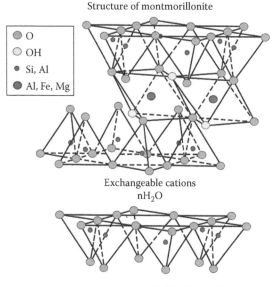

◑	O
○	OH
•	Si, Al
●	Al, Fe, Mg

Exchangeable cations
nH_2O

Modified from Grim (1962)

FIGURE 13.2
Clay structure of montmorillonite (MMT).[69]

of the silicate galleries and the exchangeable counteracting cation make it possible to introduce instead other organic or inorganic cations between the clay galleries, which can assist in the interaction of polymer chains with the clay platelets distributing them uniformly and individually separated from each other in the polymer matrix in a so-called exfoliated state.

The organic cation used for functionalizing the clay platelets by exchange of ions are, e.g., cationic surfactants including primary, secondary, tertiary, or quaternary alkyl ammonium or alkylphosphonium cations, which render the functionalized clay organophilic.

In the nonfunctionalized state, the clay would only be miscible with hydrophilic polymers like PEO (polyethylenoxide) or PVA (polyvinyl alcohol).

Some commercial producers of clay NP products are:

> Nanomer® by Nanocor: The nanoclays produced by Nanocor are MMT minerals that have been treated with compatibilizing agents, enabling them to disperse to nanoscale size in plastic resins.[70]
>
> Cloisite® and Nanofil® is produced by Southern Clay Products/ Rockwood additives. They consist of organically modified nanometer-scale, layered magnesium aluminum silicate platelets. The silicate platelets are 1 nm thick and 70–150 nm across. The platelets are surface modified with an organic chemistry treatment to allow complete dispersion into and provide miscibility with the thermoplastic systems for which they were designed.[71]
>
> BENTONE® by Elementis Specialties: Organoclays for flame, barrier, and mechanical property improvement in thermoplastic and thermoset plastics.[72]

ZnO as an NP reinforcement in PNCs is mainly used because it is adding antibacterial properties to the PNC. Besides antibacterial properties, inorganic zinc-oxide NPs show also antifungal, anticorrosion, catalytic, and UV filtering properties. Food packaging materials with ZnO NPs have been already prepared and described in a polypropylene matrix,[30,31] in an LDPE matrix,[26] in combination with Ag particles in an LDPE matrix,[73] and in a PVC matrix.[28]

Ag is used as nanoreinforcement in a multitude of polymer matrices but also because it is functionalizing the resulting polymer to be strongly antibacterial. Ag is a very effective antibiotic substance known and used already for a long time in human history. It is reported that the Phoenicians have already used Ag to preserve food and wine. Prior to development of antibiotics, Ag has been used. Milk could be kept fresh for longer time with a Silver dollar inside the milk bottle.

The number of commercially available Ag containing nanoproducts is fast growing (it was 93 in 2006 and already 313 in 2010).[74] Nano Ag is produced by either chemical or physical methods. Aqueous solutions of Ag salts can be reduced to nano Ag particles by hydrogen, tetrahydroborates, citrates, alcohols, hypophosphites, or organometallic compounds. Physical methods of Ag NP production are consisting in plasma, laser or electron beam ablation processes, sputtering or mechanical grinding of metallic silver.[75] The worldwide production of nano Ag is estimated to be already between 500 and 1200 ton/year.[76]

Ag NPs have been used mainly with renewable polymer matrices, with chitosan matrix,[42] with a chitosan/starch matrix,[45] and with a zein polymer matrix.[44] Ag has also been successfully combined with MMT by replacing the exchangeable cations with Ag cations. Such Ag MMT NPs have been applied in a matrix of PLA[13] and in zein, agar, and polyepsilon caprolactone (PCL) matrix yielding strongest antimicrobial effects in agar.[43]

Cellulose nanowhiskers, chitin whiskers, and bacterial cellulose fibers are naturally occurring polysaccharides that are derived by chemical processing from cellulose, chitin, and bacterial produced cellulose. Such naturally occurring polysaccharides are renewable, biodegradable, and biocompatible and are considered to have a great potential for being used as a raw material for food packaging applications.

Cellulose is the most abundant naturally occurring biopolymer. Cellulose can either be used as a raw material for polymer matrix in a PNC or the CNWs can be used as reinforcing nanorods in various polymer matrices like PLA,[34] carrageenan polymer,[40] amylopectin,[37] and shellac.[38] Bacterial cellulose has been used as a nanofiller in a matrix of polyhydroxy butyrate,[33] forming a biocompatible and biodegradable nanocomposite material. CNWs can be extracted from naturally occurring cellulose by mechanical, chemical, or enzymatic separation techniques.[57] The sulfate route for obtaining NPs from cellulose has been replaced by a hydrochloric acid route, which showed better composite properties.[35] A challenge for the production of PNC exists because microfibrillated cellulose (MFC) is lacking compatibility with hydrophobic polymers. This makes a pretreatment and organic functionalization of nanocellulose necessary in order to render its surface hydrophobic.

Chitin is the second most abundant naturally occurring biopolymer. In its deacetylated form (chitosan), it can be used as a polymer matrix for forming a natural nanocomposite with various nanofillers. Microfibrillated chitin can be used as nanofillers by employing nanosized chitin whiskers.[39] Chitosan-derived NPs have been used as nanofillers in a matrix of glycerol plasticized starch.[36] By using a layer-by-layer technique, a unique biodegradable nanocomposite has been produced and tested consisting of a chitosan matrix and wood CNWs derived from Eucalyptus. The authors were using the electrostatic attraction of sulfate groups on the surface of the CNWs and the amine groups at the matrix polymer as driving forces for the growth of the composite films.[77] PVA as matrix and bacterial-cellulose-derived NP have been shown to form successful composite material for food packaging applications. Edible films consisting of mango puree reinforced by cellulose nanofibers have been produced as well.[41] Mechanical properties could be considerably increased as well as the water vapor barrier properties.

TiO₂, titanium dioxide, is used as nanoreinforcement too, but it is used mainly because it adds photooxidation capabilities to the resulting polymers making it self-cleaning and antibacterial, and it changes the optical properties of resulting films. Nano-TiO₂ is used already in large amounts as a photocatalyst in a great range of products and in cosmetics.

Whey protein isolates have been converted successfully into a nanocomposite by using small amounts (0.25%) of nano-TiO₂ particles. The resulting composite films showed an exponential decrease of transmittance of visible, UVA, and UVB light relative to the TiO₂ content.[29]

The photocatalytic activity of embedded nano-TiO₂ in a matrix of starch-based polymer has been determined by the methylene blue method under UV and visible light. The resulting film has shown photocatalytic activity.[32] However, the long-term stability of such films remains to be seen since the photocatalyst will also attack the polymer matrix itself.

The self-sterilizing capability of a polymer film composed of a matrix of ethylene vinyl alcohol copolymer containing nano-TiO₂ has been tested by employing nine different microorganisms. Testing showed that an incorporation of 2–5 wt% TiO₂ proved to be most effective in killing a large range of different microorganisms.[78]

13.7 Polymer Matrices Most Commonly Used in PNC for Food Packaging

Petrochemical-based synthetic polymers like PE (LDPE/HDPE), PP, PVC, PU, PET, PETG, Polymethylmethacrylat (PMMA), and PVA are usually not or only very poorly biodegradable and are not renewable. Exceptions are polybutylene succinate (PBS), PCL, aliphatic polyesters, and polyvinyl alcohol, which are usually produced from petrochemicals-derived monomers but are biodegradable to a certain extent (Table 13.6).

The biodegradable polymer matrix PCL blended with polyhydroxybutyrate (PHB) has been processed into a PNC with MMT and kaolinite.[16] Enhanced barrier properties have been obtained against oxygen, D-limonene, and water. PCL has been mixed with PLA by melt blending and was reinforced by clay.[84] A modified Mg/Al hydrotalcite (3%) has been dispersed in PCL matrix by melt extrusion, which showed some improvement in mechanical properties, although full exfoliation has not been achieved.[85]

Naturally occurring polymers are biodegradable and biocompatible and potentially edible.

Naturally occurring biopolymers are various polysaccharides (starches, cellulose, and chitin), natural proteins (prolamines), or others of different chemistry (PHB, biopolyester, natural rubber). Natural biopolymer resources can be further processed in various ways to form commercially important

TABLE 13.6

Nanocomposite Intended for Food Packaging with Petroleum-Based Polymer Matrix That Is Not Biodegradable

Polymer Matrix	Nanoparticle	Reference	Comment	Applications
PET	MMT	[3]	Enhanced barrier properties against water, oxygen, limonene	Bottles for carbonated drinks
PA6	MMT	[10]		Enhanced barrier properties, water oxygen
LDPE	MMT	[17]	50% reduction in oxygen permeability with 3% clay	Replacement of multilayer films
PA	MMT	[79]	Migration study of toxic components of polyamide	
PP	ZnO	[30]	ZnO nanoparticles stabilized with starch	Antibacterial properties, UV photodegradation
iPP (isotactic polypropylene)	ZnO	[31]	Mechanical and oxygen barrier properties	
iPP	$CaCO_3$	[80]	Drastic reduction of permeability of oxygen, CO_2	Increase of shelf life of fresh produce
LDPE	Ag/ZnO	[73]		Antimicrobial activity
PVA	Cellulose NP	[35]	Thermal and mechanical properties	
PVC	ZnO	[28]	UV and climate resistance	
EVA	Silicate	[81]		
PE/PEG	Silicate	[9]	PE/PEG film on LDPE	50%–70% decrease of oxygen permeability
PMMA	MMT	[9]	On PE/terephthalate film, brick wall structure	Very low oxygen transmission rate
PU	MMT	[82]		
HDPE	MMT	[19,20]	Oxygen permeability reduced by 50%	
PVOH	MMT-Na	[12]		
PETG	MMT	[6]	Coupling agent maleic anhydride	
Nylon 6	Silicates	[83]	He barrier properties	Application primarily for high altitude balloon

polymers remaining biodegradable but with improved properties like chitosan, agar, zein, methylcellulose, carrageenan, soy protein isolate, whey protein isolate, amylase, and amylopectin.

Biodegradable polymers can be synthesized by polymerizing organic molecules that are derived from renewable sources to form polymers like PLA from lactide monomers or polyurethanes from polyols obtained from vegetable oils.

The mechanical properties of such biopolymers are relatively poor compared to polymers obtained from petrochemical resources and show lower stiffness and strength and are sensitive to water or absorb water readily, which lowers further their already low strength.

Improving the properties of such biopolymers by nanoreinforcement is therefore an attractive way to obtain materials with a great potential to be used for future food packaging applications.

It is estimated that the market for bioplastics will be growing until the year 2020 by 20%–25% per year.[88] The observed growth of recent years is mainly attributable to the use of blends of PLA for packaging of short-lived goods. The biodegradable nature of such short-lived products is of special importance because it opens up for the consumer an associated alternative disposal route.

PLA consists of polymerized lactic acid molecules. PLA was invented at Du Pont in 1932. The PLA polymer is derived by ionic polymerization starting from lactides that are formed from two lactic acid molecules. The lactides itself are obtained by bacterial fermentation of sugar molecules like glucose.

PLA blends have different mechanical properties than the raw PLA. Most of the blends can replace the conventional packaging plastics: PE and PP. PLA is blended frequently by PBAT, which is a biodegradable thermoplastic polymer too. Since PLA is high in strength (63 MPa) and modulus (3.4 GPa), but brittle (strain at break 3.8%) and PBAT is flexible and tough (strain at break approximately 710%), a blend of PLA and PBAT yields beneficial properties and the blend is still biodegradable.

A variety of food products needed for daily use are packaged in bags or nets. Networks for bags or foil applications must withstand sudden stresses during the filling and must have high weld strength.[86] As an added benefit, PLA films from a certain blend of PLA have a very pleasant soft finish. Examples of successful packaging applications include biobags and bubble bags. In 2002, a first commercial plant for the manufacture of PLA plastics was built by NatureWorks LLC with a capacity of 150,000 ton. In 2012, a production facility in Guben (Germany) with an annual capacity of 60,000 ton will be available.[87] Commercially available PLA is provided already by Biopearls, Guangzhou Bright China, Hisun Biomaterials, Kingfa Science & Tech., Nantong, Nature Works, and Toray Synbra.[88]

The thermal and mechanical properties of PLA/PEG (74/20) blends reinforced with MMT have been investigated. At 6% clay addition, the obtained blend showed excellent flexibility.[89] By a chloroform solution casting method,

PLA films reinforced with various amount of CNWs (1%–5%) have been prepared. At 3% CNW, the optimum in barrier properties could be obtained. Water permeability decreased by 82% and oxygen permeability by 90%.[34] By solvent casting, PLA films containing Ag-MMT have been prepared, which showed strong antibacterial properties against *Salmonella*.[13]

Agar is yielded from red algae and is produced mainly in Asia. Agar is chemically an unbranched polysaccharide consisting of a galactose polymer. Agar-based nanocomposite films have been prepared with differently functionalized MMT (Na-MMT, Cloisite 20A, Cloisite 30B). The authors could show that the unmodified Na-MMT was most compatible with an agar matrix and showed the highest property improvements. Tensile strength was increased by 18% and water vapor penetration was decreased by 24%. Only Cloisite 30B showed some antibacterial effectivity against *Listeria*.[25] The antimicrobial activity of agar reinforced with Ag-MMT has been compared with other nanocomposites containing a polymer matrix of zein and PCL. Agar-based nanocomposite had the highest antimicrobial effect, while the other samples showed no activity. The authors conclude that these differences are related to the differing water contents of the samples, which seemed to be the key parameter for antimicrobial effectiveness.[43]

Starch is a polysaccharide consisting of glucose units joined together by glycoside bonds. This polysaccharide is produced by all green plants as energy storage. It is contained in large amounts in potatoes, wheat, corn, rice, and many other vegetables. Starch is a white powder that is insoluble in cold water or alcohol. Starch consists of linear chains formed by the helical amylose and of branched structures formed by amylopectin. Depending on the plant, starch generally contains 20%–30% amylose and 70%–80% amylopectin. Starch can be modified by various chemical treatments in order to obtain the required properties for further manufacturing of starch into products. The great potential of starch and biobased nanocomposite materials for packaging is pointed out in a general overview (Table 13.7).

Amylopectin is the main component of starch (70%–80%). Typically about 1.200–6.200 D-glucose units are connected by α-1,4-glycosidic bonds. A branched structure develops by α-1,6-glycoside bonds, which occur typically every 25 glucose units.

Solvent cast amylopectin films reinforced with MFC have been prepared. The authors report depending on the type of MFC used that either the ductile behavior or the elastic modulus can be increased by rather high loadings of MFC of up to 50% The oxygen permeability was lowered too comparing with pure amylopectin films. The opacity of the reinforced films proved to be similar compared to pure amylopectin films.[37]

Cellulose is the main structural component in cell walls of plants. Wood consists of, e.g., 40%–50% cellulose, 30% lignin, 5% minerals, and 15% hemicellulose (polysaccharide with pentose as structural element). Cellulose is built from D-glucose molecules that are linked by 1,4-glucosidic links (Figure 13.3).

TABLE 13.7

Nanocomposite Food Packaging Materials Based on Starch-Derived Polymers

Nanoparticle	Reference	Comment	Application
MMT and hectorite	[90]	Glycerol plasticized thermoplastic starch	
Alpha ZrPO$_4$	[91]	Glycerol plasticized pea-starch	Mechanical property improvement
TiO$_2$	[32]	Starch + PMMA	Photocatalytic activity
MMT	[92]	Thermoplastic starch blended with PVOH	Mechanical property improvement
Chitin whiskers	[36]	Glycerol plasticized starch	
MMT	[93]		Mechanical property improvement
MMT	[23]	Influence of starch plasticizers tested	
MMT	[94]		
MMT	[64]	Glycerol plasticized starch, high amount of clay (30%) added	

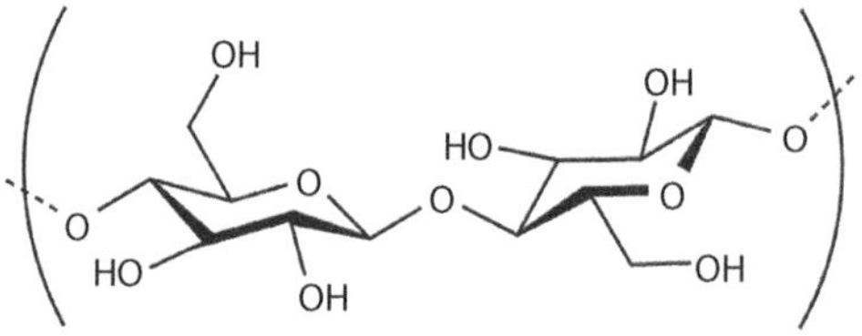

FIGURE 13.3
Chemical structure of cellulose.

The macromolecule of cellulose consists therefore of an alternating arrangement of the oxygen bridges in contrast to the structure of starch. This allows the formation of stretched chain molecules and the formation of bundles (micelles) from numerous parallel molecules. The molecules within the micelles are bound together by hydrogen bonds and as the chains overlap, the plant fibers can reach a considerable length.

Cellulose is processed in various ways to yield micro fibrillated cellulose (MFC) or cellulose whiskers with nanodimensions, which can be used as nanoreinforcement in PNC. The production methods of MFC are described in.[101]

Methyl cellulose is derived from cellulose. It is a white powder and dissolves in cold (but not in hot) water. It is used as a nondigestible, nontoxic, and nonallergenic thickener and emulsifier in various food and cosmetic products. It is synthetically produced by heating cellulose with a solution of sodium hydroxide. In a substitution reaction with methyl chloride, the OH functional groups are then replaced by OCH$_3$ groups. Different kinds of methyl cellulose can be prepared depending on the number of hydroxyl groups substituted.

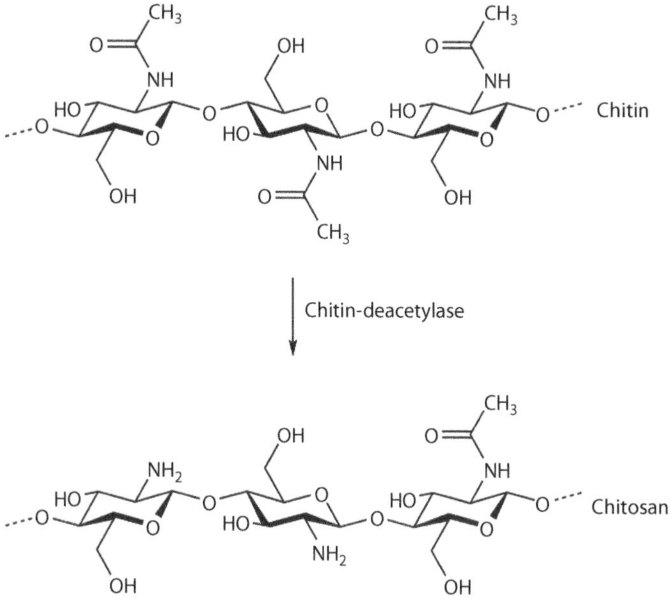

FIGURE 13.4
Synthesis and structure of chitosan.

MMT as nanoparticle and carvacrol as antibiotic agents were blended with methylcellulose and the composite was tested for antibacterial activity.[4] Chitosan NPs were blended with carboxymethyl cellulose and proved to be highly stable and having improved mechanical properties.[102]

Chitosan is a biopolymer, a naturally occurring polyaminosaccharide, derived from chitin. It consists of β-1,4-glycosidic connected N-acetylglucosamines. The resulting linear molecules consist of approximately 2000 monomer units. It was already invented in 1859 by C. Rouget from a reaction of chitin with potassium hydroxide (Figure 13.4).

Chitosan-matrix-based nanocomposites have already been developed and presented in a ternary system of hectorite and chitin whiskers[39] and showed promising results concerning mechanical properties and antibacterial effects. A solution cast film of chitosan with Ag-oxide[42] has been prepared and showed excellent antibacterial properties. By gamma irradiation of an Ag ion containing chitosan solution, a stable Ag NP solution was obtained, which was later prepared into a chitosan/starch film that showed good antibacterial and slightly improved mechanical properties.[45] By a layer-by-layer processing technique, a high-density nanocomposite chitosan film reinforced with CNW was prepared.[77] By in situ preparation, Ag NPs have been incorporated in a chitosan film with good antibacterial properties.[102] A chitosan/hectorite–clay nanocomposite film was prepared by a casting/solvent evaporation technique and showed promise for successful drug delivery applications.[11]

Carrageenans are sulfated linear polysaccharides that are extracted from red seaweeds. Carrageenans are large, highly flexible molecules forming helical structures, which give them the ability to form pseudoplastic gels. They are used in the food industry as thickeners. Carrageenans are high-molecular-weight polysaccharides made up of repeating galactose units and 3,6 anhydrogalactose (3,6-AG), both sulfated and nonsulfated. The units are joined by alternating alpha 1–3 and beta 1–4 glycosidic linkages. A carrageenan nanocomposite with 1 wt%–5 wt% CNWs produced by solution casting[40] had a water vapor reduction of 71%. Carrageenan/zein prolamine–based nanocomposite with mica produced by solvent casting[47] had significantly reduced water permeability and water uptake.

Prolamines are a group of plant storage proteins and are found in cereal grains. The prolamin in wheat is called gliadin, in barley it is called hordein, in rye it is called secalin, and in corn it is called zein. Prolamines are characterized by a high glutamine and proline content and are generally soluble only in alcohol solutions.

Zein is a plant protein extracted from corn and has a variety of industrial and food uses.[96,97] Zein can be processed into resins and bioplastic polymers, which can be extruded or rolled into a variety of plastic products.[98] Zein has been used in the manufacture of commercial packaging products like coatings for paper cups, soda bottle cap linings, adhesives, coatings, and binders. It is also used as a coating for candy and other encapsulated foods and drugs. It is classified as generally recognized as safe (GRAS) by the U.S. Food and Drug Administration.

A zein polymer matrix has been used with Ag testing their antibacterial properties by changing pH conditions at preparation.[44] Low pH conditions were revealed to be important for good nanocomposite stability.

A matrix of soy protein isolate with different types of functionalized MMT has been tested. Mechanical properties could be sufficiently increased while water vapor barrier properties remained below expectation compared to the state-of-the-art plastic used for food packaging.[22] However, the same authors found in another investigation significant improvement in both mechanical and barrier properties of a matrix of soy protein isolate with MMT obtained by melt extrusion.[7] A matrix of whey protein isolate with TiO_2 has been studied concerning its optical properties in UV light.[29] A matrix of whey protein isolate with Cloisite 30B has been tested for its optical, tensile, and water vapor barrier properties.[8]

ε-Polylysine is a natural occurring protein composed of lysine units that form a straight polymer chain. Lysine is an α-amino acid with the chemical formula $HO_2CCH(NH_2)(CH_2)_4NH_2$ and is an essential amino acid. It is produced by, e.g., *Streptomyces albulus* and inhibits the growth of many gram-positive and gram-negative bacteria. ε-Polylysine is a hygroscopic, light yellow powder, which is soluble in water, slightly soluble in ethanol, and insoluble in organic solvents. Polylysine is used as an efficient and safe food preservative with no side effects.[95] A nanocomposite based on an EP matrix

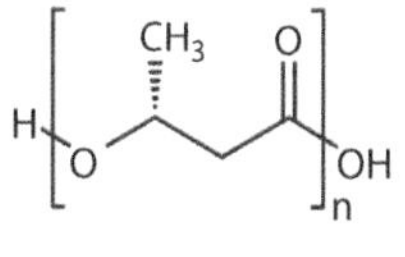

FIGURE 13.5
General structure of poly-3-hydroxybutyrate (P3HB).

with multi-walled carbon nanotubes (MWNTs) was produced and showed enhanced and effective antibacterial properties.[49]

PHB is a naturally occurring polyester and is produced by microorganisms (like *Ralstonia eutrophus* or *Bacillus megaterium*) (Figure 13.5).

P3HB is probably the most common type of PHB polyhydroxyalkanoates, but many other polymers of this class are produced by a variety of organisms: poly-4-hydroxybutyrate (P4HB), polyhydroxyvalerate (PHV), polyhydroxyhexanoate (PHH), polyhydroxyoctanoate (PHO), and their copolymers.

An increase in mechanical properties has been reported for a PHB nanocomposite with bacterial cellulose.[33] PHB polymers are very difficult to be processed successfully because of their significant instability during melt processing. A PHBV matrix was filled with different types of clay (MMT, kaolinite) and the influence of clay type on polymer degradation during processing has been investigated.[99]

A melt mixed blend of PHB and PCL filled with kaolinite proved to be better concerning melt stability than with MMT fillers.[16] The composite showed enhanced barrier properties against oxygen, D-limonene, and water.

Shellac is a resin secreted by the female lac bug on trees in the forests of India and Thailand. *Kerria lacca* is a species of scale insect of the family *Kerriidae*. Shellac is a polyester of various alcohols with hydroxyl-carbonic acids. It is sold as dry flakes, which can be dissolved in ethyl alcohol to make liquid shellac, which is used as a brush-on colorant, food glaze, and wood finish.

Shellac has been mixed with MFC and sprayed coatings of paper and paperboard have been produced.[38] The oxygen transmission rate and the water vapor transmission rate could be considerably reduced reaching values representing high barrier materials.

13.8 Smart and Active Packaging

Nanotechnology bears the promise not only to improve the physical and chemical nature of packaging foils used for food packaging but may also provide innovative solutions for high-tech food packages. Various smart sensors, intelligent tags, dynamic pricing labels, controlled release devices, food spoilage indicators, time-temperature indicators, ripeness indicators, biosensors, advanced product tracing elements, and protection measures against forgery may soon become available and could be applied on a larger scale. However, before such applications become widespread it is not only a technical and cost issue that has to be solved but it requires also some legislative and regulation measures. Safety issues like migration of substances to the food have to be investigated and food safety concerns have to be settled first.[103] The development of such smart packaging elements is presently ongoing in many laboratories worldwide.

Smart materials change their properties in a controlled quick fashion and in a predictable way as a response to outer conditions like mechanical loadings, temperature, humidity, and gaseous environment. The changing properties of such smart materials can be various like color, shape, surface appearance, conductivity, viscosity, etc. Smart materials can be realized by, e.g., PNCs containing in a transparent polymer matrix platelike NPs matching in their dimensions and interparticle distances the wavelength of the visible light. In such composite materials, colours based on a physical principle of light diffraction can be observed (a thin oily film on water or the wings of butterflies are a good demonstration for this principle). When the polymer matrix is changing its dimensions by, e.g., swelling by uptake of water, a visible color change is resulting.

The Austrian company Attophotonics is a specialist for such smart color materials and has developed a large range of sensors based on such principles, which can be printed on food packages. Such smart sensors can sense temperature changes, humidity changes, spoilage of food, and the interruption of the cooling chain.[104]

Active and intelligent packaging involves the deliberate interaction of the packaging with the food to improve the stability, shelf life, quality, and safety of the packaged food. Such techniques are, e.g., delayed oxidation techniques, controlled respiration, controlling and retarding the microbial growth and moisture migration, carbon dioxide and odor absorbers, ethylene removers, and aroma emitters. Nanotechnology and PNC materials can provide such solutions and have shown this already in the laboratory, but before widespread application will become reality production, cost has to be reduced further and safety concerns have to be settled first.[105,106]

13.9 Commercial Applications of Polymer Nanocomposite Materials Already on the Market and Used for Food Packaging

Ag in its nano form is used as a very effective antibiotic ingredient in a variety of products already on the market. Of the 1317 commercial nanoproducts listed in Woodrow Wilson's inventory of nanotechnology-based consumer products (2010) about 300 contain nano Ag.[74] Many of them are intended to be used for storing food or refrigerate food. In the sector "Cooking-items/Tableware/ Kitchenware" of Woodrow Wilson's inventory of nanotechnology-based consumer products, 12 products are listed: they are either based on polymer-Ag and nanoceramic composites providing antibiotic or nonsticking properties.

In the sector "Food storage products" of Woodrow Wilson's inventory of nanotechnology-based consumer products, 21 products are listed (Table 13.8). Refrigerators with antibacterial nanocoatings manufactured by a few companies (Korea: Daewoo, LG Electronics, Samsung; China: Haier Yu Hang) are also part of this list of 21 but are not retained in Table 13.8.

Nanocor, a subsidiary of Illinois-based AMCOL International, is a major producer of nanocomposite plastics. Barrier properties, odor absorption, improved flame and thermal resistance for making the package microwavable are claimed as being major properties of food packages based on clay nanocomposites.[109] Nylon-based nanocomposites with clay additions have been produced and show excellent barrier properties.[110] They are already used in a number of food packaging products (Table 13.9).

13.10 Risk of Nanoparticles and Nanocomposite Materials

The risk of NP is depending on the specific health hazard of the NP involved and the exposure to this hazard. The assessment of risk is especially important when NPs are used in close contact with food. Unfortunately, there are still many open questions concerning the risk of nanocomposite usage. There is a general health and environmental safety (HES) concern of NP usage but not enough secure information is available on the individual health hazard potential of specific NP. Environmental (Eco) toxicity and fate of NP when released into the environment is still largely unknown making it very difficult to assess exposure to NP. Therefore, human exposure limits and needed occupational measures are still unknown in many cases of NP. As a result of this, the consumer acceptance for using NP in food-related products is generally low, especially in Europe.

We can classify NPs roughly into three categories:

1. *Natural occurring NPs* like virus, proteins, volcanic ash, airborne ash and dust from forest fires, Sahara-dust, certain minerals (e.g., various clay minerals like MMT, sepiolite, hectorite). It can be assumed with a certain confidence that evolution has equipped humans with some counteracting bodily measures in order to cope with and survive such natural NP hazards to a certain extent as long as exposure remains limited.

2. *Involuntarily produced NPs* are generated from exhaust-gas emissions, ultrafine dust from various industrials processes, fires, cigarette smoke, barbecue grill fires, aircraft engines, and diesel engines.

TABLE 13.8

Specific Food Storage Products with Nanoingredients

Product	Nanocomponent	Producer Country	Remarks	Reference/Supplier
Adhesive for Mc Donalds burger containers	Nano-Starch	USA		Ecosynthetix [107]
Nonstick coating Al-foil from Toppits		DE		Melitta
Beer bottle with nano-modified imperm	Nano clay	USA	Used by Miller Brewing for Miller Lite, Miller Genuine Draft, and Ice House brands	Multiple manufacturers and developers (Voridian, Nanocor)
FresherLonger™ Miracle Food Storage	Nano Ag	USA		Sharper image
Nanosilver storage box	Nano Ag	China		Quan Zhou Hu Zheng Nano Technology Co., Ltd
FresherLonger™ Plastic Storage Bags	Nano Ag	USA		Sharper Image
Baby mug	Nano Ag	Korea		Baby Dream Co., Ltd
Food container (NS)	Nano Ag	Korea		A-DO global
Nano silver NS-315 water bottle	Nano Ag	Korea		A-DO global
Nano-silver salad bowl	Nano Ag	Korea		ChangMin Chemicals
Plastic beer and flavored alcoholic beverage (FAB) bottles	Nanoclay	USA	Multilayer PET bottles made with Aegis® OX barrier nylon resin	Honeywell
N-coat		USA		Constantia multifilm [108]
Nanoplastic wrap	Nano Zno	Taiwan	Properties claimed: Anti-UV, reflecting IR, sterilizing and antimold, temperature tolerance, fire-proof	SongSing nanotechnology Co., Ltd
BlueMoonGoods™ Fresh Box Food Storage Containers	Ag Nano	USA		BlueMoonGood LLC

Source: http://www.nanotechproject.org/inventories/consumer/

TABLE 13.9

General Properties of Some Commercially Available Nanoreinforced Food Packaging Materials, Their Main Application in Food Packaging and Commercial Producers

Polymer Nanocomposite	Main Properties	Food Packaging Applications	Producers	Processing/Reference
Imperm	High barrier nylon, OTR 4×, CO_2TR 2×, WVTR 2.5× improvement compared to MXD6	Carbonated soft drinks, waters, beers, and flavored alcoholic beverages, for oxygen sensitive products	ColorMatrix Corporation	Multilayer bottles, films, and thermoformed containers, also for coated paper cartons[111]
NANO-N-MXD6 (GRADE M9)	In PET bottles M9 improves CO_2TR 2× and OTR 4× compared to standard N-MXD6	Juice or beer bottles, multilayer films, containers	Mitsubishi Gas Chemical company, Nanocor	[112]
Durethan KU2-2601	Doubling of stiffness, high gloss and clarity, reduced oxygen transmission rate	Medium barrier applications	Lanxess	Barrier films, paper coating for paperboard juice containers[113]
Aegis NC	3× better OTR than nylon 6, 60% OTR improved versus PA6 on paperboard	Medium barrier bottles and films for paper/cardboard containers	Honeywell polymer	Coating or base resin, for cast or blown film. Replacement for nylon 6, coatings in paperboard juice cartons and for processed meat and cheese[114]
Aegis OX	Highly reduced oxygen transmission rate, improved clarity	High barrier beer bottles and orange juice bottles	Honeywell polymer	Coinjection molded PET bottles with oxygen barrier and D-limonene barrier[115]

Source: http://www.nanocompositech.com/commercial-nanocomposites-nanoclays.htm

There is already a well-established correlation between premature mortality (caused mainly by lung cancer and cardiovascular diseases) and particulate matter (PM2.5) associated with pollutants.[116] Also in detailed studies of fine and ultrafine particles and of particles of diesel exhaust gas, it has been shown that NP can travel to organs that are not directly exposed to the NP, i.e., NP can apparently cross the cell walls contrary to the behavior of micro-sized particles and NPs have been found in brain and testes of animals when NP has been delivered only to the lungs of the animals.[117–120]

3. Synthetic (engineered) nanoparticles (ENPs) with deliberately new properties and functionalities like metals, metal oxides, metal sulfides, carbon (CNT, MWCNT, expanded graphite, carbon black), semiconductors, polymers in NP form, nanoscale micelles, nanowhiskers. Concerning the hazards of such ENPs, there are still many more questions than answers because there is still a lack of systematic toxicological studies. However, in the last 5 years, an exponential growth of such studies can be observed (Figure 13.6).

The present focus of HES studies is concerned mainly with the nanotoxicology of ENP like CNT, fullerenes, Ag, Au, and TiO_2. Very little or even nothing can be found in the open literature on health hazards of more exotic NPs and almost nothing on exposure and exposure scenarios. Many (majority of) NP hazards and NP exposure data and scenarios are still missing. Further the results of the large number of toxicological assays, which are used, are not easily comparable with each other. The initial state of starting NPs is often undefined (agglomeration) or is even changing during exposure. A general European regulation is yet not existent. Analytical (measurements, metrics) methods to follow and assess exposure still need improvement.

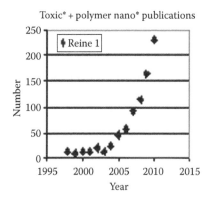

FIGURE 13.6
Exponential growth of nano-toxicological publications. (Assessed in October 2010 in ISI Web of Knowledge.)

NPs have other toxic characteristics than macroscopic materials. Individual toxicity of NP is dependent on many other properties. Toxicity of larger-scale material is generally proportional to its mass and chemistry (only). NPs are in general more toxic than an equivalent mass of larger-scale substances due to their higher surface area and due to their higher reactivity. The individual toxicity of NPs is dependent on many other physical and chemical properties:

- Surface properties: specific surface area/surface properties/surface charge of NP
- Geometrical properties: shape, porosity, size and granulometric distribution, degree of agglomeration/aggregation
- Chemical: chemical composition (purities and impurities), reactivity, functional groupings, presence of metals/redox potential, potential to generate free radicals
- Physical: solubility, crystalline structure, hydrophilicity/hydrophobicity
- Biological/medical: biopersistence, deposition site in organism
- Technical: age of particles/producer, process and source of the material used

NP distribution in the organism is differentiated unlike larger-scale materials. NP can enter the blood and nerve system and is distributed to other organs easily. NP can penetrate cell walls contrary to larger-scale particles. Once NPs have entered the cell they put oxidative stress onto the cell. NPs have "piggy back" functionality due to possibly attached toxic chemicals on the large and reactive surface area of NP and its ability of NP to penetrate cell walls. By such a mechanism, toxic chemicals can be transported inside the cell, which would otherwise be restricted to enter the cell by the cell walls acting as a barrier. The exposure to NP is most critical by inhalation (less critical by oral ingestion and by skin contact). The health hazard for each NP has therefore to be characterized and treated case by case and no general rules are unfortunately available.

Toxicological studies of nanoclay are still rather limited. Most of the studies so far have assessed the toxicity of stacked and not exfoliated and stabilized nanoclay particles. The stacked clay particles are having dimensions in the micrometer range and will behave in a toxicological assay differently than true nanosized particles.

By intratracheal instillation of MMT in rats, interstitial fibrosis was observed.[121]

Another recent study was examining the toxicological effects on human neutrophils in vitro. The test showed that MMT may potentiate infection by a direct cytotoxic effect on the neutrophils. This would make them unavailable for bacterial phagocytosis.[122] The genotoxic potential of MMT NP has been investigated in a recent study involving natural nanoclay (Cloisite Na) and an organo-modified MMT Cloisite 30B. It was assessed that the natural MMT was not genotoxic in salmonella/microsome assay and for Caco-2 cells (a human colon cancer cell line) tested in the comet assay (single-cell electrophoresis assay for the detection of DNA damage). Cloisite 30B was found to be moderately genotoxic when tested by the comet assay for genotoxicity in Caco-2 cells. The authors could show convincingly that the genotoxicity of Cloisite 30B appears to be caused by the organic modifier (methyl tallow bis 2 hydroxyethyl quaternary ammonium).[123]

For other clay types (bentonite, kaolinite), no or only weak genotoxic effects have been shown in vitro.[124,125] Toxic effects of clay particles appear to show mainly after inhalation.[126] Toxicology assessments of nanoclay in comparison with other NP have been performed by lung exposure (intratracheally instilled) to sepiolite (magnesium silicate nanoclay), titanium dioxide (ultrafine), and quartz (as reference material). Findings were that the tissue damage was increasing from nanoclay (low) over TiO_2 (low) to quartz (severe).[127]

Cytotoxicity and genotoxicity of MMT were studied in a recent study.[128] Three test systems have been used: comet assay test on Chinese Hamster ovary (CHO) cells in vitro, micronucleus (MN) in vivo, and Salmonella gene mutation assay. Findings were that no mutations, no DNA damage, and no cytotoxicity at 1000 μm/mL have been observed. The material was therefore considered to be safe for use in biomedical applications.

In contrast to these studies of nanoclay, embryo-toxicity assessment of engineered carbon NPs gave following results.[129] SWCNT (pristine) was classified as mildly embryo-toxic, COOH functionalized-SWCNT was classified as strongly embryo-toxic, nanocarbon black used as reference standard proved as not embryo-toxic. These results show very convincingly that the surface functionalization of the NP has a large influence on the results of the toxicity assessment. The authors state that in their conclusion that "All forms of SWCNT are embryo-toxic," "The functionalizing of CNT makes a great difference in its toxicity," "Proven translocation capabilities and crossing of biological barriers of SWCNT makes these findings alarming." The authors admit, "Literature data are often controversial, completely opposite to above findings and inconclusive." Findings are dependent on aggregation status, purity, size, surface chemistry, reactivity, cellular uptake, and cell types used for testing.

A cautious summary of findings on the nanotoxicity of some NP is as follows:

- Clay NPs derived from a natural source appear in general less toxic.
- The functionalizing agents appear in certain cases to be more toxic than the nanoclay itself.
- Engineered NPs appear in general more toxic (e.g., CNT) but their toxicology is strongly depending on surface chemistry characteristics of NP.
- Likelihood of intoxication is increasing in the following order: uptake by skin-intestines-lung.

It has been shown that the functionalizing compounds used to functionalize clay appear to be rather toxic. Adverse health effects of quaternary ammonium salts are well known.

Quaternary ammonium ions can cause skin and respiratory irritation, have antimicrobial activity, and are used as antimicrobials and disinfectants. They are also good against fungi, amoeba, and enveloped viruses; they act by disrupting the cell membrane of the microbes. As an example, a largely used quaternary ammonium compound like cetyltrimethylammonium bromide, which is used as a mild disinfectant and cleaning agent, has a LD50 of 410 mg/kg. The toxicity of other structural modifications of quaternary ammonium ions is different depending on the individual organic chains but is in a similar range.

In order to overcome this toxicity problem of the clay functionalizing agent, clay has been successfully functionalized by chitosan, which yields a MMT-chitosan (chito)-modified nanoclay. Chito-MMT was used for biocompatible polymers (e.g., strengthening of hydrogels). Chito-MMT has no toxicity as proven in cell culture testing (chitosan has a comparable high LD50 of 16 g/kg). XRD has also shown its ability to widen the MMT-clay galleries, which is a prerequisite to be used successfully in nanocomposite formation.[130]

When assessing the environmental impact of PNCs, the behavior of involuntarily released NP into the environment has to be taken into account. However, the fate of NP released from the nanocomposite material into the environment is still largely unknown. There is, e.g., a growing concern about what effect the increasing release of Ag into the environment will be. The release of nano Ag is coming from the already large number of Ag containing products that are used as food packaging, medical products, textiles, and as many other consumer items, which will be discarded or composted after end of life and will end up finally in the environment. Information on release and occurrence of nanosilver in the environment is very scarce. It is further not known until now whether nanosilver particles will present a threat to the environment itself. Only data from the aqueous environment, no data on nanosilver in soils are available. The present, however, still very limited state of knowledge on Ag in the environment is summarized in a recent review paper.[131] Such questions have to be answered fully before NP can be included safely into food packaging to a larger extent.

Acknowledgments

The very helpful financial and organizational contributions of the COST Action MP0701 on PNC and of the COST Action FA 901 on food packaging are thankfully acknowledged. Without the support of these COST Actions this review on PNC for food packaging would not have been possible.

References

1. CIPET-Central Institute of plastics and engineering technology/government of India, Plastics industry statistics, http://cipet.gov.in/plastics_statics.html (accessed on August 4, 2008).
2. J. Lange, Y. Wyser, Recent innovations in barrier technology for plastic packaging—A review, *Packaging Technology and Science*, 16 (2003) 149.
3. M.D. Sanchez-Garcia, E. Gimenez, J.M. Lagaron, Novel pet nanocomposites of interest in food packaging applications and comparative barrier performance with biopolyester nanocomposites, *Journal of Plastic Film and Sheeting*, 23 (2007) 133.
4. S. Tunc, O. Duman, Preparation of active antimicrobial methyl cellulose/carvacrol/montmorillonite nanocomposite films and investigation of carvacrol release, *LWT—Food Science and Technology*, 44 (2011) 465.
5. T.W. Cho, S.W. Kim, Morphologies and properties of nanocomposite films based on a biodegradable poly(ester)urethane elastomer, *Journal of Applied Polymer Science*, 121 (2011) 1622.
6. A. Ranade, N. D'Souza, B. Gnade, C. Thellen, C. Orroth, D. Froio, J. Lucciarini, J.A. Ratto, Effect of coupling agent on the dispersion of PETG montmorillonite nanocomposite films, in: I. Ovidko, C.S. Pande, R. Krishnamoorti, E. Lavernia, G. Skandan (Eds.), *Mechanical Properties of Nanostructured Materials and Nanocomposites, Materials Research Society Symposium Proceeding*, Boston, MA, Vol. 791 (2004), p. 283.
7. P. Kumar, K.P. Sandeep, S. Alavi, V.D. Truong, R.E. Gorga, Preparation and characterization of bio-nanocomposite films based on soy protein isolate and montmorillonite using melt extrusion, *Journal of Food Engineering*, 100 (2010) 480.
8. R. Sothornvit, S.I. Hong, D.J. An, J.W. Rhim, Effect of clay content on the physical and antimicrobial properties of whey protein isolate/organo-clay composite films, *LWT—Food Science and Technology*, 43 (2010) 279.
9. W.-S. Jang, I. Rawson, C.J. Grunlan, Layer-by-layer assembly of thin film oxygen barrier, *Thin Solid Films*, 516 (2008) 4819.
10. C. Poisson, J. Guerengomba, M.F. Lacrampe, P. Krawczak, B. Gupta, V. Miri, and J.M. Lefebvre, Mechanical, optical and barrier properties of PA6/nano-clay-based single-and multilayer blown films, *Polymers and Polymer Composites*, 16 (2008) 349.
11. X. Wang, Y. Du, J. Luo, B. Lin, J.F. Kennedy, Chitosan/organic rectorite nanocomposite films: Structure, characteristic and drug delivery behaviour, *Carbohydrate Polymers*, 69 (2007) 41.
12. J.C. Grunlan, A. Grigorian, C.B. Hamilton, A.R. Mehrabi, Effect of clay concentration on the oxygen permeability and optical properties of a modified poly(vinyl alcohol), *Journal of Applied Polymers Science*, 93 (2004) 1102.
13. M.A. Busolo, P. Fernandez, M.J. Ocio, J.M. Lagaron, Novel silver-based nanoclay as an antimicrobial in polylactic acid food packaging coatings, *Food Additives and Contaminants: Part A—Chemistry Analysis Control Exposure & Risk Assessment*, 27 (2010) 1617.

14. J. Vartiainen, T. Tammelin, J. Pere, U. Tapper, K. Nattinen, A. Harlin, M. Tuominen, Bio-hybrid nanocomposite coatings from polysaccharides and nanoclay, in: B.L. Lu (Ed.), *Proceedings of the 17th IAPRI World Conference on Packaging*, Tianjin, China (2010), p. 54.

15. M.A. Priolo, D. Gamboa, J.C. Grunlan, Transparent clay-polymer nano brick wall assemblies with tailorable oxygen barrier, *ACS Applied Materials and Interfaces*, 2 (2010) 312.

16. M.D. Sanchez-Garcia, E. Gimenez, J.M. Lagaron, Morphology and barrier properties of nanobiocomposites of poly(3-hydroxybutyrate) and layered silicates, *Journal of Applied Polymers Science*, 108 (2008) 2787.

17. S. Dadbin, M. Noferesti, M. Frounchi, Oxygen barrier LDPE/LLDPE/organo-clay nano-composite films for food packaging, *Macromolecular Symposia*, 274 (2008) 227.

18. J.M. Lagaron, L. Cabedo, D. Cava, J.L. Feijoo, R. Gavara, E. Gimenez, Improving packaged food quality and safety. Part 2: Nanocomposites, *Food Additives and Contaminants*, 22 (2005) 994.

19. M.A. Osman, J.E.P. Rupp, U.W. Suter, Effect of non-ionic surfactants on the exfoliation and properties of polyethylene-layered silicate nanocomposites, *Polymer*, 46 (2005) 8202.

20. M.A. Osman, J.E.P. Rupp, U.W. Suter, Gas permeation properties of polyethylene-layered silicate nanocomposites, *Journal of Materials Chemistry*, 15 (2005) 1298.

21. S.A. Jang, Y.J. Shin, Y.B. Seo, K.B. Song, Effects of various plasticizers and nanoclays on the mechanical properties of red algae film, *Journal of Food Science*, 76 (2011) 30.

22. P. Kumar, K.P. Sandeep, S. Alavi, V.D. Truong, R.E. Gorga, Effect of type and content of modified montmorillonite on the structure and properties of bio-nanocomposite films based on soy protein isolate and montmorillonite, *Journal of Food Science*, 75 (2010) 46.

23. X. Tang, S. Alavi, T.J. Herald, Barrier and mechanical properties of starch-clay nanocomposite films, *Cereal Chemistry*, 85 (2008) 433.

24. M. Sirousazar, V. Yari, B.F. Achachlouei, J. Arsalani, Y. Mansoori, Polypropylene/montmorillonite nanocomposites for food packaging, *E-Polymers*, 027 (2007).

25. J.W. Rhim, S.B. Lee, S.I. Hong, Preparation and characterization of agar/clay nanocomposite films: The effect of clay type, *Journal of Food Science*, 76 (2011) 40.

26. A. Emamifar, M. Kadivar, M. Shahedi, S. Soleimanian-Zad, Effect of nanocomposite packaging containing Ag and ZnO on inactivation of *Lactobacillus plantarum* in orange juice, *Food Control*, 22 (2011) 408.

27. A. Kubacka, C. Serrano, M. Ferrer, H. Launsdorf, P. Bielecki, M.L. Cerrada, M. Fernandez-Garcia, High-performance dual-action polymer-TiO$_2$ nanocomposite films via melting processing, *Nano Letters*, 7 (2007) 2529.

28. X. Li, W. Chen, Y. Xing, V. Yun, P. Zhang, Effect of ZnO nanoparticles on the UV light fastness and climate resistance of PVC film, in: Z. Jiang, C.L. Zhang (Eds.), Manufacturing Science and Engineering, PTS 1-5, *Advanced Materials Research*, 97–101 (2010) 2197.

29. Y. Li, Y. Jiang, F. Liu, F. Ren, G. Zhao, X. Leng, Fabrication and characterization of TiO(2)/whey protein isolate nanocomposite film, *Food Hydrocolloids*, 25 (2011) 1098.

30. S. Chandramouleeswaran, S.T. Mhaske, A.A. Kathe, P.V. Varadarajan, V. Prasad, N. Vigneshwaran, Functional behaviour of polypropylene/ZnO-soluble starch nanocomposites, *Nanotechnology*, 18 (2007).

31. N. Lepot, M.K. Van Bael, H. Van den Rul, J. D'Haen, R. Peeters, D. Franco, J. Mullens, Influence of incorporation of ZnO nanoparticles and biaxial orientation on mechanical and oxygen barrier properties of polypropylene films for food packaging applications, *Journal of Applied Polymer Science*, 120 (2011) 1616.

32. Y.H. Yun, K.J. Hwang, Y.J. Wee, S.D. Yoon, Synthesis, physical properties, and characterization of starch-based blend films by adding nano-sized TiO(2)/poly(methylmetacrylate-co-acrylamide), *Journal of Applied Polymer Science*, 120 (2011) 1850.

33. Z. Cai, G. Yang, Optical nanocomposites prepared by incorporating bacterial cellulose nanofibrils into poly(3-hydroxybutyrate), *Materials Letters*, 65 (2011) 182.

34. M.D. Sanchez-Garcia, A. Lopez-Rubio, J.M. Lagaron, Natural micro and nanobiocomposites with enhanced barrier properties and novel functionalities for food biopackaging applications, *Trends in Food Science & Technology*, 21 (2010) 528.

35. J. George, A.S. Bawa, Siddaramaiah, Synthesis and characterization of bacterial cellulose nanocrystals and their PVA nanocomposites, in: J.H. Lee (Ed.), Multi-Functional Materials and Structures III, PTS 1 and 2, *Advanced Materials Research*, 123–125 (2010) 383.

36. P.R. Chang, R. Jian, J. Yu, X. Ma, Fabrication and characterisation of chitosan nanoparticles/plasticised-starch composites, *Food Chemistry*, 120 (2010) 736.

37. D. Plackett, H. Anturi, M. Hedenqvist, M. Ankerfors, M. Gallstedt, T. Lindstrom, I. Siro, Physical properties and morphology of films prepared from microfibrillated cellulose and microfibrillated cellulose in combination with amylopectin, *Journal of Applied Polymer Science*, 117 (2010) 3601.

38. E.L. Hult, M. Iotti, M. Lenes, Efficient approach to high barrier packaging using microfibrillar cellulose and shellac, *Cellulose*, 17 (2010) 575.

39. X. Li, B. Ke, X. Shi, Y. Du, Cooperative performance of chitin whisker and rectorite fillers on chitosan films, *Carbohydrate Polymers*, 85 (2011) 747.

40. M. Dolores Sanchez-Garcia, L. Hilliou, J.M. Lagaron, Morphology and water barrier properties of nanobiocomposites of k/i-hybrid carrageenan and cellulose nanowhiskers, *Journal of Agricultural and Food Chemistry*, 58 (2010) 12847.

41. H.M.C. Azeredo, L.H.C. Mattoso, D. Wood, T.G. Williams, R.J. Avena-Bustillos, T.H. McHugh, Nanocomposite edible films from mango puree reinforced with cellulose nanofibers, *Journal of Food Science*, 74 (2009) 31.

42. S. Tripathi, G.K. Mehrotra, P.K. Dutta, Chitosan-silver oxide nanocomposite film: Preparation and antimicrobial activity, *Bulletin of Materials Science*, 34 (2011) 29.

43. A.L. Incoronato, G. Buonocore, A. Conte, M. Lavorgna, M.A. Del Nobile, Active systems based on silver-montmorillonite nanoparticles embedded into bio-based polymer matrices for packaging applications, *Journal of Food Protection*, 73 (2010) 2256.

44. B. Zhang, Y. Luo, Q. Wang, Development of silver-zein composites as a promising antimicrobial agent, *Biomacromolecules*, 11 (2010) 2366.

45. R. Yoksan, S. Chirachanchai, Silver nanoparticle-loaded chitosan-starch based films: Fabrication and evaluation of tensile, barrier and antimicrobial properties, *Materials Science and Engineering C: Materials for Biological Applications*, 30 (2010) 891.

46. R. Gupta, G.U. Kulkarni, Removal of organic compounds from water by using a gold nanoparticle-poly(dimethylsiloxane) nanocomposite foam, *ChemSusChem*, 4 (2011) 737.

47. M.D. Sanchez-Garcia, L. Hilliou, J.M. Lagaron, Nanobiocomposites of carrageenan, zein, and mica of interest in food packaging and coating applications, *Journal of Agricultural and Food Chemistry*, 58 (2010) 6884.

48. L. Bastarrachea, S. Dhawan, S.S. Sablani, J.H. Mah, D.H. Kang, J. Zhang, J. Tang, Biodegradable poly(butylene adipate-co-terephthalate) films incorporated with nisin: Characterization and effectiveness against *Listeria innocua*, *Journal of Food Science*, 75 (2010) 215.

49. J. Zhou, X. Qi, Multi-walled carbon nanotubes/epsilon-polylysine nanocomposite with enhanced antibacterial activity, *Letters in Applied Microbiology*, 52 (2011) 76.

50. NANOCOMPOSITECH, Metal/Polymer Nanocomposites: Brief Outline, http://www.nanocompositech.com/review-metal-nanocomposite.htm (accessed on August 20, 2012).

51. C. Jeffrey Brinker, W. George Scherer (Eds.), *Sol Gel Science. The Physics and Chemistry of Sol-Gel Processing*. Academic Press, Boston, MA (1990).

52. V. Uskovic, M. Drofenik, Synthesis of materials within reverse micelles, *Surface Review and Letters*, 12 (2005) 239–277.

53. M. Trunec, Dispersion of Nanoparticles in Solvents, Polymer Solutions and Melts. Principles and Possibilities, Oral presentation at *COST MP0701 Workshop*, Novi Sad, Serbia (2010).

54. W. Han, Y.J. Yu, N.T. Li, L.B. Wang, Application and safety assessment for nanocomposite materials in food packaging, *Chinese Science Bulletin*, 56 (2011) 1216.

55. J. Ahmed, S.K. Varshney, Polylactides—Chemistry, properties and green packaging technology: A review, *International Journal of Food Properties*, 14 (2011) 37.

56. M. Jamshidian, E.A. Tehrany, M. Imran, M. Jacquot, S. Desobry, Poly-lactic acid: Production, applications, nanocomposites, and release studies, *Comprehensive Reviews in Food Science and Food Safety*, 9 (2010) 552.

57. I. Siro, D. Plackett, Microfibrillated cellulose and new nanocomposite materials: A review, *Cellulose*, 17 (2010) 459.

58. A. Arora, G.W. Padua, Review: Nanocomposites in food packaging, *Journal of Food Science*, 75 (2010) R43.

59. H.M.C. de Azeredo, Nanocomposites for food packaging applications, *Food research International*, 42 (2009) 1240.

60. C. Andersson, New ways to enhance the functionality of paperboard by surface treatment—A review, *Packaging Technology and Science*, 21(2008) 339.

61. R. Zhao, P. Torley, P.J. Halley, Emerging biodegradable materials: Starch- and protein-based bio-nanocomposites, *Journal of Materials Science*, 43 (2008) 3058.

62. J.W. Rhim, Potential use of biopolymer-based nanocomposite films in food packaging applications, *Food Science and Biotechnology*, 16 (2007) 691.

63. A. Sorrentino, G. Gorrasi, V. Vittoria, Potential perspectives of bio-nanocomposites for food packaging applications, *Trends in Food Science & Technology*, 18 (2007) 84.

64. H.M. Wilhelm, M.R. Sierakowski, G.P. Souza, F. Wypych, Starch films reinforced with mineral clay, *Carbohydrate Polymers*, 52 (2003) 101.
65. A. Okada, M. Kawasumi, A. Usuki, J. Kojima, T. Kurauchi, O. Kamigaito, Nylon 6-clay hybrid, *Materials Research Society Symposium*, P171 (1990) 45.
66. A. Usuki, J. Kojima, M. Kawasumi, A. Okada, Y. Fukushima, T. Kurauchi, O. Kamigaito, Synthesis of nylon 6 clay hybrid, *Journal of Materials Research*, 8 (1993) 1179.
67. M. Huang, J. Yu, Structure and properties of thermoplastic corn starch/montmorillonite biodegradable composites, *Journal of Applied Polymer Science*, 99 (2006) 170.
68. J.T. Kloprogge, S. Komarneni, J.E. Amonette, Synthesis of smectite clay minerals: A critical review, *Clays and Clay Minerals*, 47(1999) 529.
69. US Department of the Interior, US Geological Survey, U.S. Geological Survey Open-File Report 01-041, http://pubs.usgs.gov/of/2001/of01–041/htmldocs/clays/smc.htm (accessed on August 20, 2012).
70. Nanocor, Technical Data G-100 (12/04) General Information About Nanomer® Nanoclay, http://www.nanocor.com/tech_sheets/G100.pdf (accessed on August 20, 2012).
71. Rockwood, Cloisite® and Nanofil® additives, http://www.rockwoodadditives.com/nanoclay/ (accessed on August 20, 2012).
72. Elementis Specialties, Plastics-main page, http://www.elementis.com/esweb/esweb.nsf/pages/plastics?opendocument (accessed on August 20, 2012).
73. A. Emamifar, M. Kadivar, M. Shahedi, S. Soleimanian-Zad, Preparation and evaluation of nanocomposite LDPE films containing Ag and ZnO for food-packaging applications, in: X. Yi, L. Mi (Eds.), Materials and Manufacturing Technology, PTS 1 and 2, *Advanced Materials Research*, 129–131 (2010) 1228.
74. The Project on Emerging Nanotechnologies, Consumer Products – An inventory of nanotechnology-based consumer products currently on the market, http://www.nanotechproject.org/inventories/consumer/ (accessed on August 20, 2012).
75. M.E. Balezin, O.R. Timoshenkova, S.Yu. Sokovnin, Using Nanosecond Electron Beam to Produce a Silver Nanopowder, http://www.ispc-conference.org/ispcproc/papers/127.pdf (accessed on August 20, 2012).
76. N. Müller, Nanoparticles in the environment—Risk assessment based on exposure-modelling, Diploma thesis, ETH Zurich, Department of Environmental Sciences, Empa—Material Science and Technology, St Gallen, Switzerland (2007).
77. J.P. de Mesquita, C.L. Donnici, F.V. Pereira, Biobased nanocomposites from layer-by-layer assembly of cellulose nanowhiskers with chitosan, *Biomacromolecules*, 11 (2010) 473.
78. M.L. Cerrada, C. Serrano, M. Sanchez-Chaves, M. Fernandez-Garcia, F. Fernandez-Martin, A. de Andres, R.J. Jimenez Rioboo, A. Kubacka, M. Ferrer, Self-sterilized EVOH-TiO$_2$ nanocomposites: Interface effects on biocidal properties, *Advanced Functional Materials*, 18 (2008) 1949.
79. D.A. Pereira de Abreu, J. Manuel Cruz, I. Angulo, P. Paseiro Losada, Mass transport studies of different additives in polyamide and exfoliated nanocomposite polyamide films for food industry, *Packaging Technology and Science*, 23 (2010) 59.
80. M. Avella, G. Bruno, M.E. Errico, G. Gentile, N. Piciocchi, A. Sorrentino, M.G. Volpe, Innovative packaging for minimally processed fruits, *Packaging Technology and Science*, 20 (2007) 325.

81. D. Dimonie, C. Radovici, R.M. Coserea, S. Garea, M. Teodorescu, The polymer molecular weight and silicate treatment influence upon the morphology of nanocomposites for food packaging, *Revue Roumanie de Chemie*, 53 (2008) 1017.

82. C.E. Corcione, P. Prinari, D. Cannoletta, G. Mensitieri, A. Maffezzoli, Synthesis and characterization of clay-nanocomposite solvent-based polyurethane adhesives, *International Journal of Adhesion and Adhesives*, 28 (2008) 91.

83. A. Ranade, N.A. D'Souza, B. Gnade, A. Dharia, Nylon-6 and montmorillonite-layered silicate (MLS) nanocomposites, *Journal of Plastic Film & Sheeting*, 19 (2003) 271.

84. L. Cabedo, J.L. Feijoo, M.P. Villanueva, J.M. Lagaron, E. Gimenez, Optimization of biodegradable nanocomposites based on aPLA/PCL blends for food packaging applications, *Macromolecular Symposia*, 233 (2006) 191.

85. R. Pucciariello, L. Tammaro, V. Villani, V. Vittoria, New nanohybrids of poly(epsilon-caprolactone) and a modified Mg/Al hydrotalcite: Mechanical and thermal properties, *Journal of Polymer Science Part B—Polymer Physics*, 45 (2007) 945.

86. K. van de Velde, P. Kiekens, Biopolymers: Overview of several Properties and Consequences on their Applications, *Polymer Testing*, 21(4) (2002) 433–442.

87. H.-J. Müller, CEO von Pyramid Technologies, Biopolymer-Produktion in Guben startet später, http://www.tbia.or.th/uploads/attach/1251427724.pdf (accessed on August 20, 2012).

88. E. Hans-Josef, A. Siebert-Raths, *Technische Biopolymere*. Hanser-Verlag, München, Germany (2009), p. 293.

89. J. Ahmed, S.K. Varshney, R. Auras, S.W. Hwang, Thermal and rheological properties of L-polylactide/polyethylene glycol/silicate nanocomposites films, *Journal of Food Science*, 75 (2010) N97.

90. B.Q. Chen, J.R.G. Evans, Thermoplastic starch-clay nanocomposites and their characteristics, *Carbohydrate Polymers*, 61 (2005) 455.

91. H. Wu, C. Liu, J. Chen, P.R. Chang, Y. Chen, D.P. Anderson, Structure and properties of starch/alpha-zirconium phosphate nanocomposite films, *Carbohydrate Polymers*, 77 (2009) 358.

92. K. Majdzadeh-Ardakani, B. Nazari, Improving the mechanical properties of thermoplastic starch/poly(vinylalcohol)/clay nanocomposites, *Composites Science and Technology*, 70 (2010) 1557.

93. K. Majdzadeh-Ardakani, A.H. Navarchian, F. Sadeghi, Optimization of mechanical properties of thermoplastic starch/clay nanocomposites, *Carbohydrate Polymers*, 79 (2010) 547.

94. M. Avella, J.J. De Vlieger, M.E. Errico, S. Fischer, P. Vacca, M.G. Volpe, Biodegradable starch/clay nanocomposite films for food packaging applications, *Food Chemistry*, 93 (2005) 467.

95. E. Curylo, S.C. Khor, C. Mayo, M.G. Ruda, G. Parks, Isolation and Antimicrobial Potential Of Epsilon-Poly-L-Lysine- a Major Qualifying Project Report: submitted to the Faculty of the Worcester Polytechnic Institute, http://www.wpi.edu/Pubs/E-project/Available/E-project-042408-104838/unrestricted/ePLMQPFINAL.pdf (accessed on August 20, 2012).

96. J.W. Lawton, Zein: A history of processing and use, *Cereal Chemistry*, 79 (1) 1–18.

97. A. Gennadios (Ed.), *Handbook on Protein-Based Films and Coatings*. CRC Press, Boca Raton, FL (2001).

98. J.W. Jr. Lawton, Plasticizers for zein: Their effect on tensile properties and water absorption of zein films, *Cereal Chemistry*, 12 (2004) 1–5.

99. L. Cabedo, D. Plackett, E. Gimenez, J.M. Lagaron, Studying the degradation of polyhydroxybutyrate-co-valerate during processing with clay-based nanofillers, *Journal of Applied Polymer Science*, 112 (2009) 3669.

100. R. Yoksan, S. Chirachanchai, Silver nanoparticles dispersing in chitosan solution: Preparation by gamma-ray irradiation and their antimicrobial activities, *Materials Chemistry and Physics*, 115 (2009) 296.

101. K.L. Spence, R.A. Venditti, O.J. Rojas, Y. Habibi, J.J. Pawlak, The effect of chemical composition on microfibrillar cellulose films from wood pulps: Water interactions and physical properties for packaging applications, *Cellulose*, 17 (2010) 835.

102. M.R. de Moura, M.V. Lorevice, L.H.C. Mattoso, V. Zucolotto, Highly stable, edible cellulose films incorporating chitosan nanoparticles, *Journal of Food Science*, 76 (2011) S25.

103. D. Restuccia, U.G. Spizzirri, O.I. Parisi, G. Cirillo, M. Curcio, F. Iemma, F. Puoci, G. Vinci, N. Picci, New EU regulation aspects and global market of active and intelligent packaging for food industry applications, *Food Control*, 21 (2010) 425.

104. Attophotonics Biosciences GmbH, main page, http://www.attophotonics.com (accessed on August 20, 2012).

105. C.D. Papaspyrides, Nanotechnology and food contact materials, In: A. D'Amore, D. Acierno, L. Grassia (Eds.), *Fifth International Conference on Times of Polymers TOP and Composites AIP Conference Proceedings*, Ischia, Italy, Vol. 1255 (2010), p. 234.

106. L. Xu, Y. Liu, R. Bai, C. Chen, Applications and toxicological issues surrounding nanotechnology in the food industry, *Pure and Applied Chemistry*, 82 (2010) 349.

107. Ecosynthetix, Sustainable Polymers for Planet Earth-main page, http://www.ecosynthetix.com/ecosphere.html (accessed on August 20, 2012).

108. Multifilm Packaging Corporation, N-Coat multifilm, http://www.multifilm.com/products_N-Coat.htm (accessed on August 20, 2012).

109. Nanocor a new operating subsidiary of AMCOL International Corporation, main page, http://www.nanocor.com (accessed on August 20, 2012).

110. Nanocompositech, Commercial Nanocomposites and Nanoclays, http://www.nanocompositech.com/commercial-nanocomposites-nanoclays.htm (accessed on August 20, 2012).

111. Nanocor, Technical bulletin, Imperm® Grade 103-Pet Multilayer Barrier Bottles, http://www.nanocor.com/tech_sheets/I103.pdf (accessed on August 20, 2012).

112. Mitsubishi Gas Chemical Co and Nanocor, Multilayer Containers Featuring Nano-Nylon Mxd6 Barrier Layers with Superior Performance and Clarity, http://www.nanocor.com/tech_papers/NOVAPACK03.pdf (accessed on August 20, 2012).

113. Nanocor, Commercial Product Lines, http://www.nanocor.com/products.asp (accessed on August 20, 2012).

114. Honeywell, Aegis® barrier nylon resins, http://www51.honeywell.com/sm/aegis/ (accessed on August 20, 2012).

115. Honeywell, Aegis® OXCE Barrier Nylon Resin, http://www51.honeywell.com/sm/aegis/products-n2/aegis-ox.html (accessed on August 20, 2012).

116. I. Annesi-Maesano, F. Forastiere, N. Kunzli, B. Brunekref, Particulate matter, science and EU policy, *European Respiratory Journal*, 29 (2007) 428.

117. K. Donaldson, L. Tran, L.A. Jimemez et al., Combustion derived nanoparticles: A review of their toxicology following inhalation exposure. *Particle and Fibre Toxicology*, 21 (2005) 10.

118. J.T. Kwon, D.S. Kim, A. Minai-Tehrani et al., Inhaled fluorescent magnetic nanoparticles induced extramedullary hematopoiesis in the spleen of mice, *Journal of Occupational Health*, 51(2009) 423.
119. N. Watanabe, Y. Oonuki, Inhalation of diesel engine affects spermatogenesis in growing male rats, *Environmental Health Perspectives*, 107 (1999) 539.
120. G. Chen, G. Song, L. Jiang et al., Short term effects of ambient gaseous pollutants and particulate matter on daily mortality in Shanghai, *China Journal of Occupational Health*, 50 (2008) 41.
121. A.R. Elmore, F.A. Andersen, Final report on the safety assessment of aluminium silicate, calcium silicate, magnesium aluminium silicate magnesium silicate magnesium trisilicate sodium magnesium silicate zirconium silicate attapulgite, bentonite, fullers earth, hectorite, kaolin, lithium magnesium silicate, lithium magnesium sodium silicate, montmorillonite, pyrophyllite, and zeolite, *International Journal of Toxicology*, 22 (2003) 37.
122. S.H. Dougherty, V.D. Fiegel, R.D. Nelson, G.T. Rodeheaver, R.L. Simmons, Effects of soil infection potentiating factors on neutrophils in vitro, *American Journal of Surgery*, 150 (1985) 306.
123. A.K. Sharma, B. Schmidt, H. Frandsen, N.R. Jacobsen, E.H. Larsen, M. Binderup, Genotoxicity of unmodified and organo-modified montmorillonite, *Mutation Research*, 700 (2010) 18.
124. N. Gao, M.J. Keane, T. Ong, W.E. Wallace, Effects of simulated pulmonary surfactant on the cytotoxicity and DNA damaging activity of respirable quartz and Koalin, *Journal of Toxicology and Environmental Health—Part A*, 60 (2000) 153.
125. S. Geh, T.M. Shi, B. Shoukouhi, R.P.F. Shins, L. Armbruster, A.W. Rettenmeier, E. Dopp, Genotoxic potential of bentonite particles with different quartz content and chemical modification in human lung fibroblasts. *Inhalation Toxicology*, 18 (2006) 405.
126. M.I. Carretero, C.S.F. Gomes, F. Tateo, Clays and human health, in: F. Bergaya, B.K.G. Theng, G. Lagaly (Eds.), *Handbook of Clay Science*. Elsevier, Amsterdam, the Netherlands (2006), p. 717.
127. D.B. Warheit, C.M. Sayes, S.R. Frame, K.L. Reed, Pulmonary exposures to sepiolite nanoclay particulates in rat: Resolution following multinucleate giant cell formation, *Toxicology Letters*, 192 (2010) 286.
128. P.R. Li, J.C. Wei, Y.F. Chiu, H.L. SU, F.C. Peng, J.J. Linn, Evaluation on cytotoxicity and genotoxicity of the exfoliated silicate nanoclay, *Applied Materials & Interfaces*, 2 (2010) 1608.
129. A. Pietroiusti, Embryonic stem Cell test (EST), Classification of embryo-toxicity, in: Embryo toxicity of Engineered Nanoparticles, *Proceedings of COST Workshop on Modern Polymeric Materials for Environmental Applications*, Vol. 4(2), Krakow, Poland, December 1–4 (2010).
130. A. Almagableh, P.R. Mantena, A. Alostaz, W. Liu, L.T. Drzal, Chitosan modified nanoclay poly (AMPS) nanocomposite hydrogel with improved strength, *Express Polymer Letters*, 3 (2009) 724.
131. S.W.P. Wijnhoven, W.J.G.M. Peijnenburg, C.A. Herberts, W.I. Hagens, A.G. Oomen, E.H.W. Heugens, B. Roszek, J. Bisschops, I. Gosens, D. Van De Meent, S. Dekkers, W.H. de Jong, M. Van Zijverden, A.J.A.M. Sips, R.E. Geertsma, Nano-silver a review of available data and knowledge gaps in human and environmental risk assessment, *Nanotoxicology*, 3 (2009) 109.

Index

A

Active/bioactive nanomaterials
  Alicyclobacillus acidoterrestris, 262
  antibacterial and antifungal
    activity, 263
  antimicrobial packaging
    system, 261–262
  coated asparagus, 262
  effects, sterile silver-coated
    refrigerators, 262–263
  Lactobacillus plantarum, 262
  LDPE nanocomposite films, 261
  nano-active-packaging, 264
  oxygen scavenger films, 263
  substances, 260–261
Active packaging (AP), 32
AFM, *see* Atomic force microscopy
    (AFM)
Antibacterial properties, PNC
  improvements
    by Ag NP, 343
    by cellulose and chitin NP, 343
    by clay NP, 342
    by ZnO NP and TiO$_2$ NP, 342
  quaternary amines use, 341
AP, *see* Active packaging (AP)
Applications, PNFP
  active, 10–12
  communication and containment
    function, 6
  improved
    advantages, 10
    barrier properties, 7
    EVOH, 10
    layer-by-layer assembly, 8
    morphology, 8
    nanoclay composites, 10
    nanofillers, 7
    nanoparticles, 7
    oxygen and water vapor
      permeability, 7, 8
    oxygen permeability, linear
      HDPE, 7, 9

polymers and clay fillers, 8
ultrahigh barrier properties, 8, 9
intelligent/smart, 12–13
mechanical and thermal properties,
    materials, 7
polymer nanotechnology, 7
principal, 6
Atomic force microscopy (AFM)
  description, 94
  phase imaging, 97

B

Bacterial cellulose nanowhiskers
    (BCNWs)
  anisotropic biohybrid fiber
    yarns, 211–212
  electrospun PLA fibers
    reinforcement, 212–213
  EVOH fibers reinforcement,
    209–210
  nanofabrication
    aspect ratio (L/D), 199
    crystallinity index, 200
    disruption and digestion, 201
    MFC, 199
    morphology, 200
    TEM, 200, 201
    thermal stability, 201
    vegetal microfibrillated
      cellulose, 199–200
    XRD, 200
  vs. plant cellulose nanowhiskers,
    202–203
Barrier coatings
  diamond-like carbon films, 129–132
  food and beverage packaging, 122
  metal coatings, 122
  nanocomposite functional coatings,
    132–135
  oxygen transition rates, polymers,
    122, 123
  SiO$_x$ films, *see* SiO$_x$ films